FUNDAMENTALS OF OBJECT TRACKING

Kalman filter, particle filter, IMM, PDA, ITS, random sets ... The number of useful object tracking methods is exploding. But how are they related? How do they help to track everything from aircraft, missiles and extra-terrestrial objects to people and lymphocyte cells? How can they be adapted to novel applications? *Fundamentals of Object Tracking* tells you how.

Starting with the generic object tracking problem, it outlines the generic Bayesian solution. It then shows systematically how to formulate the major tracking problems – maneuvering, multi-object, clutter, out-of-sequence sensors – within this Bayesian framework and how to derive the standard tracking solutions. This structured approach makes very complex object tracking algorithms accessible to the growing number of users working on real-world tracking problems and supports them in designing their own tracking filters under their unique application constraints. The book concludes with a chapter on issues critical to the successful implementation of tracking algorithms, such as track initialization and merging.

SUBHASH CHALLA is a Professor and Senior Principal Researcher at the NICTA, Victoria Research Laboratory at the University of Melbourne. He is also the co-founder and CEO of SenSen Networks Pty Ltd, a leading video business intelligence solutions company.

MARK R. MORELANDE is a Senior Research Fellow in the Melbourne Systems Laboratory at the University of Melbourne.

DARKO MUŠICKI is a Professor in the Department of Electronic Systems Engineering at Hanyang University in Ansan, Republic of Korea.

ROBIN J. EVANS is a Professor of Electrical Engineering at the University of Melbourne and the Director of NICTA, Victoria Research Laboratory.

William of Ockham

Frustra fit per plura, quod fieri potest per pauciora
(It is vain to do with more what can be done with less)

FUNDAMENTALS OF OBJECT TRACKING

SUBHASH CHALLA

National ICT Australia (NICTA), University of Melbourne, Australia

MARK R. MORELANDE

University of Melbourne, Australia

DARKO MUŠICKI

Hanyang University, Ansan, Republic of Korea

ROBIN J. EVANS

National ICT Australia (NICTA), University of Melbourne, Australia

CAMBRIDGE
UNIVERSITY PRESS

CAMBRIDGE
UNIVERSITY PRESS

University Printing House, Cambridge CB2 8BS, United Kingdom

One Liberty Plaza, 20th Floor, New York, NY 10006, USA

477 Williamstown Road, Port Melbourne, VIC 3207, Australia

314-321, 3rd Floor, Plot 3, Splendor Forum, Jasola District Centre, New Delhi - 110025, India

79 Anson Road, #06-04/06, Singapore 079906

Cambridge University Press is part of the University of Cambridge.

It furthers the University's mission by disseminating knowledge in the pursuit of education, learning and research at the highest international levels of excellence.

www.cambridge.org
Information on this title: www.cambridge.org/9780521876285

© S. Challa, M. R. Morelande, D. Mušicki and R. J. Evans 2011

First published 2011

A catalogue record for this publication is available from the British Library

Library of Congress Cataloging in Publication data
Fundamentals of object tracking / Subhash Challa. . . [et al.].
p. cm.
Includes index.
ISBN 978-0-521-87628-5 (hardback)
1. Linear programming.   2. Programming (Mathematics)   I. Challa, Sudha, 1953–
QA402.5.F86   2011
519.7 – dc22   2011008595

ISBN 978-0-521-87628-5 Hardback

Contents

Preface

Tracking the paths of moving objects is an activity with a long history. People in ancient societies used to track moving prey to hunt and feed their kith and kin, and invented ways to track the motion of stars for navigation purposes and to predict seasonal changes in their environments. Object tracking has been an essential technology for human survival and has significantly contributed to human progress.

In recent times, there has been an explosion in the use of object tracking technology in non-military applications. Object tracking algorithms have become an essential part of our daily lives. For example, GPS-based navigation is a daily tool of humankind. In this application a group of artificial satellites in outer space continuously locate the vehicles people drive and the object tracking algorithms within the GPS perform self-localization and enable us to enjoy a number of location-based services, such as finding places of interest and route planning. Similarly, tracking of objects is used in a wide variety of contexts, such as airspace surveillance, satellite and space vehicle tracking, submarine and whale tracking and intelligent video surveillance. They are also used in autonomous robot navigation using lasers, stereo cameras and other proximity sensors, radiosonde-enabled balloon tracking for accurate weather predictions, and, more recently, in the study of cell biology to study cell fate under different chemical and environmental influences by tracking many kinds of cells, including lymphocyte and stem cells through multiple generations of birth and death.

This book is an introduction to the fascinating field of object tracking and provides a solid foundation to the collection of diverse algorithms developed over the past 60 years by academics, scientific researchers and engineers. Historically, advances in the field of object tracking were a result of the systematic extension of methods that worked under severely restrictive ideal-world conditions to less restrictive real-world conditions. The advances were often a result of inspired innovations by scientists incorporating either more descriptive object dynamics models and/or sensor measurement models and associated statistical methods/approximations to deal with them. This led to a repertoire of extremely valuable, sometimes apparently unrelated techniques and tools to solve real-world object tracking problems. Most of the books on object tracking present the methods as a collection of these diverse algorithms.

However, all these techniques have firm foundations in recursive Bayesian logic. They can be formulated and derived in a Bayesian probabilistic framework. The aim of this book is to present such a unifying approach along with the latest advances in efficient computational algorithms for real time implementation. We have designed this book to provide a thorough understanding of object tracking to the growing number of engineers and researchers working on or contemplating working on real-world object tracking problems and to students in university programs in engineering and statistics.

In Chapter 1, we introduce the generic object tracking problem and propose a generic Bayesian solution. We refer to an object being tracked as a target, to use the common engineering language, and develop a general approach to solve the problems considered in the book. At the end of the chapter we briefly review the major tracking algorithms used by practitioners in the field. All object tracking algorithms use estimation or filtering as a core component. The Kalman filter, extended Kalman filter, unscented Kalman filter, point mass filter and particle filter are examples of algorithms developed to solve several generic estimation and filtering problems. In Chapter 2, these algorithms are derived as approximations to the recursive form of Bayes' theorem. These algorithms also form the basis of tracking a single target that is moving at approximately constant speed. Tracking a maneuvering target has led to a rich literature on filtering, including the generalized pseudo-Bayesian filter, the interacting multiple model filter, and many others. These approaches and their Bayesian foundations are considered in Chapter 3.

A unique feature of tracking arises because the sensor measurements from which track estimates must be extracted contain "false" detections. A plethora of techniques aimed at addressing this problem are described in the literature. The nearest neighbor filter, probabilistic data association filter and the like are derived in Chapter 4. Most of these techniques assume that the target exists. However, in reality, there can be uncertainty about whether a target exists or not. A class of object tracking algorithms introduces the concept of the existence of objects as a random jump Markov process and estimates its probability from the available observations. Integrated track splitting (ITS) and its derivatives (e.g., integrated probabilistic data association) fall into this category and we introduce them in Chapter 5. These algorithms are applicable for both single- and multiple-object tracking.

The task of the multiple-object tracker is not only to estimate the state of each object, but also to provide an estimate of the number of objects. The number of objects can be inferred from ITS- or IPDA-type algorithms (see Chapter 5); the process is not part of the algorithm itself. The randomly varying finite set gives a method to model the multi-object tracking problem in such a way that the tracker can estimate the number of objects along with the state of each invidual object. Random-set-based modeling is introduced in Chapter 6. The resulting Bayesian random-set-based trackers and their approximations, such as probability hypothesis density (PHD) and cardinalized PHD, are introduced in the chapter. It also demonstrates how the random set formalism leads to object-existence-based filters, such as the IPDA and joint IPDA algorithms introduced in Chapter 5.

Most target tracking systems use filtering and/or prediction techniques that use the current sensor measurements to obtain best estimates at the current time or into the future. However, significant situational awareness can be gained by improving the estimates of the past target states using current measurements. Such an approach is traditionally referred to as smoothing. Chapter 7 presents the Bayesian foundations of smoothing as well as certain approximations. The smoothing framework of this chapter is extended in Chapter 8 to deal with out-of-sequence measurements. Finally, Chapter 9 introduces key tricks and methods in implementing object tracking algorithms, e.g., track initiation, merging and termination, and provides insights into how object tracking methods are realized in real applications. The chapter gives a systematic framework to design these practical object tracking methods.

A useful way of assessing an algorithm's performance in the real world is to compare its performance with the best achievable performance. Indicators of best achievable performance include the Cramér–Rao bound (CRB), Barankin bound (BB) and Weiss–Weinstein bound (WWB), and others. We present the performance bounds for all the real-world scenarios in their respective chapters.

The object tracking algorithms described in the book are at times elaborate and mathematically and notationally extensive due to the realistic engineering situations they address. We provide illustrative examples on how to use the methods presented in each chapter on representative real-world problems, and these are also expected to help the reader with a better understanding of these algorithms. We hope that the unification of all the object tracking approaches under the Bayesian probabilistic paradigm makes it possible for any reader at the first-year university level in statistics to comprehend the material and master the techniques.

We are grateful to the many colleagues and students who have helped us through the years in developing some of the material in this book. Specifically, we wish to mention Dr. Khalid Aboura and Dr. Rajib Chakravorty, who helped us in pulling together some of the elements of this book. We also wish to thank Prof. Nozer Singpurwalla for introducing us to the editors of Cambridge University Press and encouraging us to undertake the mammoth task of writing this book. We would like to express our deepest gratitude to all of these wonderful people and also to our families, who had to sacrifice a lot to help transform this book from a figment of our imagination into reality.

1

Introduction to object tracking

Object/target tracking refers to the problem of using sensor measurements to determine the location, path and characteristics of objects of interest. A sensor can be any measuring device, such as radar, sonar, ladar, camera, infrared sensor, microphone, ultrasound or any other sensor that can be used to collect information about objects in the environment. The typical objectives of object tracking are the determination of the number of objects, their identities and their states, such as positions, velocities and in some cases their features. A typical example of object/target tracking is the radar tracking of aircraft. The object tracking problem in this context attempts to determine the number of aircraft in a region under surveillance, their types, such as military, commercial or recreational, their identities, and their speeds and positions, all based on measurements obtained from a radar.

There are a number of sources of uncertainty in the object tracking problem that render it a highly non-trivial task. For example, object motion is often subject to random disturbances, objects can go undetected by sensors and the number of objects in the field of view of a sensor can change randomly. The sensor measurements are subject to random noises and the number of measurements received by a sensor from one look to the next can vary and be unpredictable. Objects may be close to each other and the measurements received might not distinguish between these objects. At times, sensors provide data when no object exists in the field of view. As we will see in later chapters of this book, the problems in object tracking can be classified according to the various types of uncertainties involved.

In this chapter, we introduce Bayes' rule, a deceptively simple yet extremely powerful tool from statistical inference, which facilitates recursive reasoning and estimation in the presence of uncertainty. This, in conjuction with the Chapman–Kolmogorov theorem, provides the foundation used to derive the object tracking algorithm presented in this book.

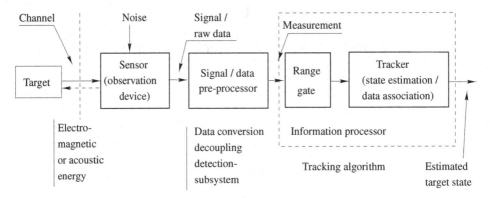

Figure 1.1 Typical object tracking system.

1.1 Overview of object tracking problems

The typical object tracking problem is essentially a state estimation problem where the object states to be estimated from noisy corrupted and false measurements are kinematic states such as position, velocity and acceleration. The tracking system consists of an object or objects to be tracked, a sensor which measures some aspect of the object, a signal processor and an information processor, as shown in Figure 1.1.

There are a number of applications for object tracking. Some of these are reviewed here.

1.1.1 Air space monitoring

An important tracking problem is the tracking of aircraft using radar (Krause, 1995), such as for air traffic control. Radar tracking is also used in military surveillance systems, where the problem involves identifying aircraft, their type, identity, speed, location, and the likely intentions of the object in order to determine, for example, if the object is a threat. Radar comes with a wide variety of measurement capabilities ranging from simple range measurements to high-resolution imaging. Radar uses reflected radio waves to measure the direction, distance and radial speed of the detected object. A radar transmitter emits electromagnetic waves, which are reflected by the object and detected by a receiver (Figure 1.2). The measured data are used to extract tracks, which are often displayed along with the object reflections on a display screen.

The tracking of aircraft using radar data is made particularly difficult because of uncertainties in the origin of the measurements. Birds, animals, vegetation, terrain, clouds, sea, rain, snow and signals generated by other radars create radar signal noise, known as clutter. Clutter consists of those detections which are not

Figure 1.2 Radar sensors.

the objects of interest. The problem of determining which detections belong to the object of interest and should be used to extract the track of the object is called the data association problem and is discussed in Chapter 4 and subsequent chapters.

Aircraft display maneuvering behaviors. Tracking such objects requires elaborate solutions that can adaptively change the model of the dynamics being used to represent the object behavior. Chapter 3 presents several approaches for dealing with maneuvering object tracking and discusses efficient algorithms to solve this problem.

Very often, air surveillance is conducted in areas where a large number of often closely spaced aircraft are present. This leads to the multi-object tracking problem (see, for example, Hwang *et al.*, 2004). A major difficulty in such situations is deciding which measurement belongs to which object in an environment where the radar produces false measurements and furthermore does not always produce a measurement for every object. This generalizes the single object tracking problem and involves consideration of more complex probabilistic models. Chapters 5 and 6 introduce solutions to multi-object tracking.

1.1.2 Video surveillance

The use of digital-video-based surveillance is growing significantly. Video surveillance is now instrumental in implementing security at airports, buildings, banks, department stores, casinos, railway stations, highways, streets, stadiums, crowd gathering places and all government institutions. In practically all sectors of society, video surveillance is used as a means to increase public safety and security and deter criminal acts. With the proliferation of high-speed broad band wired/wireless networks, many institutions now deploy large networks of cameras for surveillance. A typical large building in a major city owns a large network of cameras deployed at major entrances, floors, large gathering places, elevators, hallways, labs and offices. While such surveillance systems allow trained security guards to visually monitor the protected areas, intelligent software is needed to make use of the huge amount of information collected by the cameras. An increasingly large part of research and development is devoted to intelligent visual surveillance software (Olsen and Brill, 1997; Wren *et al.*, 1997; Boult, 1998; Lipton *et al.*, 1998; Tan *et al.*, 1998; Collins *et al.*, 2000; Haritaoglu *et al.*, 2004).

To detect and track a person or vehicle in video images, and furthermore to infer their behavior, such as unusual, loitering or even criminal behavior, one must solve non-trivial problems (Hu *et al.*, 2004). For example, the US military supported the Visual Surveillance and Monitoring (VSAM) project (Collins *et al.*, 2000) in 1997. The purpose of VSAM was to enable an operator to monitor behavior over complex areas such as battlefields. Another project, called the Human Identification at a Distance (HID) project, followed in 2000. HID aimed to develop a full range of multi-modal surveillance technologies for detecting and identifying humans from a distance. Some intelligent systems have been implemented for crowd estimation. Real-time systems for crowd estimation have been implemented based on existing closed circuit television (CCTV) at railway stations in London (Davies *et al.*, 1995) and Genova (Regazzoni *et al.*, 1993; Regazzoni and Tesei, 1996).

The tracking algorithm often breaks down when people or objects move in unusual ways. Other problems hinder traditional tracking in the visual context. Distraction occurs when some motion or light comes across the object and creates distorted measurements. Obstruction is another case when an object appears between the camera and another object. The measurements are not produced for the hidden object, which results in the termination of the track. In case both tracks do not terminate, then, as the two people separate, it is not easy to associate the new measurements with the objects.

A myriad of other problems not usually found in radar tracking situations lead to difficulties in establishing a robust approach for visual tracking. Interesting research challenges result. For example the probabilistic data association filter

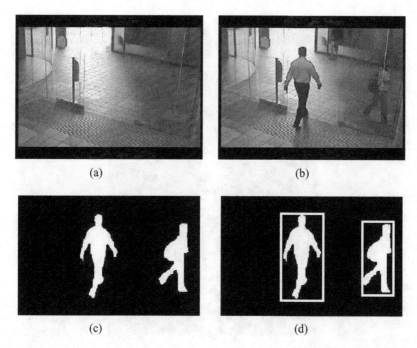

(a)   (b)

(c)   (d)

Figure 1.3 Sensor measurement extraction in video images: (a) background image, (b) background and foreground, (c) blobs, (d) bounding boxes as sensor measurements.

(PDAF) and the joint probabilistic data association filter (JPDAF) described in later chapters are excellent methods for dealing with data association problems in tracking. These are problems in which the measurements come with significant noise, or clutter. Rasmussen and Hager (2001) make good use of these filters by adapting them to visual tracking problems.

1.1.3 Weather monitoring

In order to provide weather forecasts, weather bureaus use several techniques. One technique is to track weather balloons, which provide information on high-altitude wind velocities, pressure, humidity and temperature. Weather bureaus release 50 to 70 balloons each day with the release time spaced throughout the day. In extreme weather conditions, the number of releases increases because more information is needed for accurate weather prediction. In order to get weather-related parameters at different levels of atmosphere, each weather balloon needs to be tracked.

A simple approach is to track the balloon using ground-based radars. But this approach has limitations as the initial trajectory is not in the line-of-sight of the radar. So an operator needs to track the balloon manually until it reaches a specific

Figure 1.4 Balloon for weather monitoring (NASA/courtesy of nasaimages.org).

height from which the ground-based radar can track it automatically. A second approach is to use GPS devices in the balloon. But as the balloon is unrecoverable, the cost of the device adds to the cost of the whole process. Moreover, as this process is repeated throughout the day, using a GPS device becomes extremely costly.

A clever solution is to build a system that tracks the initial stages of the balloon trajectory and then hands it over to the ground radar once the balloon reaches a specified height. In order to do so, a radiosonde device is attached to the balloon. This device sends a synchronization pulse to three base stations. The location of the balloon is estimated by a filter that works on the principle of triangulation using the time-of-arrival of the synchronization pulse in three base stations. Once the balloon crosses a pre-specified height, a ground-based radar is locked onto the balloon and tracks it further into the atmosphere.

1.1.4 Cell biology

In studies of humans, animals, plants and insects, medical researchers and pathologists routinely study birth and death rates and the movement of biological cells. In immunology, the organism's immune response is correlated with the life cycle

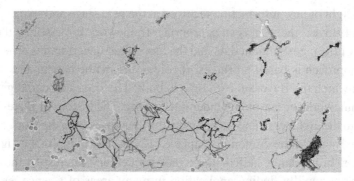

Figure 1.5 Tracking of lymphocyte cells.

of lymphocytes. The parameters of interest are the division/birth and death time of each generation of cells. In fertility studies, the interesting parameter is the velocity or shape of sperm cells. In the case of anti-inflammatory diseases, researchers look for the speed and acceleration of lymphocyte cells (Figure 1.5). Medical researchers carefully prepare the cells and take sequential images at regular intervals. In some cases, the images can be collected over several days. These images are then examined manually. The process of going through the set of images and finding out the required parameters, i.e., speed, division and death, of each cell is both time-consuming and prone to error and in some cases is almost impractical. The problem can be formulated as the problem of tracking cells over a sequence of images. The events of cell division and death are observed by looking at the track initiation and termination probabilities. The process is not only faster and reliable but also enables the discovery of previously unobserved phenomena. The key concept is to define the "data association" parameters for cell identification and tracking and then using the domain knowledge of cell immunology to model the "division/birth" and "death" of cells and integrate those into the recursive Bayesian solutions.

1.2 Bayesian reasoning with application to object tracking

The Bayesian approach is a well-developed probabilistic and statistical theory that can be applied to the modeling and solution of problems encountered in object tracking. Bayesian methods are widely used in the statistical and engineering communities, and engineers have long made use of them in solving a large variety of problems.

In most object tracking problems, like the ones introduced in Section 1.1, measurements from sensors are received in a sequential manner over time. At each measurement stage, new estimates of the object state are made by combining new information with current estimates of the object state. These latest estimates are updated in light of new information received at the next stage and so on. The recursive form of Bayes' theorem is an appropriate framework for handling the sequential nature of the measurement data and the associated uncertainties. The Bayesian recursive approach is based on Bayes' theorem, the mathematical and conceptual basis upon which the Bayesian paradigm is built. The Bayesian paradigm is a theoretical framework for reasoning under uncertainty. The history of Bayes' theorem goes back to the eighteenth and early nineteenth centuries with the work of Thomas Bayes (1764) and Pierre-Simon de Laplace (1812). The theorem was ignored for a long time. In the latter half of the twentieth century interest in the Bayesian approach grew rapidly finding application in many areas of science, engineering and statistical inference. Stigler (1986) provides an historical account of Bayesian thinking.

1.2.1 Bayes' theorem

Bayes' theorem is the encapsulation of a philosophy to consistently reconcile past and current information by exploiting the conditional probability concepts of probability theory. It may be stated as follows: given two related events $\mathbf{x}$ and $\mathbf{y}$, the conditional probability of event $\mathbf{x}$ given observation of event $\mathbf{y}$ is

$$p(\mathbf{x}|\mathbf{y}) = \frac{p(\mathbf{x}, \mathbf{y})}{p(\mathbf{y})}. \tag{1.1}$$

It is equal to the joint probability of events $\mathbf{x}$ and $\mathbf{y}$, $p(\mathbf{x}, \mathbf{y})$, normalized by the unconditional probability of event $\mathbf{y}$, $p(\mathbf{y})$. Using (1.1) twice, Bayes' theorem can be rewritten as

$$p(\mathbf{x}|\mathbf{y}) = \frac{p(\mathbf{y}|\mathbf{x})p(\mathbf{x})}{p(\mathbf{y})}. \tag{1.2}$$

Various aspects of the object tracking problem, e.g., the number of objects and their states, can be modeled as events $\mathbf{x}$ and many types of sensor outputs, e.g., radar returns or infrared images, as events $\mathbf{y}$. Bayes' theorem in (1.2) can then be applied to obtain the conditional probabilities $p(\mathbf{x}|\mathbf{y})$ as a probabilistic answer to the object tracking problem.

Interpretation of Bayes' theorem

Let $\mathbf{x}$ be a random variable of interest. In object tracking, $\mathbf{x}$ is often the state of the object under consideration. Let $\mathbf{y}$ be a measurement related to $\mathbf{x}$. Given knowledge of $\mathbf{y}$, we want to update our existing knowledge about $\mathbf{x}$. Since $\mathbf{x}$ is a random variable, and using a probabilistic approach in dealing with uncertainty, we represent our knowledge about $\mathbf{x}$ with a probability distribution or a probability density function (pdf) $p(x)$, depending on whether x takes discrete values or is continuous. $p(\mathbf{x})$ is a function of $\mathbf{x}$ that attributes probability values to $\mathbf{x}$. The problem at the core of most tracking problems is how to update $p(\mathbf{x})$ to take into account the new information $\mathbf{y}$. The answer is provided in (1.2) by $p(\mathbf{x}|\mathbf{y})$, the conditional distribution of $\mathbf{x}$ *given* $\mathbf{y}$. Viewing $\mathbf{x}$ as a random variable that can take a range of values, we can see that for a fixed $\mathbf{y}$, $p(\mathbf{x})$ is altered by $p(\mathbf{y}|\mathbf{x})$ to become $p(\mathbf{x}|\mathbf{y})$. We can ignore the effect of $p(\mathbf{y})$, since it is the same for all $\mathbf{x}$ values, once $\mathbf{y}$ is known. It is this, theoretically sound, alteration of $p(\mathbf{x})$ for each possible value $\mathbf{x}$ that makes us change our probability representation of $\mathbf{x}$, from $p(\mathbf{x})$ to $p(\mathbf{x}|\mathbf{y})$. It is done with a re-weighting of the initial probability distribution with the term $p(\mathbf{y}|\mathbf{x})$ for each $\mathbf{x}$ value. This deceptively simple operation is at the heart of many complex algorithms for estimating parameters of dynamic processes. The term $\mathcal{L}(\mathbf{x}) = p(\mathbf{y}|\mathbf{x})$ is known as the *likelihood function* when viewed as a function of $\mathbf{x}$. We write

$$p(\mathbf{x}|\mathbf{y}) \propto p(\mathbf{y}|\mathbf{x})p(\mathbf{x})$$

to mean that $p(\mathbf{x}|\mathbf{y})$ is proportional to the product of $p(\mathbf{y}|\mathbf{x})$ and $p(\mathbf{x})$. The term $p(\mathbf{y})$ is a constant when viewed as a function of $\mathbf{x}$ and is called the normalizing constant or normalization factor which ensures that $p(\mathbf{x}|\mathbf{y})$ as a function of $\mathbf{x}$ sums up or integrates to 1. The initial distribution $p(\mathbf{x})$ is called the *prior distribution*. The new distribution of $\mathbf{x}$, $p(\mathbf{x}|\mathbf{y})$, is called the *posterior distribution*.

Consider the single object whose state can be represented by the one-dimensional vector $\mathbf{x}$. Suppose that given all past history, our current knowledge about $\mathbf{x}$ is contained in the probability density function

$$p(\mathbf{x}) = \frac{1}{\Sigma_0 \sqrt{2\pi}} e^{-\frac{(x-x_0)^2}{2\Sigma_0^2}}.$$

This means that we think that $\mathbf{x}$ is near x_0, with an estimation error quantified by Σ_0. We may have arrived at this normal distribution (Gaussian distribution) through some previous calculations. We write $p(\mathbf{x}) = N(\mathbf{x}; x_0, \Sigma_0^2)$. The bigger Σ_0 the less certain we are about $\mathbf{x}$. Assume that a sensor produces an observation $\mathbf{y}$ about $\mathbf{x}$. Knowing that the sensor output has measurement error, we write the following equation:

$$\mathbf{y} = \mathbf{x} + \mathbf{w}, \quad p(\mathbf{w}) = N(\mathbf{w}; 0, \Sigma^2).$$

The measurement error $\mathbf{w}$ is modeled probabilistically using a Gaussian distribution with zero mean and variance Σ^2. Using the basic properties of the Gaussian distribution (see Hogg and Craig, 1995), the likelihood function $\mathcal{L}(x) = p(\mathbf{y}|\mathbf{x})$

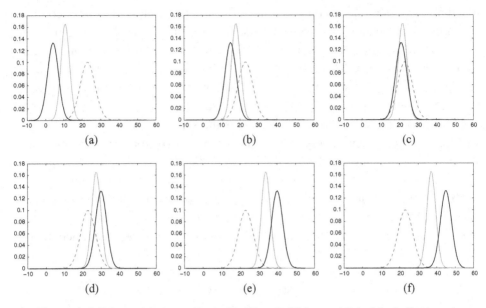

Figure 1.6 Prior $p(\mathbf{x})$ (grey line), likelihood $\mathcal{L}(\mathbf{x}) = p(\mathbf{y}|\mathbf{x})$ (black line) and posterior $p(\mathbf{x}|\mathbf{y})$ (grey line) for data values (a) $\mathbf{y} = 4$, (b) $\mathbf{y} = 15$, (c) $\mathbf{y} = 21$, (d) $\mathbf{y} = 30$, (e) $\mathbf{y} = 40$, (f) $\mathbf{y} = 45$.

becomes

$$p(\mathbf{y}|\mathbf{x}) = \frac{1}{\Sigma\sqrt{2\pi}}e^{-\frac{(\mathbf{y}-\mathbf{x})^2}{2\Sigma^2}}.$$

Following Bayes' theorem, we can easily show that $p(x|y) = N(\mathbf{x}; \hat{\mathbf{x}}, \hat{\boldsymbol{\Sigma}}^2)$, where

$$\hat{\mathbf{x}} = \left(\frac{\mathbf{x}_0}{\Sigma_0^2} + \frac{\mathbf{y}}{\Sigma^2}\right)\left[\frac{1}{\Sigma_0^2} + \frac{1}{\Sigma^2}\right]^{-1},$$

$$\hat{\boldsymbol{\Sigma}}^2 = \left[\frac{1}{\Sigma_0^2} + \frac{1}{\Sigma^2}\right]^{-1},$$

where $p(\mathbf{x})$ is the prior distribution and $p(\mathbf{x}|\mathbf{y})$ is the posterior distribution. Figure 1.6 shows how the likelihood function, which differs for different values of the sensor measurement, affects the prior to produce a posterior distribution. $\mathbf{x}_0 = 23$, $\boldsymbol{\Sigma}_0 = 4$ and $\boldsymbol{\Sigma} = 3$ are assumed to be the parameter values in this example. In Figure 1.6(a), $\mathbf{y} = 4$ leads to an estimate $\hat{\mathbf{x}} = 10.84$ with a posterior variance $\hat{\boldsymbol{\Sigma}}^2 = 5.76$. As the value of $\mathbf{y}$ increases, the likelihood function drifts to the right, pulling with it the posterior distribution. Table 1.1 lists the estimates of $\mathbf{x}$ for different values of $\mathbf{y}$. This result is intuitive, and Bayes' theorem provides the equation that computes the posterior distribution.

Table 1.1 *Mean estimate of* **x***, with variance, for different measurements* **y***.*

Estimators for different data	y = 4	y = 15	y = 21	y = 30	y = 40	y = 45
$\hat{\mathbf{x}}$	10.84	17.88	21.72	27.48	33.88	37.08
$\hat{\mathbf{\Sigma}}^2$	5.76	5.76	5.76	5.76	5.76	5.76

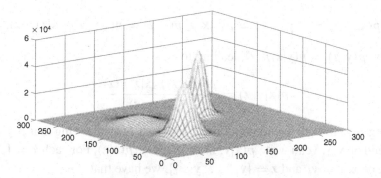

Figure 1.7 Illustration of the prior and posterior.

Figure 1.7 illustrates an example of a prior distribution in two dimensional space with two successive posterior distributions, in the case of a recursive estimation solution. The two dimensions considered are the location coordinates of the object on a plane. The flatter curve is the prior distribution. It represents the prior knowledge about the vector of interest **x**. When the first set of data is received, the posterior distribution is computed via Bayes' theorem. In this example, it is sharper and more concentrated around the estimates than the prior distribution. Yet, it is not as sharp, or as focused, as the second posterior distribution. This second distribution is computed after more data are received.

In many industrial, engineering, economic and other settings, one repetitively receives measurements. Assume that at times $t_1, t_2, \ldots, t_k, \ldots$, we observe $\mathbf{y}_1, \mathbf{y}_2, \ldots, \mathbf{y}_k, \ldots$. For example, while testing a machine for its reliability, one observes the times of failures of the machine. Similarly when sampling for quality control, we obtain values for the number of defects in batches of products tested for quality. In object tracking, the **y** values are sensor measurements about the state of the object. Treating the data one value at a time, and each time treating the old posterior distribution as the new prior distribution, creates a recursive process:

- at time 0, before any measurement value arrives, $p(\mathbf{x})$ represents our knowledge of **x**;
- at time t_1, $p(\mathbf{x}|\mathbf{y}_1) \propto p(\mathbf{y}_1|\mathbf{x})p(\mathbf{x})$;

- at time t_2, $p(\mathbf{x}|\mathbf{y}_1, \mathbf{y}_2) \propto p(\mathbf{y}_2|\mathbf{y}_1, \mathbf{x})p(\mathbf{x}|\mathbf{y}_1)$;
- at time t_k, $p(\mathbf{x}|\mathbf{y}_1, \mathbf{y}_2, \dots, \mathbf{y}_{k-1}, \mathbf{y}_k) \propto p(\mathbf{y}_k|\mathbf{y}_{k-1}, \dots, \mathbf{y}_1, \mathbf{x})p(\mathbf{x}|\mathbf{y}_1, \dots, \mathbf{y}_{k-1})$.

Applying (1.1) in the case of three random events $\mathbf{x}$, $\mathbf{y}$ and $\mathbf{z}$,

$$p(\mathbf{x}|\mathbf{y}, \mathbf{z}) = \frac{p(\mathbf{x}, \mathbf{y}, \mathbf{z})}{p(\mathbf{y}, \mathbf{z})}.$$

Switching the order in which we write the events,

$$p(\mathbf{x}, \mathbf{y}, \mathbf{z}) = p(\mathbf{y}, \mathbf{x}, \mathbf{z}) = p(\mathbf{y}|\mathbf{x}, \mathbf{z})p(\mathbf{x}, \mathbf{z}) = p(\mathbf{y}|\mathbf{x}, \mathbf{z})p(\mathbf{x}|\mathbf{z})p(\mathbf{z}).$$

Similarly, $p(\mathbf{y}, \mathbf{z}) = p(\mathbf{y}|\mathbf{z})p(\mathbf{z})$, leading to

$$p(\mathbf{x}|\mathbf{y}, \mathbf{z}) = \frac{p(\mathbf{y}|\mathbf{x}, \mathbf{z})p(\mathbf{x}|\mathbf{z})p(\mathbf{z})}{p(\mathbf{y}|\mathbf{z})p(\mathbf{z})}.$$

Canceling $p(\mathbf{z})$ in both the numerator and denominator and noting that $p(\mathbf{y}|\mathbf{z})$ does not depend on $\mathbf{x}$, we can write $p(\mathbf{x}|\mathbf{y}, \mathbf{z}) \propto p(\mathbf{y}|\mathbf{x}, \mathbf{z})p(\mathbf{x}|\mathbf{z})$. For each $k = 1, 2, \dots$, letting $\mathbf{x}$ be $\mathbf{x}$, $\mathbf{y} = \mathbf{y}_k$ and $\mathbf{z} = \{\mathbf{y}_1, \dots, \mathbf{y}_{k-1}\}$, we have that

$$p(\mathbf{x}|\mathbf{y}_1, \mathbf{y}_2, \dots, \mathbf{y}_{k-1}, \mathbf{y}_k) \propto p(\mathbf{y}_k|\mathbf{y}_{k-1}, \dots, \mathbf{y}_1, \mathbf{x})p(\mathbf{x}|\mathbf{y}_1, \dots, \mathbf{y}_{k-1}),$$

or rearranging terms,

$$p(\mathbf{x}|\mathbf{y}_k, \{\mathbf{y}_1, \dots, \mathbf{y}_{k-1}\}) \propto p(\mathbf{y}_k|\mathbf{x}, \{\mathbf{y}_1, \dots, \mathbf{y}_{k-1}\})p(\mathbf{x}|\{\mathbf{y}_1, \dots, \mathbf{y}_{k-1}\}).$$

A recursive Bayesian solution is applied by taking the posterior distribution $p(\mathbf{x}|\mathbf{y}_1, \dots, \mathbf{y}_{k-1})$ at time t_{k-1} and using it as a prior distribution to compute the posterior distribution at time t_k, $p(\mathbf{x}|\mathbf{y}_1, \dots, \mathbf{y}_{k-1}, \mathbf{y}_k)$. It is this recursive approach that is used in the remainder of this book to derive tracking algorithms. In object tracking, $\mathbf{x}$ changes with time, and therefore another step is introduced to derive the posterior distribution. Nevertheless, it is this idea of using a posterior at one stage to derive the posterior at the next stage that is used in the general Bayesian solution to object tracking. The optimal Bayesian solution for object tracking in such recursive form will be derived in the next section.

1.2.2 Application to object tracking

In object tracking, the complete probabilistic knowledge of the object states is described by the joint pdf $p(\mathbf{S}^k) = p(\mathbf{S}_k, \mathbf{S}_{k-1}, \dots, \mathbf{S}_0)$, where $\mathbf{S}_k$ denotes a generic object state that represents the single- or multiple-object states, or the number of objects, or the identity of an object/objects, or a combination of these at time t_k. Often, we will refer to time t_k as simply time k, or stage k. Let

$\mathbf{y}^k = (\mathbf{y}_1, \mathbf{y}_2, \ldots, \mathbf{y}_k)$, where $\mathbf{y}_i$ is the measurement at time i, be the sensor outputs. When a physical relationship governing $\mathbf{y}^k$ and the object states $\mathbf{S}^k$ is known and the knowledge of the object states is available in the form of a probability density function $p(\mathbf{S}^k)$, then Bayes' theorem provides a fundamentally sound method to update the knowledge of these states. Bayes' theorem (1.2) derives $p(\mathbf{S}^k|\mathbf{y}^k)$ as

$$p(\mathbf{S}^k|\mathbf{y}^k) = \frac{p(\mathbf{y}^k|\mathbf{S}^k)p(\mathbf{S}^k)}{p(\mathbf{y}^k)}. \tag{1.3}$$

The conditional probability distribution $p(\mathbf{S}^k|\mathbf{y}^k)$ is the posterior distribution of the object state. The term $p(\mathbf{y}^k|\mathbf{S}^k)$ is the likelihood function, which is the probability that the observed sequence of measurements is $\mathbf{y}^k$ given that the underlying state sequence is $\mathbf{S}^k$. The second term in the numerator $p(\mathbf{S}^k)$ is the prior distribution of $\mathbf{S}^k$. The denominator $p(\mathbf{y}^k)$ is the normalization factor that ensures that the resultant probability distribution $p(\mathbf{S}^k|\mathbf{y}^k)$ satisfies the axioms of probability and sums up to 1. The normalization factor can be computed using the following integral equation:

$$p(\mathbf{y}^k) = \int_{\mathbf{S}_k} \cdots \int_{\mathbf{S}_0} p(\mathbf{y}^k|\mathbf{S}_k, \ldots, \mathbf{S}_0)p(\mathbf{S}_k, \ldots, \mathbf{S}_0)d\mathbf{S}_k \cdots d\mathbf{S}_0.$$

We are using $\int$ to represent a generic integral which has all the operations of summation and conventional integrals as appropriate. For example, if $\mathbf{S}_k$ is a hybrid state consisting of a continuous state $\mathbf{x}_k = \{x_1, \ldots, x_i\}$ and a discrete state $\mathbf{r}_k = \{r_1, \ldots, r_j\}$, $\mathbf{S}_k = \{x_1, \ldots, x_i, r_1, \ldots, r_j\}$, the integral $\int$ is meant as

$$\int_{\mathbf{S}_k} (.) \, d\mathbf{S}_k = \int_{x_1} \int_{x_2} \cdots \int_{x_i} \sum_{r_1} \cdots \sum_{r_j} (.) \, dx_1 dx_2 \cdots dx_i.$$

The application of Bayes' theorem in the context of object tracking leads to a process where the updated knowledge of the states of the objects can be obtained by multiplying the prior joint density with the likelihood function and re-normalizing it with the normalizing factor. In theory (1.3) provides a solution for computing the distribution of $\mathbf{S}^k$ based on the knowledge of the sequence of measurements $\mathbf{y}^k$. However, the measurements are received in a sequence over time. At each stage of the measurement process, the distribution of the object state can be updated. This is done to ensure that all information is used as soon as received and that the most accurate estimations and predictions are made in the object tracking problem. The Bayesian solution of (1.3) can be further expanded to provide a recursive solution. In it, the conditional distribution $p(\mathbf{S}^k|\mathbf{y}^k)$ is obtained at time k, after receiving the last measurement $\mathbf{y}_k$, by updating the conditional distribution of the object state $p(\mathbf{S}^{k-1}|\mathbf{y}^{k-1})$ assessed at time $k - 1$. Recall that

$\mathbf{y}^k = (\mathbf{y}_1, \mathbf{y}_2, \ldots, \mathbf{y}_k)$, where $\mathbf{y}_i$ is the measurement at time i. It can be written as $\mathbf{y}^k = (\mathbf{y}_k, \mathbf{y}^{k-1})$. Using (1.1) we can rewrite the terms in the solution equation (1.3) as

$$p(\mathbf{y}^k|\mathbf{S}^k) = p(\mathbf{y}_k, \mathbf{y}^{k-1}|\mathbf{S}^k) = p(\mathbf{y}_k|\mathbf{y}^{k-1}, \mathbf{S}^k)p(\mathbf{y}^{k-1}|\mathbf{S}^k),$$

$$p(\mathbf{S}^k) = p(\mathbf{S}_k, \mathbf{S}^{k-1}) = p(\mathbf{S}_k|\mathbf{S}^{k-1})p(\mathbf{S}^{k-1}),$$

$$p(\mathbf{y}^k) = p(\mathbf{y}_k, \mathbf{y}^{k-1}) = p(\mathbf{y}_k|\mathbf{y}^{k-1})p(\mathbf{y}^{k-1}).$$

From the *causality principle*, it is clear that the measurements at time $k - 1$ do not depend on object states at times $\geq k$. Thus $p(\mathbf{y}^{k-1}|\mathbf{S}^k) = p(\mathbf{y}^{k-1}|\mathbf{S}^{k-1})$. This leads to a simplification of the likelihood function:

$$p(\mathbf{y}^k|\mathbf{S}^k) = p(\mathbf{y}_k|\mathbf{y}^{k-1}, \mathbf{S}^k)p(\mathbf{y}^{k-1}|\mathbf{S}^{k-1}).$$

Substituting into the Bayes' rule of (1.3) gives

$$p(\mathbf{S}^k|\mathbf{y}^k) = \frac{p(\mathbf{y}_k|\mathbf{y}^{k-1}, \mathbf{S}^k)p(\mathbf{y}^{k-1}|\mathbf{S}^{k-1})p(\mathbf{S}_k|\mathbf{S}^{k-1})p(\mathbf{S}^{k-1})}{p(\mathbf{y}_k|\mathbf{y}^{k-1})p(\mathbf{y}^{k-1})}.$$

Regrouping the right-hand side of the above equation, the Bayes' solution can be written as

$$p(\mathbf{S}^k|\mathbf{y}^k) = \frac{p(\mathbf{y}_k|\mathbf{y}^{k-1}, \mathbf{S}^k)p(\mathbf{S}_k|\mathbf{S}^{k-1})}{p(\mathbf{y}_k|\mathbf{y}^{k-1})} \left\{ \frac{p(\mathbf{y}^{k-1}|\mathbf{S}^{k-1})p(\mathbf{S}^{k-1})}{p(\mathbf{y}^{k-1})} \right\}.$$

Recognizing that the term in the braces is the prior joint pdf of the object state, i.e.,

$$\frac{p(\mathbf{y}^{k-1}|\mathbf{S}^{k-1})p(\mathbf{S}^{k-1})}{p(\mathbf{y}^{k-1})} = p(\mathbf{S}^{k-1}|\mathbf{y}^{k-1}),$$

the recursive form of the Bayesian solution is

$$p(\mathbf{S}^k|\mathbf{y}^k) = \frac{p(\mathbf{y}_k|\mathbf{y}^{k-1}, \mathbf{S}^k)p(\mathbf{S}_k|\mathbf{S}^{k-1})}{p(\mathbf{y}_k|\mathbf{y}^{k-1})} p(\mathbf{S}^{k-1}|\mathbf{y}^{k-1}).$$

Under the assumption that the measurements at a given time depend only on the object states at the corresponding time and are conditionally independent of measurements taken at other times, i.e., the measurements at time k are independent of the measurements at times $\leq k - 1$, and that they depend only on the current states of objects via $\mathbf{S}_k$ and not on its entire state sequence, the measurement likelihood $p(\mathbf{y}_k|\mathbf{y}^{k-1}, \mathbf{S}^k)$ simplifies to $p(\mathbf{y}_k|\mathbf{S}_k)$.

In object tracking, it is usual to make simplifying assumptions in order to model adequately the problem at hand in a probabilistic framework. Many real-world systems obey the **Markov** property where the present state does not depend on previous states, given the last state. In object tracking this translates to S_k depending only on S_{k-1} and not on $(S_{k-2}, \ldots, S_0)$. Therefore, $p(S_k|S^{k-1}) = p(S_k|S_{k-1})$, and we obtain the recursive form of the Bayesian solution for the object tracking problem:

Recursive Bayesian solution

$$p(S^k|y^k) = \frac{p(y_k|S_k)}{p(y_k|y^{k-1})} p(S_k|S_{k-1}) p(S^{k-1}|y^{k-1}). \qquad (1.4)$$

The recursive Bayesian solution (1.4) provides the posterior conditional distribution $p(S^k|y^k)$ at time k, after receiving the last measurement y_k. In object tracking, it is important to know the sequence of objects and their states and also to know the number of objects and their states at a particular time k. This knowledge can be derived from $p(S^k|y^k)$ by integrating out objects and their states as follows:

$$p(S_k|y^k) = \int_{S_{k-1}} \cdots \int_{S_0} p(S^k|y^k) dS_0 \cdots dS_{k-1}$$

$$= \int_{S_{k-1}} \cdots \int_{S_0} p(S_k, S_{k-1}, \ldots, S_0|y^k) dS_0 \cdots dS_{k-1}.$$

Using (1.4), the above equation becomes

$$p(S_k|y^k) = \frac{p(y_k|S_k)}{p(y_k|y^{k-1})} \int_{S_{k-1}} \cdots \int_{S_0} p(S_k|S_{k-1}) p(S^{k-1}|y^{k-1}) dS_0 \cdots dS_{k-1}$$

$$= \frac{p(y_k|S_k)}{p(y_k|y^{k-1})} \int_{S_{k-1}} p(S_k|S_{k-1}) \int_{S_{k-2}} \cdots \int_{S_0} p(S^{k-1}|y^{k-1}) dS_0 \cdots dS_{k-1}.$$

The inner integrals $\int_{S_{k-2}} \cdots \int_{S_0} p(S^{k-1}|y^{k-1}) dS_0 \cdots dS_{k-2}$ simplify to $p(S_{k-1}|y^{k-1})$ since $p(S^{k-1}|y^{k-1}) = p(S_{k-1}, S^{k-2}|y^{k-1})$, leading to the posterior pdf of the object state S_k conditioned on the available measurement data y^k:

The state conditional density

$$p(S_k|y^k) = \frac{1}{p(y_k|y^{k-1})} p(y_k|S_k) \int_{S_{k-1}} p(S_k|S_{k-1}) p(S_{k-1}|y^{k-1}) dS_{k-1}. \qquad (1.5)$$

It is this conditional density that most object tracking algorithms compute or seek to approximate. The integral $\int_{\mathbf{S}_{k-1}} p(\mathbf{S}_k|\mathbf{S}_{k-1}) p(\mathbf{S}_{k-1}|\mathbf{y}^{k-1}) d\mathbf{S}_{k-1}$ is the **Chapman–Kolmogorov** equation. The solution of this integral gives the predicted state of the elements of $\mathbf{S}_k$, given all the measurements up to time $k-1$ and the state at time $k-1$. Upon receipt of measurement $\mathbf{y}_k$ at time k, the predicted state is corrected by likelihood factor $p(\mathbf{y}_k|\mathbf{S}_k)$ and re-normalized. Solving the recursive relation in (1.5) is at the core of solving object tracking problems. Different object tracking problems differ in their problem requirements and manifest in different forms of the likelihood function $p(\mathbf{y}_k|\mathbf{S}_k)$ and the transition density $p(\mathbf{S}_k|\mathbf{S}_{k-1})$, and pose different challenges in solving this recursion.

1.3 Recursive Bayesian solution for object tracking

1.3.1 The generalized object dynamics equation

Before applying Bayes' theorem, a significant effort must be spent on modeling the dynamics of the processes which describe the problem. In object tracking this corresponds to developing object dynamical models and sensor measurement models.

Object motion is usually described by characteristics such as position, velocity, acceleration and other kinematic components, like angular velocity and its rate of change. These kinematic components constitute the object state. Using simple definitions like the rate of change of position is equal to velocity, it is possible to relate one component of object state to the other. Such models can be elegantly derived using the state space approach in either discrete or continuous time. The ideas and concepts in this book all refer to discrete-time. For example, in the case of the single-object tracking problem, we may consider the state $\mathbf{S}_k$ to represent only the position $\mathbf{x}_k$ of a single object. In the state-space approach, the components of the state are grouped as a single vector called the state vector $\mathbf{x}_k$. An example state vector of the object is

$$\mathbf{x}_k = [\text{position, velocity, acceleration}]_k^T.$$

The dimension of the object state is equal to the number of components of the vector and in this case it is 3. Another example of the object state vector is

$$\mathbf{x}_k = [\text{position, velocity, acceleration, angular velocity, angular accelaration}]_k^T,$$

where the dimension of the state vector is 5. Thus the object state is a finite-dimensional vector of those kinematic components of the object motion that explain its motion.

Many physical processes, such as the motion of an object, which is subject to random disturbances, and whose state can be represented by a finite-dimensional vector, can be modeled using a vector difference equation. Suppose $\mathbf{S}_k$ is an n-dimensional state vector at time k and $\mathbf{v}_k$ is the m-dimensional vector of random disturbances ($m \leq n$). We can write a difference equation

$$\mathbf{S}_k = \mathbf{g}(\mathbf{S}_{k-1}, \mathbf{v}_k), \tag{1.6}$$

where $\mathbf{g}$ is a real, in general non-linear, n-vector function, which, we suppose, is a twice continuously differentiable function of its arguments. The disturbance $\mathbf{v}_k$ is often called the random noise input to the system.

1.3.2 The generalized sensor measurement equation

Sensors are devices which observe aspects of the object state and the measurements made by sensors are used to make inferences about the entire object state. Measurement or sensor models enable one to determine the likelihood function $p(\mathbf{y}_k|\mathbf{S}_k, \mathbf{y}^{k-1})$. Hence one of the primary properties of the sensor model is that it must be a function of the object state. Moreover, as presented earlier, the Bayes' recursion is based on another sensor measurement characteristic, i.e., the measurements originated from the object state at a particular time are conditionally independent of measurements from other times. Sensor models of the form

$$\mathbf{y}_k = \mathbf{l}(\mathbf{S}_k, \mathbf{w}_k), \tag{1.7}$$

where $\mathbf{w}_k$ is a white noise (measurement error), satisfy the conditions needed for a recursive Bayesian solution.

1.3.3 Generalized object state prediction and conditional densities

In the case where the state and measurement equations do satisfy (1.6) and (1.7), and under certain regularity conditions the solution (1.5) can be further expanded to obtain the object state prediction and conditional densities. The state prediction density is $p(\mathbf{S}_k|\mathbf{y}^{k-1})$ at time k. It represents knowledge about the state $\mathbf{S}_k$ given all measurements up to time $k - 1$. The state conditional density, or posterior pdf of $\mathbf{S}_k$, is $p(\mathbf{S}_k|y^k)$, and represents the updated knowledge at time k after receiving the measurement $\mathbf{y}_k$. In complex tracking problems the state dynamics equation may not be captured in an equation of the form of (1.6), or it may not satisfy some regularity conditions. Similarly, the measurement equation in some tracking problems may not be in the form of (1.7) with the regularity conditions met. Even in these cases, the solution to the object tracking problem is derived from the recursive solutions (1.4) and (1.5). The derivation of the solution differs for each problem.

1.3.4 Generalized object state prediction and update

The generalized object state transition density

Using the object dynamics model of (1.6) we derive the transition density $p(\mathbf{S}_k|\mathbf{S}_{k-1})$ needed in the Bayes' solution (1.5). If the random input is absent from (1.6), we have an ordinary difference equation and we speak of $\mathbf{S}_k$ as its solution. However, in the presence of the random input term, we refer to (1.6) as a stochastic difference equation (SDE). In using the SDE, it is the pdf of $\mathbf{S}_k$ which is of interest. If the probability law of $\mathbf{v}_k$ is arbitrary, little can be said about the dynamic system (1.6). However, by modeling $\mathbf{v}_k$ as a white noise sequence, we can make the following observation. Since $\mathbf{S}_{k-1}$, $\mathbf{S}_k$ depends only on $\mathbf{v}_k$, which is independent of $S_{k-2}, \ldots, S_0$, the solution of (1.6) is a Markov sequence. It is well known that the Markov sequence is defined using its transition density and an initial condition for its states. In the context of the SDE in (1.6), the transition density is $p(\mathbf{S}_k|\mathbf{S}_{k-1})$ and the initial condition is $p(\mathbf{S}_{k-1})$.

Assume that the SDE in (1.6) can be solved for $\mathbf{v}_k$. In other words, for a given $\mathbf{S}_{k-1}$, $\mathbf{g}(\mathbf{S}_{k-1}, \cdot)$ has an inverse $\mathbf{g}^{-1}$ which is continuously differentiable. Then given $\mathbf{S}_{k-1}$, the pdf of $\mathbf{S}_k$, the transition density $p(\mathbf{S}_k|\mathbf{S}_{k-1})$ is

$$p(\mathbf{S}_k|\mathbf{S}_{k-1}) = p_{\mathbf{v}_k}(\mathbf{g}^{-1}(\mathbf{S}_k, \mathbf{S}_{k-1})) \left| \frac{\partial \mathbf{g}^{-1}}{\partial \mathbf{S}_k} \right|. \tag{1.8}$$

Here it is assumed that $\mathbf{g}^{-1}(\cdot)$ exists (i.e., it is assumed $m = n$). If $m < n$, the above direct derivation of the transition density is not possible. However, by partitioning $\mathbf{S}_k$ into

$$\mathbf{S}_k^T = [\mathbf{S}_k^{(1)T} \quad \mathbf{S}_k^{(2)T}],$$

and with the aid of the implicit function theorem (Kudryavtsev, 2001), (1.6) can be put in the form of $\mathbf{S}_k^{(1)} = \mathbf{g}^{(1)}(\mathbf{S}_{k-1}, \mathbf{w}_k)$ and $\mathbf{S}_k^{(2)} = \mathbf{g}^{(2)}(\mathbf{S}_{k-1}, \mathbf{S}_k^{(1)})$ and the transition density can be derived as follows:

$$p(\mathbf{S}_k|\mathbf{S}_{k-1}) = p(\mathbf{S}_k^{(1)}, \mathbf{S}_k^{(2)}|\mathbf{S}_{k-1})$$
$$= p(\mathbf{S}_k^{(2)}|\mathbf{S}_k^{(1)}, \mathbf{S}_{k-1}) p(\mathbf{S}_k^{(1)}|\mathbf{S}_{k-1}).$$

Since $\mathbf{S}_k^{(2)}$ depends only on $\mathbf{S}_k^{(1)}$ and $\mathbf{S}_k$, the first term in the above equation is given by the following dirac delta function:

$$p(\mathbf{S}_k^{(2)}|\mathbf{S}_k^{(1)}, \mathbf{S}_{k-1}) = \delta(\mathbf{S}_k^{(2)} - \mathbf{g}^{(2)}(\mathbf{S}_{k-1}, \mathbf{S}_k^{(1)})).$$

The second term depends on $\mathbf{w}_k$ and is given by

$$p(\mathbf{S}_k^{(1)}|\mathbf{S}_{k-1}) = p_{\mathbf{v}_k}(\mathbf{g}^{(1)^{-1}}(\mathbf{S}_k, \mathbf{S}_{k-1})) \left| \frac{\partial \mathbf{g}^{(1)^{-1}}}{\partial \mathbf{S}_k^{(1)}} \right|.$$

Thus the transition density when $m < n$ is given by

$$p(\mathbf{S}_k|\mathbf{S}_{k-1}) = \delta(\mathbf{S}_k^{(2)} - \mathbf{g}^{(2)}(\mathbf{S}_{k-1}, \mathbf{S}_k^{(1)})) p_{\mathbf{v}_k}(\mathbf{g}^{(1)^{-1}}(\mathbf{S}_k, \mathbf{S}_{k-1})) \left| \frac{\partial \mathbf{g}^{(1)^{-1}}}{\partial \mathbf{S}_k^{(1)}} \right|. \quad (1.9)$$

The generalized object state prediction density

Substituting the state transition density $p(\mathbf{S}_k|\mathbf{S}_{k-1})$ from either (1.8) or (1.9), we obtain the state prediction density

$$p(\mathbf{S}_k|\mathbf{y}^{k-1}) = \int_{\mathbf{S}_{k-1}} p(\mathbf{S}_k|\mathbf{S}_{k-1}) p(\mathbf{S}_{k-1}|\mathbf{y}^{k-1}) d\mathbf{S}_{k-1},$$

where $p(\mathbf{S}_{k-1}|\mathbf{y}^{k-1})$ is the prior pdf of the object state, or the state conditional density at time $k - 1$. It is assumed known at time $k - 1$, having been derived recursively, based on all measurements $\mathbf{y}^{k-1}$ up to time $k - 1$.

1.3.5 Generalized object state filtering

The likelihood function

The sensor model (1.7) satisfies the two measurement related assumptions required by (1.5). The whiteness property of $\mathbf{w}_k$ is what gives the measurements, $\mathbf{y}_k$, the property of conditional independence and the fact that $\mathbf{l}(\cdot)$ is a function of $\mathbf{S}_k$ enforces the property that the measurements at time k depend on the object state at time k. To infer about $\mathbf{S}_k$ using $\mathbf{y}_k$, the function $\mathbf{l}(\cdot)$ must be invertible. Under these assumptions the likelihood function in (1.5) can be derived by treating (1.7) as a transformation of the random variable $\mathbf{w}_k$:

$$p(\mathbf{y}_k|\mathbf{S}_k) = p_{\mathbf{w}_k}(\mathbf{l}^{-1}(\mathbf{y}_k, \mathbf{S}_k)) \left| \frac{\partial \mathbf{l}^{-1}}{\partial \mathbf{y}_k} \right|. \quad (1.10)$$

The normalization factor

The normalization factor is given by:

$$p(\mathbf{y}_k|\mathbf{y}_{k-1}) = \int_{\mathbf{S}_k} p(\mathbf{y}_k|\mathbf{S}_k) p(\mathbf{S}_k|\mathbf{y}^{k-1}) d\mathbf{S}_k.$$

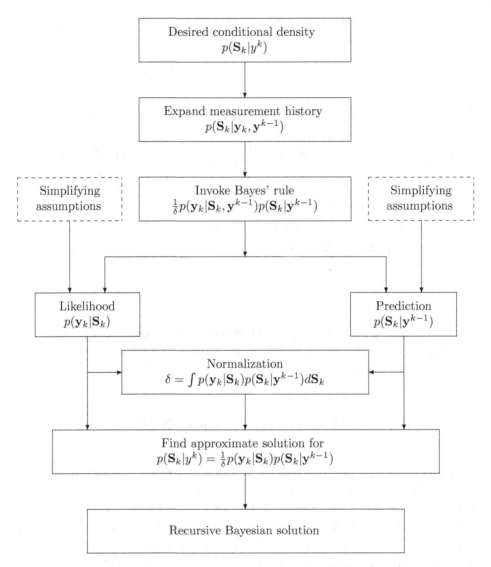

Summary of the recursive Bayesian framework for object tracking.

Substituting for the likelihood function $p(\mathbf{y}_k|\mathbf{S}_k)$ from (1.10) and for the predicted density $p(\mathbf{S}_k|\mathbf{y}^{k-1})$, the normalization factor becomes

$$p(\mathbf{y}_k|\mathbf{y}_{k-1}) = \int_{\mathbf{S}_k} p_{\mathbf{w}_k}(\mathbf{l}^{-1}(\mathbf{y}_k, \mathbf{S}_k)) \left| \frac{\partial \mathbf{l}^{-1}}{\partial \mathbf{y}_k} \right| p(\mathbf{S}_k|\mathbf{y}^{k-1}) d\mathbf{S}_k. \qquad (1.11)$$

The generalized object state conditional density

The conditional density of the general single-object tracking problem can be obtained by substituting the transition density of (1.8) or (1.9), the likelihood function of (1.10) and the normalizing factor of (1.11) into the recursive Bayes' formula in (1.5), and simplifying it.

1.3.6 Generalized object state estimates

In object tracking, the expected value $\hat{\mathbf{S}}_k$ of the object state $\mathbf{S}_k$ is often used as an estimate of the object state at time k. To assess the accuracy of this estimate, the covariance matrix $\mathbf{P}_{k|k}$ of $\mathbf{S}_k$ is computed. These are the first two moments of the conditional density (1.5) of the object state $\mathbf{S}_k$ at time k,

$$p(\mathbf{S}_k|\mathbf{y}^k) = \frac{1}{p(\mathbf{y}_k|\mathbf{y}^{k-1})} p(\mathbf{y}_k|\mathbf{S}_k) \int_{\mathbf{S}_{k-1}} p(\mathbf{S}_k|\mathbf{S}_{k-1}) p(\mathbf{S}_{k-1}|\mathbf{y}^{k-1}) d\mathbf{S}_{k-1}.$$

$p(\mathbf{S}_k|\mathbf{y}^k)$ is obtained after all the terms have been computed in the above equation, using (1.8) or (1.9), (1.10) and (1.11). The track estimate is $\hat{\mathbf{S}}_k = E(\mathbf{S}_k|\mathbf{y}^k)$,

$$\hat{\mathbf{S}}_k = \int_{\mathbf{S}_k} \mathbf{S}_k \, p(\mathbf{S}_k|\mathbf{y}^k) d\mathbf{S}_k.$$

The covariance matrix is $\mathbf{P}_{k|k} = E((\mathbf{S}_k - \hat{\mathbf{S}}_k)(\mathbf{S}_k - \hat{\mathbf{S}}_k)^T|\mathbf{y}^k)$,

$$\mathbf{P}_{k|k} = \int_{\mathbf{S}_k} (\mathbf{S}_k - \hat{\mathbf{S}}_k)(\mathbf{S}_k - \hat{\mathbf{S}}_k)^T p(\mathbf{S}_k|\mathbf{y}^k) d\mathbf{S}_k.$$

1.4 Summary

This chapter introduced the various important areas where object tracking forms the core part. It has discussed the application of Bayesian reasoning in object tracking problems. The recursive form of the Bayesian formulation is presented as a useful format that can be applied to any object tracking problem and will be used in later chapters.

2

Filtering theory and non-maneuvering object tracking

In this chapter we introduce and derive several well-known Bayesian algorithms for recursive state estimation. These algorithms, known as filters, result from the general Bayesian recursive equation (1.5) presented in Chapter 1. Well-known filters like the Kalman filter (KF), the extended Kalman filter (EKF), the unscented Kalman filter (UKF), the point mass filter (PMF) and the particle filter (PF) are developed as approximations to the optimal Bayesian filter under different modeling constraints. These filtering algorithms can also be viewed as algorithms for the single non-maneuvering object tracking problem.

2.1 The optimal Bayesian filter

The goal of all Bayesian estimation problems is to compute the posterior density of the parameter of interest. The posterior density encapsulates the information contained in the measurements and can be used to compute optimal parameter estimates. In object tracking the parameter of interest is the dynamic random variable referred to as the object state. The optimal Bayesian filter is a procedure for computing the posterior density of the object state in a recursive manner. Estimation problems in the Bayesian framework are defined by the prior density and the likelihood function of the parameter of interest. In the context of object tracking the prior density is determined by the object dynamic equation and the likelihood can be found from the measurement equation.

2.1.1 Object dynamics and sensor measurement equations

Object dynamics equation

Let $\mathbf{x}_k \in \mathbb{R}^{n_x}$ denote the object state at time k. The object dynamics are often adequately modeled using a stochastic difference equation of the form

$$\mathbf{x}_k = \mathbf{g}(\mathbf{x}_{k-1}, \mathbf{v}_k), \tag{2.1}$$

where $\mathbf{g} : \mathbb{R}^{n_x} \times \mathbb{R}^{n_v} \to \mathbb{R}^{n_x}$ is assumed to be a twice continuously differentiable function of its arguments. The disturbance $\mathbf{v}_k$ is a random noise input to the system which is used to account for modeling errors. The stochastic difference equation (2.1) defines the prior density for the object state.

Sensor measurement equation

Let $\mathbf{y}_k \in \mathbb{R}^{n_y}$ denote the observed measurement at time k. Most of the sensors used for object tracking can be adequately described using sensor models of the form

$$\mathbf{y}_k = \mathbf{l}(\mathbf{x}_k, \mathbf{w}_k), \tag{2.2}$$

where $\mathbf{l} : \mathbb{R}^{n_x} \times \mathbb{R}^{n_w} \to \mathbb{R}^{n_y}$ is assumed to be a twice continuously differentiable function of its arguments. The random variable $\mathbf{w}_k$ models the measurement error. The sensor measurement equation (2.2) establishes the quantitative relationship between the object state $\mathbf{x}_k$ and the observation $\mathbf{y}_k$. The likelihood function $p(\mathbf{y}_k|\mathbf{x}_k)$ is derived from the sensor measurement equation.

When the functions $\mathbf{g}$ and $\mathbf{l}$ of (2.1) and (2.2) are linear and the random variables $\mathbf{v}_k$ and $\mathbf{w}_k$ are Gaussian, the posterior density of $\mathbf{x}_k$ is Gaussian and can be found using the Kalman filter. More generally, approximations or numerical techniques are required.

Prediction and filtering

The transition density function $p(\mathbf{x}_k|\mathbf{x}_{k-1})$ derived from the object dynamics equation, along with the likelihood function $p(\mathbf{y}_k|\mathbf{x}_k)$ derived from the sensor measurement equation, are used for the recursive estimation of the conditional density function as follows:

$$p(\mathbf{x}_k|\mathbf{y}^k) = \frac{1}{p(\mathbf{y}_k|\mathbf{y}^{k-1})} p(\mathbf{y}_k|\mathbf{x}_k) \int_{\mathbf{x}_{k-1}} p(\mathbf{x}_k|\mathbf{x}_{k-1}) p(\mathbf{x}_{k-1}|\mathbf{y}^{k-1}) \, d\mathbf{x}_{k-1}, \tag{2.3}$$

where:

- $p(\mathbf{x}_{k-1}|\mathbf{y}^{k-1})$ is the posterior density at time $k-1$;
- $p(\mathbf{y}_k|\mathbf{y}^{k-1})$ is the normalization factor.

At time $k-1$, after processing measurement set $\mathbf{y}_{k-1}$, $p(\mathbf{x}_{k-1}|\mathbf{y}^{k-1})$ represents the probabilistic knowledge about the object state. It is referred to as the *prior probability density function* or *prior distribution* in the Bayesian recursive solution. After $\mathbf{y}_k$ is observed, the conditional density $p(\mathbf{x}_k|\mathbf{y}^k)$ of the object state at time k given all the observed sensor measurements $\mathbf{y}^k = (\mathbf{y}_1, \ldots, \mathbf{y}_k)$, is obtained using 2.3. It is known as the *posterior probability density function* of the object state. We also refer to it at times as the *conditional distribution* or the *filtering distribution*. The recursion (2.3) computes the posterior density of the object state in two stages:

1. The *prediction step* takes the conditional density at the previous stage $p(\mathbf{x}_{k-1}|\mathbf{y}^{k-1})$ through the transition density to form the predicted density via the Chapman–Kolmogorov equation (CKE),

$$p(\mathbf{x}_k|\mathbf{y}^{k-1}) = \int p(\mathbf{x}_k|\mathbf{x}_{k-1})p(\mathbf{x}_{k-1}|\mathbf{y}^{k-1})\, d\mathbf{x}_{k-1}. \qquad (2.4)$$

The predicted density $p(\mathbf{x}_k|\mathbf{y}^{k-1})$ encapsulates the current knowledge about $\mathbf{x}_k$ gathered up to and including time $k - 1$, before incorporating the new information $\mathbf{y}_k$. The predicted density can be used to infer and predict in any problem involving $\mathbf{x}_k$. In object tracking, the problem of interest is the value of $\mathbf{x}_k$, and the predicted mean of $\mathbf{x}_k$ is obtained from $p(\mathbf{x}_k|\mathbf{y}^{k-1})$ as an estimate, along with the accuracy value for that estimate in the form of the covariance matrix of $\mathbf{x}_k$.

2. The *filtering step* uses the new data $\mathbf{y}_k$ through the likelihood function $p(\mathbf{y}_k|\mathbf{x}_k)$ to form the filtering distribution $p(\mathbf{x}_k|\mathbf{y}^k)$ in (2.3). The filtering distribution or posterior distribution of $\mathbf{x}_k$ contains all the information about $\mathbf{x}_k$ given all the received measurements $\mathbf{y}^k$. The posterior mean and covariance matrix of $\mathbf{x}_k$ can be computed from the filtering density.

2.1.2 The optimal non-maneuvering object tracking filter recursion

The transition density

Most object tracking algorithms generally impose an additive noise assumption on the object dynamics equation, yielding an equation of the form

$$\mathbf{x}_k = \mathbf{f}(\mathbf{x}_{k-1}) + \mathbf{v}_k. \qquad (2.5)$$

Under such an additive noise assumption, the inverse of the function $\mathbf{g}$ of the object dynamics equation (2.1) is $\mathbf{g}^{-1}(\mathbf{x}_k, \mathbf{x}_{k-1}) = \mathbf{x}_k - \mathbf{f}(\mathbf{x}_{k-1})$. From the basic result of transformation of random variables, the transition density function is

$$p(\mathbf{x}_k|\mathbf{x}_{k-1}) = p_{\mathbf{v}_k}(\mathbf{g}^{-1}(\mathbf{x}_k, \mathbf{x}_{k-1})) \left| \nabla_{\mathbf{x}_k}\mathbf{g}^{-1}(\mathbf{x}_k, \mathbf{x}_{k-1}) \right| = p_{\mathbf{v}_k}(\mathbf{x}_k - \mathbf{f}(\mathbf{x}_{k-1})). \qquad (2.6)$$

The likelihood function

Most object tracking algorithms impose an additive assumption on the measurement noise, yielding the measurement equation of the form

$$\mathbf{y}_k = \mathbf{h}(\mathbf{x}_k) + \mathbf{w}_k. \qquad (2.7)$$

Under such an additive noise assumption, the inverse of the function $\mathbf{l}$ of the measurement equation (2.2) is $\mathbf{l}^{-1}(\mathbf{y}_k, \mathbf{x}_k) = \mathbf{y}_k - \mathbf{h}(\mathbf{x}_k)$. From the basic result of transformation of random variables, the likelihood function is

$$p(\mathbf{y}_k|\mathbf{x}_k) = p_{\mathbf{w}_k}(\mathbf{l}^{-1}(\mathbf{y}_k, \mathbf{x}_k)) \left| \nabla_{\mathbf{y}_k}\mathbf{l}^{-1}(\mathbf{y}_k, \mathbf{x}_k) \right| = p_{\mathbf{w}_k}(\mathbf{y}_k - \mathbf{h}(\mathbf{x}_k)). \qquad (2.8)$$

The filter recursion

Substituting the transition density (2.6) and the likelihood function (2.8) in (2.3), the posterior density $p(\mathbf{x}_k|\mathbf{y}^k)$ of the object state (or filtering density) is

$$p(\mathbf{x}_k|\mathbf{y}^k) = \frac{p_{\mathbf{w}_k}(\mathbf{y}_k - \mathbf{h}(\mathbf{x}_k)) \int p_{\mathbf{v}_k}(\mathbf{x}_k - \mathbf{f}(\mathbf{x}_{k-1})) p(\mathbf{x}_{k-1}|\mathbf{y}^{k-1}) \, d\mathbf{x}_{k-1}}{\int p_{\mathbf{w}_k}(\mathbf{y}_k - \mathbf{h}(\mathbf{x}_k)) p(\mathbf{x}_k|\mathbf{y}^{k-1}) d\mathbf{x}_k}, \qquad (2.9)$$

where $p(\mathbf{x}_k|\mathbf{y}^{k-1}) = \int p_{\mathbf{v}_k}(\mathbf{x}_k - \mathbf{f}(\mathbf{x}_{k-1})) p(\mathbf{x}_{k-1}|\mathbf{y}^{k-1}) \, d\mathbf{x}_{k-1}$.

Equation (2.9) summarizes, in mathematical form, the Bayesian approach to non-maneuvering object tracking. This is the starting point for the various Bayesian tracking algorithms which have been proposed in the literature. Differences in these algorithms arise due to the different assumptions used in the evaluation of (2.9). Some of the commonly used assumptions include:

A1 The object dynamics and measurement equations are linear:

$$\mathbf{x}_k = \mathbf{F}\mathbf{x}_{k-1} + \mathbf{v}_k, \qquad (2.10)$$

$$\mathbf{y}_k = \mathbf{H}\mathbf{x}_k + \mathbf{w}_k. \qquad (2.11)$$

A2 $\mathbf{v}_k$ and $\mathbf{w}_k$ are white, uncorrelated, Gaussian noise sequences with zero mean and covariance $\mathbf{Q}_k$ and $\mathbf{R}_k$ respectively.

A3 The posterior density of the object state $p(\mathbf{x}_{k-1}|\mathbf{y}^{k-1})$ at time $k - 1$ is Gaussian with mean $\hat{\mathbf{x}}_{k-1|k-1}$ and covariance $\mathbf{P}_{k-1|k-1}$.

In the following sections these assumptions will be used as required to derive several of the more prominent methods for evaluating (2.9).

2.2 The Kalman filter

2.2.1 Derivation of the Kalman filter

Assumptions of the Kalman filter

The KF can be derived by assuming **A1**, **A2** and **A3**, i.e., the object dynamic and measurement equations are linear/Gaussian and the posterior density at time $k - 1$ is Gaussian.

Example 2.1 Consider an object moving in one dimension at a fairly constant velocity, without any maneuvering. Let x_k be its one-dimensional position, and $\dot{x}_k$ be its velocity, at time t_k, $k = 1, 2, \ldots$. The two-dimensional vector $\mathbf{x}_k = [x_k, \dot{x}_k]^T$ represents the state of the object. The object dynamics equation can be found in a

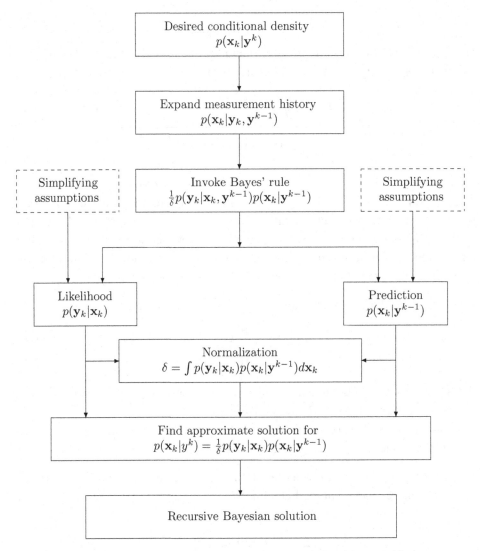

Summary of the recursive Bayesian framework for object tracking.

straightforward way. Assuming the velocity of the object motion is constant, the position x_k of the object at time t_k can be written in terms of the position and velocity at time t_{k-1}:

$$x_k = x_{k-1} + \dot{x}_{k-1} T,$$

where $T = t_k - t_{k-1}$ is the interval between measurements and is assumed to be the same for all k. Also, since the object velocity is assumed constant $\dot{x}_k = \dot{x}_{k-1}$. The assumption of a perfectly constant object velocity will rarely be met in practice. Departures from constant velocity motion can be allowed for by the introduction of additive process noise. We thus have the equation (2.10) for the object dynamics with

$$\mathbf{F} = \begin{bmatrix} 1 & T \\ 0 & 1 \end{bmatrix}.$$

The sensor provides measurements of the object position embedded in zero-mean additive Gaussian noise at times $t_k = 1, 2, \ldots$. The measurement equation is then given by (2.11) with

$$\mathbf{H} = \begin{bmatrix} 1 & 0 \end{bmatrix}.$$

This example introduces typical, yet simple, forms of the object dynamics and measurement equations. In most tracking problems, the dimensions of the object state and the measurement vectors are higher. For example, the state vector may consist of position, velocity and acceleration in three orthogonal directions leading to a nine-dimensional state vector.

Gaussian densities will appear regularly in the derivations so it will be convenient to use the notation

$$N(\mathbf{z}; \boldsymbol{\mu}, \boldsymbol{\Sigma}) = \frac{1}{\sqrt{|2\pi \boldsymbol{\Sigma}|}} \exp\left[-(\mathbf{z} - \boldsymbol{\mu})^T \boldsymbol{\Sigma}(\mathbf{z} - \boldsymbol{\mu})/2\right].$$

The Kalman filter will be derived using the following theorem.

Theorem 2.1 *(Gaussian product) For* $\mathbf{x}_1, \boldsymbol{\mu}_1 \in \mathbb{R}^{d_1}$, $\mathbf{H} \in \mathbb{R}^{d_2 \times d_1}$, $\mathbf{x}_2 \in \mathbb{R}^{d_2}$ *and positive definite matrices* $\mathbf{P}_1, \mathbf{P}_2$:

$$N(\mathbf{x}_2; \mathbf{Hx}_1, \mathbf{P}_2)N(\mathbf{x}_1; \boldsymbol{\mu}_1, \mathbf{P}_1) = N(\mathbf{x}_2; \mathbf{H}\boldsymbol{\mu}_1, \mathbf{P}_3)N(\mathbf{x}_1; \boldsymbol{\mu}, \mathbf{P}),$$

where

$$\mathbf{P}_3 = \mathbf{HP}_1\mathbf{H}^T + \mathbf{P}_2,$$

$$\boldsymbol{\mu} = \boldsymbol{\mu}_1 + \mathbf{K}(\mathbf{x}_2 - \mathbf{H}\boldsymbol{\mu}_1),$$

$$\mathbf{P} = \mathbf{P}_1 - \mathbf{KHP}_1,$$

with $\mathbf{K} = \mathbf{P}_1\mathbf{H}^T\mathbf{P}_3^{-1}$.

Proof The method of proving the above theorem is relatively well known, being first shown in Ho and Lee (1964) and later appearing in a number of texts. □

The transition and prediction densities

Using (2.6), and since $p_{\mathbf{v}_k}$ is a Gaussian density with zero mean and covariance $\mathbf{Q}_k$, the transition density is

$$p(\mathbf{x}_k | \mathbf{x}_{k-1}) = N(\mathbf{x}_k; \mathbf{Fx}_{k-1}, \mathbf{Q}_k).$$

By assumption, the posterior density of the object state at time $k-1$ is Gaussian so that $p(\mathbf{x}_{k-1}|\mathbf{y}^{k-1}) = N(\mathbf{x}_k; \hat{\mathbf{x}}_{k-1|k-1}, \mathbf{P}_{k-1|k-1})$. The predicted density is

$$p(\mathbf{x}_k|\mathbf{y}^{k-1}) = \int p(\mathbf{x}_k|\mathbf{x}_{k-1}) p(\mathbf{x}_{k-1}|\mathbf{y}^{k-1}) \, d\mathbf{x}_{k-1}$$

$$= \int N(\mathbf{x}_k; \mathbf{F}\mathbf{x}_{k-1|k-1}, \mathbf{Q}_k) N(\mathbf{x}_k; \hat{\mathbf{x}}_{k-1|k-1}, \mathbf{P}_{k-1|k-1}) \, d\mathbf{x}_{k-1}. \quad (2.12)$$

The predicted density can be obtained by using Theorem 2.1 to solve the above integral directly.[1] Applying Theorem 2.1 to (2.12) gives

$$p(\mathbf{x}_k|\mathbf{y}^{k-1}) = \int N(\mathbf{x}_k; \hat{\mathbf{x}}_{k|k-1}, \mathbf{P}_{k|k-1})$$

$$\times N(\mathbf{x}_{k-1}; \hat{\mathbf{x}}_{k-1|k-1} + \mathbf{G}_k(\mathbf{x}_k - \hat{\mathbf{x}}_{k|k-1}), \mathbf{M}_k)) \, d\mathbf{x}_{k-1}$$

$$= N(\mathbf{x}_k; \hat{\mathbf{x}}_{k|k-1}, \mathbf{P}_{k|k-1}), \quad (2.13)$$

where $\mathbf{G}_k = \mathbf{P}_{k-1|k-1}\mathbf{F}^T\mathbf{P}_{k|k-1}^{-1}$, $\mathbf{M}_k = \mathbf{P}_{k-1|k-1} - \mathbf{G}_k\mathbf{F}\mathbf{P}_{k-1|k-1}$ and

$$\hat{\mathbf{x}}_{k|k-1} = \mathbf{F}\hat{\mathbf{x}}_{k-1|k-1},$$

$$\mathbf{P}_{k|k-1} = \mathbf{F}\mathbf{P}_{k-1|k-1}\mathbf{F}^T + \mathbf{Q}_k.$$

Equation (2.13) shows that the prediction density is Gaussian if the posterior density at time $k-1$ is Gaussian and the object dynamic equation is linear/Gaussian.

This operation denotes the standard Kalman filter prediction, and its pseudo-function is

$$\left[\hat{\mathbf{x}}_{k|k-1}, \mathbf{P}_{k|k-1}\right] = \mathrm{KF_P}\left[\hat{\mathbf{x}}_{k-1|k-1}, \mathbf{P}_{k-1|k-1}, \mathbf{F}, \mathbf{Q}\right],$$

defined by

$$\hat{\mathbf{x}}_{k|k-1} = \mathbf{F}\hat{\mathbf{x}}_{k-1|k-1},$$

$$\mathbf{P}_{k|k-1} = \mathbf{F}\mathbf{P}_{k-1|k-1}\mathbf{F}^T + \mathbf{Q}.$$

The likelihood function and normalization factor

Using (2.8), and since $p_{\mathbf{w}_k}(\cdot)$ is a Gaussian density with zero mean and covariance $\mathbf{R}_k$, the likelihood function is $p(\mathbf{y}_k|\mathbf{x}_k) = N(\mathbf{y}_k; \mathbf{H}\mathbf{x}_k, \mathbf{R}_k)$. Using Theorem 2.1, the

---

[1] Most of the books that deal with Kalman filtering and tracking either use orthogonal projections in conjunction with the Graham–Schmidt orthogonalization process, as used by Kalman himself in his celebrated paper (Kalman, 1960), or use the principle of linear transformation and linear combination of random variables to show that the predicted density is a Gaussian density. The approach presented in this book is to solve the CKE directly. In the single-dimension case the solution of the CKE is straightforward. However, its solution in the multi-dimensional case is more involved. We have adopted the multi-dimensional CKE solution presented by Challa and Koks (2004) in the derivation of the KF.

normalization factor in the Bayes' recursion can be found as

$$
\begin{aligned}
p(\mathbf{y}_k | \mathbf{y}^{k-1}) &= \int p(\mathbf{y}_k | \mathbf{x}_k) p(\mathbf{x}_k | \mathbf{y}^{k-1}) \, d\mathbf{x}_k \\
&= \int N(\mathbf{y}_k; \mathbf{H}\mathbf{x}_k, \mathbf{R}_k) N(\mathbf{x}_k; \hat{\mathbf{x}}_{k|k-1}, \mathbf{P}_{k|k-1}) \, d\mathbf{x}_k \\
&= N(\mathbf{y}_k; \hat{\mathbf{y}}_{k|k-1}, \mathbf{S}_k),
\end{aligned}
$$

where $\hat{\mathbf{y}}_{k|k-1} = \mathbf{H}\hat{\mathbf{x}}_{k|k-1}$ and $\mathbf{S}_k = \mathbf{H}\mathbf{P}_{k|k-1}\mathbf{H}^T + \mathbf{R}_k$.

This operation returns the mean and covariance of the object measurement predicted pdf, and its pseudo-function is

$$
\left[\hat{\mathbf{y}}_{k|k-1}, \mathbf{S}_k \right] = \mathrm{MP} \left[\hat{\mathbf{x}}_{k|k-1}, \mathbf{P}_{k|k-1}, \mathbf{H}, \mathbf{R} \right].
$$

The measurement prediction is defined by

$$
\begin{aligned}
\hat{\mathbf{y}}_{k|k-1} &= \mathbf{H}\hat{\mathbf{x}}_{k|k-1}, \\
\mathbf{S}_k &= \mathbf{H}\mathbf{P}_{k|k-1}\mathbf{H}^T + \mathbf{R}.
\end{aligned}
$$

The conditional density

The posterior density $p(\mathbf{x}_k | \mathbf{y}^k)$ is obtained by substituting the predicted density, the likelihood function and the normalization factor in the Bayesian solution (2.3):

$$
p(\mathbf{x}_k | \mathbf{y}^k) = \frac{N(\mathbf{y}_k; \mathbf{H}\mathbf{x}_k, \mathbf{R}_k) N(\mathbf{x}_k; \hat{\mathbf{x}}_{k|k-1}, \mathbf{P}_{k|k-1})}{N(\mathbf{y}_k; \mathbf{H}\hat{\mathbf{x}}_{k|k-1}, \mathbf{S}_k)}.
$$

Using Theorem 2.1 once again gives

$$
p(\mathbf{x}_k | \mathbf{y}^k) = N(\mathbf{x}_k; \hat{\mathbf{x}}_{k|k}, \mathbf{P}_{k|k}), \tag{2.14}
$$

where

$$
\begin{aligned}
\hat{\mathbf{x}}_{k|k} &= \hat{\mathbf{x}}_{k|k-1} + \mathbf{P}_{k|k-1}\mathbf{H}^T \mathbf{S}_k^{-1} (\mathbf{y}_k - \hat{\mathbf{y}}_{k|k-1}), \\
\mathbf{P}_{k|k} &= \mathbf{P}_{k|k-1} - \mathbf{P}_{k|k-1}\mathbf{H}^T \mathbf{S}_k^{-1} \mathbf{H}\mathbf{P}_{k|k-1}.
\end{aligned}
$$

Thus, given a Gaussian prediction density and a linear/Gaussian measurement equation, the posterior density is also Gaussian. In Bayesian terminology, the Gaussian prediction density is said to be conjugate to the Gaussian likelihood.

Conditional density calculation is a standard procedure which applies the measurement $\mathbf{y}$ to the a priori probability density function represented by its mean $\hat{\mathbf{x}}_{k|k-1}$ and covariance $\mathbf{P}_{k|k-1}$, with the measurement matrix $\mathbf{H}$ and the measurement noise covariance matrix $\mathbf{R}$ being the parameters. The outputs are the mean $\hat{\mathbf{x}}_{k|k}$ and covariance $\mathbf{P}_{k|k}$ of the a posteriori probability density function. The

Kalman filter estimation pseudo-function is

$$\left[\hat{\mathbf{x}}_{k|k}, \mathbf{P}_{k|k}\right] = \text{KF}_\text{E}\left[\mathbf{y}, \hat{\mathbf{x}}_{k|k-1}, \mathbf{P}_{k|k-1}, \mathbf{H}, \mathbf{R}\right].$$

The Kalman filter estimation is defined by

$$\left[\hat{\mathbf{y}}_{k|k-1}, \mathbf{S}_k\right] = \text{MP}\left[\hat{\mathbf{x}}_{k|k-1}, \mathbf{P}_{k|k-1}, \mathbf{H}, \mathbf{R}\right],$$
$$\mathbf{K}_k = \mathbf{P}_{k|k-1}\mathbf{H}^T\mathbf{S}_k^{-1},$$
$$\hat{\mathbf{x}}_{k|k} = \hat{\mathbf{x}}_{k|k-1} + \mathbf{K}_k\left(\mathbf{y} - \hat{\mathbf{y}}_{k|k-1}\right),$$
$$\mathbf{P}_{k|k} = (\mathbf{I} - \mathbf{K}_k\mathbf{H})\mathbf{P}_{k|k-1},$$

where $\mathbf{I}$ denotes the identity matrix.

2.2.2 The Kalman filter equations

An important feature of the optimal Bayesian recursion under linear/Gaussian assumptions on the dynamic and measurement equations is that a Gaussian posterior density at time $k - 1$ will produce a Gaussian posterior density at time k. This is extremely convenient since it means that, given a Gaussian prior at time zero and assuming linear/Gaussian dynamic and measurement equations for all k, the posterior density at each k can be represented compactly and precisely by a mean and a covariance matrix. The method used to recursively compute the posterior mean and covariance matrix as measurements are acquired is known as the **Kalman filter**. The recursion is given by Algorithm 1.

Algorithm 1 Kalman filter equations

1: Compute the predicted mean and covariance matrix:

$$\hat{\mathbf{x}}_{k|k-1} = \mathbf{F}\hat{\mathbf{x}}_{k-1|k-1},$$
$$\mathbf{P}_{k|k-1} = \mathbf{F}\mathbf{P}_{k-1|k-1}\mathbf{F}^T + \mathbf{Q}_k.$$

2: Compute the predicted measurement, innovation covariance matrix and Kalman gain:

$$\hat{\mathbf{y}}_{k|k-1} = \mathbf{H}\mathbf{x}_{k|k-1},$$
$$\mathbf{S}_k = \mathbf{H}\mathbf{P}_{k|k-1}\mathbf{H}^T + \mathbf{R}_k,$$
$$\mathbf{K}_k = \mathbf{P}_{k|k-1}\mathbf{H}^T\mathbf{S}_k^{-1}.$$

3: Compute the posterior mean and covariance matrix:

$$\hat{\mathbf{x}}_{k|k} = \hat{\mathbf{x}}_{k|k-1} + \mathbf{K}_k(\mathbf{y}_k - \hat{\mathbf{y}}_{k|k-1}),$$
$$\mathbf{P}_{k|k} = \mathbf{P}_{k|k-1} - \mathbf{K}_k\mathbf{H}_k\mathbf{P}_{k|k-1}.$$

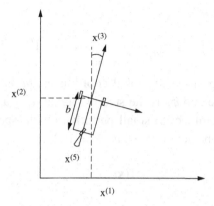

Figure 2.1 Planar model of the robot.

2.3 The extended Kalman filter

If any of the assumptions **A1–A3** do not hold, exact computation of the posterior density becomes impossible, with the exception of a few special cases (Daum, 1986). The object tracking problem then amounts to developing an accurate and computationally feasible approximation to the exact Bayesian recursion.

In many object tracking applications it is reasonable to assume Gaussian process and measurement noises. Further, for computational reasons it is desirable, and often surprisingly accurate, to adopt a Gaussian approximation to the posterior distribution since it can then be specified by a mean and a covariance matrix. However, the assumption of linear dynamic and measurement equations is often impossible. There has therefore been a strong focus on developing algorithms for use with non-linear dynamic and/or measurement equations. The most well known of these algorithms is the extended KF (EKF). In this section, the EKF will be derived as approximation to the exact Bayesian recursion under assumptions **A2** and **A3**.

Example 2.2: Robot navigation
Roumeliotis and Bekey (1997) compare the performance of an EKF to a dead-reckoning system for robot navigation. Dead-reckoning is a term used in robot navigation for the technique of localization of a robot or an autonomous vehicle. In the exploration of the planet Mars, NASA use a robot that performs scientific tasks, navigating in an autonomous manner to accomplish the tasks. The robot needs to know its exact position and orientation. To do so, the robot has on-board sensors that provide several measures. Roumeliotis and Bekey (1997) make the basic assumption that the robot moves on a plane. They assume that the vehicle moves at a constant speed and steering angle. The kinematic model used is that of a three-wheeled planar vehicle which has the steering at the rear wheel (see Figure 2.1).

The object (robot) state is

$$\mathbf{x}_k = \begin{bmatrix} x_k & y_k & \theta_k & s_k & \phi_k \end{bmatrix}^T,$$

where (x_k, y_k) is the position in global coordinates, θ_k is the orientation, s_k is the speed of the vehicle and ϕ_k is the steering angle. It is assumed that the robot moves with a velocity subject to small perturbations in speed and heading. The object motion can then be decribed by, for $k = 1, 2, \ldots,$

$$\mathbf{x}_k = \mathbf{f}(\mathbf{x}_{k-1}) + \mathbf{v}_k, \tag{2.15}$$

where

$$\mathbf{f}(\mathbf{x}_{k-1}) = \begin{bmatrix} x_{k-1} - T s_{k-1} \sin(\theta_{k-1} + \phi_{k-1}) \\ y_{k-1} + T s_{k-1} \cos(\theta_{k-1} + \phi_{k-1}) \\ \theta_{k-1} - T s_{k-1} \tan(\phi_{k-1})/b \\ s_{k-1} \\ \phi_{k-1} \end{bmatrix}, \tag{2.16}$$

where $T = t_k - t_{k-1}$ is the time interval between measurements, assumed constant for all k, and b is the distance between the wheel axes.

The measurements are provided by a sun sensor which gives the absolute orientation of the vehicle according to the sun position, two encoders mounted on the front wheels which measure the velocity of each wheel and a potentiometer mounted on the rear wheel which provides its steering angle. The measurement equation is

$$\mathbf{y}_k = \mathbf{H}\mathbf{x}_k + \mathbf{w}_k,$$

where $\mathbf{w}_k$ is the measurement error and

$$\mathbf{H} = \begin{bmatrix} 0 & 0 & 1 & 0 & 0 \\ 0 & 0 & 0 & 1 & 0 \\ 0 & 0 & 0 & 0 & 1 \end{bmatrix}.$$

2.3.1 Linear filter approximations

Linear approximation of the object dynamics equation

If the function $\mathbf{f}$ of the dynamic equation (2.5) is non-linear, the derivation of the prediction density via the Chapman–Kolmogorov equation is non-trivial, even if the process noise $\mathbf{v}_k$ is assumed to be Gaussian. An approximation in closed-form can be obtained by considering the Taylor series expansion of $\mathbf{f}(\mathbf{x}_{k-1})$ at $\hat{\mathbf{x}}_{k-1|k-1}$:

$$\mathbf{f}(\mathbf{x}_{k-1}) = \mathbf{f}(\hat{\mathbf{x}}_{k-1|k-1})$$

$$+ \sum_{m=1}^{\infty} \frac{1}{m!} \left(\bigotimes_{i=1}^{m} \nabla_{\mathbf{x}^T} \right) \mathbf{f}(\mathbf{x}) \Bigg|_{\mathbf{x}=\hat{\mathbf{x}}_{k-1|k-1}} \left(\bigotimes_{i=1}^{m} (\mathbf{x}_{k-1} - \hat{\mathbf{x}}_{k-1|k-1}) \right),$$

where, for a vector $\mathbf{b} = [b_1, \ldots, b_q]^T$, $\nabla_{\mathbf{b}} = [\partial/\partial b_1, \ldots, \partial/\partial b_q]^T$ and the symbol $\otimes$ denotes the Kronecker product. The EKF proceeds under the assumption that $\mathbf{f}(\mathbf{x}_{k-1})$ can be accurately approximated by only the zeroth- and first-order terms of the Taylor series expansion. Thus,

$$\mathbf{f}(\mathbf{x}_{k-1}) \approx \hat{\mathbf{f}}(\mathbf{x}_{k-1}) = \mathbf{f}(\hat{\mathbf{x}}_{k-1|k-1}) + \mathbf{F}_k(\mathbf{x}_{k-1} - \hat{\mathbf{x}}_{k-1|k-1}), \qquad (2.17)$$

where

$$\mathbf{F}_k = \nabla_{\mathbf{x}^T} \mathbf{f}(\mathbf{x})|_{\mathbf{x}=\hat{\mathbf{x}}_{k-1|k-1}}.$$

Linear approximation of the measurement equation

A non-linear $\mathbf{h}$ in the measurement equation (2.7) prevents derivation of the normalization factor and posterior density. Once again a first-order Taylor series approximation can be used to provide an analytic approximation. The Taylor series expansion of $\mathbf{h}(\mathbf{x}_k)$ about the point $\hat{\mathbf{x}}_{k|k-1}$ is given by

$$\mathbf{h}(\mathbf{x}_k) = \mathbf{h}(\hat{\mathbf{x}}_{k|k-1}) + \sum_{m=1}^{\infty} \frac{1}{m!} \left(\bigotimes_{i=1}^{m} \nabla_{\mathbf{x}^T} \right) \mathbf{h}(\mathbf{x}) \Bigg|_{\mathbf{x}=\hat{\mathbf{x}}_{k|k-1}} \left(\bigotimes_{i=1}^{m} (\mathbf{x}_k - \hat{\mathbf{x}}_{k|k-1}) \right).$$

The first-order approximation to this series expansion of $\mathbf{h}(\mathbf{x}_k)$ is

$$\mathbf{h}(\mathbf{x}_k) \approx \hat{\mathbf{h}}(\mathbf{x}_k) = \mathbf{h}(\hat{\mathbf{x}}_{k|k-1}) + \mathbf{H}_k(\mathbf{x}_k - \hat{\mathbf{x}}_{k|k-1}), \qquad (2.18)$$

where

$$\mathbf{H}_k = \nabla_{\mathbf{x}^T} \mathbf{h}(\mathbf{x})|_{\mathbf{x}=\hat{\mathbf{x}}_{k|k-1}}.$$

With the functions $\mathbf{f}$ and $\mathbf{h}$ replaced by $\hat{\mathbf{f}}$ and $\hat{\mathbf{h}}$, respectively, the derivation of the EKF is similar to that of the KF. This will be demonstrated in the following sections.

The transition and prediction densities

Using the linearized approximation (2.17), the transition density can be written as

$$p(\mathbf{x}_k|\mathbf{x}_{k-1}) = p_{\mathbf{v}_k}(\mathbf{x}_k - \mathbf{f}(\mathbf{x}_{k-1}))$$

$$\approx p_{\mathbf{v}_k}(\mathbf{x}_k - \hat{\mathbf{f}}(\mathbf{x}_{k-1}))$$

$$= p_{\mathbf{v}_k}(\mathbf{x}_k - \mathbf{F}_k\mathbf{x}_{k-1} - \boldsymbol{\epsilon}_{\mathbf{f}}(\hat{\mathbf{x}}_{k-1|k-1})), \tag{2.19}$$

where $\boldsymbol{\epsilon}_{\mathbf{f}}(\mathbf{x}) = \mathbf{f}(\mathbf{x}) - \mathbf{F}_k\mathbf{x}$. Since the process noise is Gaussian by assumption **A1**, the linearized approximation to the transition density is

$$p(\mathbf{x}_k|\mathbf{x}_{k-1}) = N(\mathbf{x}_k; \mathbf{F}_k\mathbf{x}_{k-1} + \boldsymbol{\epsilon}_{\mathbf{f}}(\hat{\mathbf{x}}_{k-1|k-1}), \mathbf{Q}_k). \tag{2.20}$$

Under assumption **A3**, the posterior pdf of the object state at time $k-1$ is

$$p(\mathbf{x}_{k-1}|\mathbf{y}^{k-1}) = N(\mathbf{x}_{k-1}; \hat{\mathbf{x}}_{k-1|k-1}, \mathbf{P}_{k-1|k-1}). \tag{2.21}$$

The prediction density can be found by substituting (2.20) and (2.21) into the CKE (2.4) and using Theorem 2.1:

$$p(\mathbf{x}_k|\mathbf{y}^{k-1})$$

$$= \int N(\mathbf{x}_k; \mathbf{F}_k\mathbf{x}_{k-1} + \boldsymbol{\epsilon}_{\mathbf{f}}(\hat{\mathbf{x}}_{k-1|k-1}), \mathbf{Q}_k)N(\mathbf{x}_{k-1}; \hat{\mathbf{x}}_{k-1|k-1}, \mathbf{P}_{k-1|k-1})\, d\mathbf{x}_{k-1}$$

$$= N(\mathbf{x}_k; \hat{\mathbf{x}}_{k|k-1}, \mathbf{P}_{k|k-1}), \tag{2.22}$$

where

$$\hat{\mathbf{x}}_{k|k-1} = \mathbf{F}_k\hat{\mathbf{x}}_{k-1|k-1} + \boldsymbol{\epsilon}_{\mathbf{f}}(\hat{\mathbf{x}}_{k-1|k-1})$$

$$= \mathbf{F}_k\hat{\mathbf{x}}_{k-1|k-1} + \mathbf{f}(\hat{\mathbf{x}}_{k-1|k-1}) - \mathbf{F}_k\hat{\mathbf{x}}_{k-1|k-1}$$

$$= \mathbf{f}(\hat{\mathbf{x}}_{k-1|k-1}),$$

$$\mathbf{P}_{k|k-1} = \mathbf{F}_k\mathbf{P}_{k-1|k-1}\mathbf{F}_k^T + \mathbf{Q}_k.$$

The likelihood function and normalization factor

Using the linear approximation in (2.18), we can derive the likelihood function,

$$p(\mathbf{y}_k|\mathbf{x}_k) = p_{\mathbf{w}_k}(\mathbf{y}_k - \mathbf{h}(\mathbf{x}_k))$$

$$\approx p_{\mathbf{w}_k}(\mathbf{y}_k - \mathbf{h}(\hat{\mathbf{x}}_{k|k-1}) - \mathbf{H}_k(\mathbf{x}_k - \hat{\mathbf{x}}_{k|k-1}))$$

$$= p_{\mathbf{w}_k}(\mathbf{y}_k - \mathbf{H}_k\mathbf{x}_k - \boldsymbol{\epsilon}_{\mathbf{h}}(\hat{\mathbf{x}}_{k|k-1})), \tag{2.23}$$

where $\boldsymbol{\epsilon}_{\mathbf{h}}(\mathbf{x}) = \mathbf{h}(\mathbf{x}) - \mathbf{H}_k\mathbf{x}$. By assumption **A2**, the measurement noise is Gaussian with zero mean and covariance matrix R_k so that the linearized likelihood is

$$p(\mathbf{y}_k|\mathbf{x}_k) = N(\mathbf{y}_k; \mathbf{H}_k\mathbf{x}_k + \epsilon_{\mathbf{h}}(\hat{\mathbf{x}}_{k|k-1}), \mathbf{R}_k). \tag{2.24}$$

Using the predicted density (2.22), the linearized likelihood (2.24) and Theorem 2.1, the normalization factor is

$$p(\mathbf{y}_k|\mathbf{y}^{k-1}) = \int N(\mathbf{y}_k; \mathbf{H}_k\mathbf{x}_k + \epsilon_{\mathbf{h}}(\hat{\mathbf{x}}_{k|k-1}), \mathbf{R}_k)N(\mathbf{x}_k; \hat{\mathbf{x}}_{k|k-1}, \mathbf{P}_{k|k-1})\, d\mathbf{x}_k$$

$$= N(\mathbf{y}_k; \hat{\mathbf{y}}_{k|k-1}, \mathbf{S}_k), \tag{2.25}$$

where

$$\hat{\mathbf{y}}_{k|k-1} = \mathbf{H}_k\hat{\mathbf{x}}_{k|k-1} + \epsilon_{\mathbf{h}}(\hat{\mathbf{x}}_{k|k-1})$$

$$= \mathbf{H}_k\hat{\mathbf{x}}_{k|k-1} + \mathbf{h}(\hat{\mathbf{x}}_{k|k-1}) - \mathbf{H}_k\hat{\mathbf{x}}_{k|k-1}$$

$$= \mathbf{h}(\hat{\mathbf{x}}_{k|k-1}),$$

$$\mathbf{S}_k = \mathbf{H}_k\mathbf{P}_{k|k-1}\mathbf{H}_k^T + \mathbf{R}_k.$$

The conditional density

Substituting the predicted density (2.22), the linearized likelihood function (2.24) and the normalization factor (2.25) into Bayes' rule (2.3) and using Theorem 2.1 gives the Gaussian conditional density,

$$p(\mathbf{x}_k|\mathbf{y}^k) = N(\mathbf{x}_k; \hat{\mathbf{x}}_{k|k}, \mathbf{P}_{k|k}),$$

where

$$\hat{\mathbf{x}}_{k|k} = \hat{\mathbf{x}}_{k|k-1} + \mathbf{P}_{k|k-1}\mathbf{H}_k^T\mathbf{S}_k^{-1}(\mathbf{y}_k - \hat{\mathbf{y}}_{k|k-1}),$$

$$\mathbf{P}_{k|k} = \mathbf{P}_{k|k-1} - \mathbf{P}_{k|k-1}\mathbf{H}_k^T\mathbf{S}_k^{-1}\mathbf{H}_k\mathbf{P}_{k|k-1}.$$

2.3.2 The extended Kalman filter equations

The EKF equations are summarized in Algorithm 2.

A comparison of Algorithms 1 and 2 shows that the KF and the EKF have the same structure. Both algorithms represent the posterior density by a Gaussian, the mean and covariance matrix of which are computed recursively via prediction and then corrected using the most recent measurement. An important difference is that the posterior covariance matrix in the EKF is data dependent for non-linear dynamic and/or measurement equations. This is because the posterior covariance matrix uses Jacobians, $\mathbf{F}_k$ and $\mathbf{H}_k$, evaluated at object state estimates which depend on the measurements.

Algorithm 2 Extended Kalman filter

1: Compute the Jacobian of $\mathbf{f}$ at $\hat{\mathbf{x}}_{k-1|k-1}$:

$$\mathbf{F}_k = \nabla_{\mathbf{x}^T} \mathbf{f}(\mathbf{x})|_{\mathbf{x}=\hat{\mathbf{x}}_{k-1|k-1}}.$$

2: Compute predicted mean and covariance matrix:

$$\hat{\mathbf{x}}_{k|k-1} = \mathbf{f}(\hat{\mathbf{x}}_{k-1|k-1}),$$

$$\mathbf{P}_{k|k-1} = \mathbf{F}_k \mathbf{P}_{k-1|k-1} \mathbf{F}_k^T + \mathbf{Q}_k.$$

3: Compute the Jacobian of $\mathbf{h}$ at $\hat{\mathbf{x}}_{k|k-1}$:

$$\mathbf{H}_k = \nabla_{\mathbf{x}^T} \mathbf{h}(\mathbf{x})|_{\mathbf{x}=\hat{\mathbf{x}}_{k|k-1}}.$$

4: Compute the predicted measurement, innovation covariance matrix and Kalman gain:

$$\hat{\mathbf{y}}_{k|k-1} = \mathbf{h}(\mathbf{x}_{k|k-1}),$$

$$\mathbf{S}_k = \mathbf{H}_k \mathbf{P}_{k|k-1} \mathbf{H}_k^T + \mathbf{R}_k,$$

$$\mathbf{K}_k = \mathbf{P}_{k|k-1} \mathbf{H}_k^T \mathbf{S}_k^{-1}.$$

5: Compute the posterior mean and covariance matrix:

$$\hat{\mathbf{x}}_{k|k} = \hat{\mathbf{x}}_{k|k-1} + \mathbf{K}_k (\mathbf{y}_k - \hat{\mathbf{y}}_{k|k-1}),$$

$$\mathbf{P}_{k|k} = \mathbf{P}_{k|k-1} - \mathbf{K}_k \mathbf{H}_k \mathbf{P}_{k|k-1}.$$

The EKF is a simple and elegant non-linear filtering approximation which performs well in many applications. However, there are drawbacks involved in the use of the EKF:

- The EKF is prone to divergence whereby the posterior mean approximation departs significantly from the true state. A primary cause of this divergence is an optimistic approximation of the posterior covariance matrix.
- The computations of the Jacobian matrices are non-trivial in most instances. These computations can lead to difficulties in implementing the filter in practice.

2.4 The unscented Kalman filter

The unscented Kalman filter (UKF) is an alternative to the EKF which shares its computational simplicity while avoiding the need to derive and compute Jacobians and achieving greater accuracy in many scenarios. The basis of the UKF is the unscented transformation (UT), which is a method for approximating the moments of a non-linearly transformed random variable. We begin by describing the UT and then show how it is used in the UKF.

2.4.1 The unscented transformation

Consider a random variable $\mathbf{x} \in \mathbb{R}^n$ and a non-linear transformation $\mathbf{g} : \mathbb{R}^n \to \mathbb{R}^m$. A problem that arises in many applications is the computation of the moments of the random variable $\mathbf{z} = \mathbf{g}(\mathbf{x})$. For instance, the mean of $\mathbf{z}$ may be written as

$$E(\mathbf{z}) = \int g(\mathbf{x}) p(\mathbf{x}) \, d\mathbf{x}. \tag{2.26}$$

In many cases of practical interest the integral (2.26) cannot be evaluated in closed-form. The UT is a numerical procedure for approximating the integral in such cases. The UT uses a weighted sum of function values evaluated at non-uniformly spaced points in the domain. This is in the same spirit as the well-known Gauss quadrature used for approximation of integrals with one-dimensional integrands. The collection of points at which the function $\mathbf{g}$ is evaluated are usually called sigma points. Let $\mathcal{X}^1, \ldots, \mathcal{X}^s$ denote a collection of s sigma points and $w^1, \ldots, w^s$ denote the corresponding weights. The basic requirement of these weighted sigma points is that the sample mean and covariance matrix match the true mean and covariance matrix, i.e.,

$$1/s \sum_{i=1}^{s} w^i \mathcal{X}^i = E(\mathbf{x}), \tag{2.27}$$

$$1/s \sum_{i=1}^{s} w^i [\mathcal{X}^i - E(\mathbf{x})][\mathcal{X}^i - E(\mathbf{x})]^T = \text{cov}(\mathbf{x}). \tag{2.28}$$

Let $\mathcal{Z}^i = \mathbf{g}(\mathcal{X}^i)$, $i = 1, \ldots, s$ denote the value of $\mathbf{g}$ at the ith sigma point. The UT approximation to the mean of $\mathbf{z}$ is

$$\widehat{E(\mathbf{z})} = 1/s \sum_{i=1}^{s} w^i \mathcal{Z}^i. \tag{2.29}$$

It can be seen that the UT approximation to the mean is structurally similar to a random sampling-based approximation to the mean. For example, in importance sampling $\mathbf{x}^i \sim q$ for an importance density q and the mean is approximated by

$$1/s \sum_{i=1}^{s} \frac{p(\mathbf{x}^i)}{q(\mathbf{x}^i)} g(\mathbf{x}^i) \left/ \sum_{j=1}^{s} \frac{p(\mathbf{x}^j)}{q(\mathbf{x}^j)} \right. . \tag{2.30}$$

The difference between importance sampling and the UT is that the sigma points used by the UT are selected according to a deterministic rule which ensures they satisfy the requirements (2.27) and (2.28). This means that greater accuracy can be achieved with only a few sigma points. In general, the approximation $\widehat{E(z)}$ is accurate to second-order in the sense that the Taylor series expansions of $\widehat{E(z)}$ and $E(z)$ match up to second-order.

Other moments of z can be approximated in a similar way. Of particular interest for non-linear filtering are the covariance matrix of z and the cross-covariance between x and z. The UT approximations of these quantities are

$$\widehat{\text{cov}(z)} = 1/s \sum_{i=1}^{s} w^i [Z^i - \widehat{E(z)}][Z^i - \widehat{E(z)}]^T, \tag{2.31}$$

$$\widehat{\text{cov}(x, z)} = 1/s \sum_{i=1}^{s} w^i [X^i - E(x)][Z^i - \widehat{E(z)}]^T. \tag{2.32}$$

The moment approximations $\widehat{\text{cov}(z)}$ and $\widehat{\text{cov}(x, z)}$ are also accurate up to second order.

There are many ways of selecting the sigma points in order to satisfy the requirements (2.27) and (2.28). The method originally proposed in Julier *et al.* (2000) selects $s = 2n + 1$ sigma points in the following manner. Define the matrix square root A of $(n + \kappa)\Sigma$, where κ is a pre-determined parameter, such that $AA^T = (n + \kappa)\Sigma$. The sigma points are given by, for $i = 1, \ldots, 2n + 1$,

$$X^i = \mu + \sigma^i, \tag{2.33}$$

where $\sigma^1 = 0_{n,1}$,

$$\begin{bmatrix} \sigma^2 & \cdots & \sigma^{n+1} \end{bmatrix} = A, \tag{2.34}$$

and $\sigma^{n+i+1} = -\sigma^{i+1}$, $i = 1, \ldots, n$. The corresponding weights are

$$w^i = \begin{cases} \kappa/(n + \kappa), & i = 1, \\ 0.5/(n + \kappa), & i = 2, \ldots, 2n + 1. \end{cases} \tag{2.35}$$

It can be verified that the first two sample moments of the sigma points match the moments of x, as required in (2.27) and (2.28). In particular, the sample mean is

$$\mu + \begin{bmatrix} \sigma^1 & \cdots & \sigma^{2n+1} \end{bmatrix} \begin{bmatrix} w^1 \\ \vdots \\ w^{2n+1} \end{bmatrix} = \mu + 0.5(A - A)/(n + \kappa) = \mu, \tag{2.36}$$

and the sample covariance matrix is

$$
\begin{bmatrix} \sigma^1 & \cdots & \sigma^{2n+1} \end{bmatrix}
\begin{bmatrix} w^1 & & \\ & \ddots & \\ & & w^{2n+1} \end{bmatrix}
\begin{bmatrix} \sigma^{1\prime} \\ \vdots \\ \sigma^{2n+1\prime} \end{bmatrix}
$$

$$
= \frac{1}{n+\kappa} \begin{bmatrix} \mathbf{0}_{n,1} & \mathbf{A} & -\mathbf{A} \end{bmatrix}
\begin{bmatrix} \kappa & & & \\ & 0.5 & & \\ & & \ddots & \\ & & & 0.5 \end{bmatrix}
\begin{bmatrix} \mathbf{0}_{1,n} \\ \mathbf{A}^T \\ -\mathbf{A}^T \end{bmatrix}
$$

$$
= \frac{1}{2(n+\kappa)} \begin{bmatrix} \mathbf{0}_{n,1} & \mathbf{A} & -\mathbf{A} \end{bmatrix}
\begin{bmatrix} \mathbf{0}_{1,n} \\ \mathbf{A}^T \\ -\mathbf{A}^T \end{bmatrix}
$$

$$
= \mathbf{A}\mathbf{A}^T / (n+\kappa) = \Sigma. \tag{2.37}
$$

The parameter κ determines the spread of the sigma points. It is usually desired to keep the sigma points concentrated relatively close about the mean μ so that only function values in interesting parts of the parameter space are considered in moment approximations of $\mathbf{g}(\mathbf{x})$. A small spread of sigma points requires κ such that $n + \kappa$ is small. For large, or even moderate, n this can be achieved only by choosing negative values of κ. If $\kappa < 0$, then, since $n + \kappa > 0$, the first weight $w^1 = \kappa/(n+\kappa)$ is negative and positive-definiteness of the covariance matrix approximation $\widehat{\mathrm{cov}(\mathbf{z})}$ is no longer guaranteed. This led to the proposal of the scaled UT, which permits a close concentration of sigma points without the potential for a negative covariance matrix. Other methods of sigma point selection use different numbers of sigma points. The simplex sigma point scheme reduces computational expense by using only $s = n + 2!$ sigma points. Alternatively, an improvement in accuracy can be obtained by matching higher-order moments using $s = 2n^2 + 1$ sigma points (Lerner, 2002). A common element of all these sigma point selection schemes is that the number of sigma points increases with the dimension n of the variable undergoing the non-linear transformation.

2.4.2 The unscented Kalman filter algorithm

The UT can be used as the basis of a non-linear filtering approximation which has the same form as the EKF but is derived in a quite different manner. We assume **A2** and **A3**.

The posterior PDF of $\mathbf{x}_k$ can be written using Bayes' rule as

$$
p(\mathbf{x}_k | \mathbf{y}^k) = p(\mathbf{x}_k, \mathbf{y}_k | \mathbf{y}^{k-1}) / p(\mathbf{y}_k | \mathbf{y}^{k-1}). \tag{2.38}
$$

The UKF approximates the joint density of the state $\mathbf{x}_k$ and measurement $\mathbf{y}_k$ conditional on the measurement history $\mathbf{y}^{k-1}$ is approximated by a Gaussian density,

$$p(\mathbf{x}_k, \mathbf{y}_k | \mathbf{y}^{k-1}) = N\left(\begin{bmatrix} \mathbf{x}_k \\ \mathbf{y}_k \end{bmatrix}; \begin{bmatrix} \hat{\mathbf{x}}_{k|k-1} \\ \hat{\mathbf{y}}_{k|k-1} \end{bmatrix}, \begin{bmatrix} \mathbf{P}_{k|k-1} & \mathbf{\Psi}_k \\ \mathbf{\Psi}_k^T & \mathbf{S}_k \end{bmatrix} \right). \tag{2.39}$$

This implies that

$$p(\mathbf{y}_k | \mathbf{y}^{k-1}) = N(\mathbf{y}_k; \hat{\mathbf{y}}_{k|k-1}, \mathbf{S}_k). \tag{2.40}$$

In general, the moments appearing in (2.39) and (2.40) cannot be computed exactly and so must be approximated. The UT is used for this purpose.

Predicted state statistics

The predicted mean and covariance matrix of $\mathbf{x}_k$ are moments of the transformation

$$\mathbf{x}_k = \mathbf{f}(\mathbf{x}_{k-1}) + \mathbf{v}_k, \tag{2.41}$$

where the statistics of $\mathbf{x}_{k-1}$ given $\mathbf{y}^{k-1}$ are available by **A3**. The predicted mean and covariance matrix can be written as

$$E(\mathbf{x}_k | \mathbf{y}^{k-1}) = E(\mathbf{f}(\mathbf{x}_{k-1}) + \mathbf{v}_k | \mathbf{y}^{k-1}) = E(\mathbf{f}(\mathbf{x}_{k-1}) | \mathbf{y}^{k-1}), \tag{2.42}$$

$$\text{cov}(\mathbf{x}_k | \mathbf{y}^{k-1}) = \text{cov}(\mathbf{f}(\mathbf{x}_{k-1}) + \mathbf{v}_k | \mathbf{y}^{k-1}) = \text{cov}(\mathbf{f}(\mathbf{x}_{k-1}) | \mathbf{y}^{k-1}) + \text{cov}(\mathbf{v}_k). \tag{2.43}$$

Approximations to the predicted mean and covariance matrix are to be obtained using the UT. As described in Section 2.4.1, sigma points $\mathcal{X}_{k-1}^1, \ldots, \mathcal{X}_{k-1}^s$ and weights $w^1, \ldots, w^s$ are selected to match the mean and covariance matrix of $p(\mathbf{x}_{k-1} | \mathbf{y}^{k-1})$. With $\mathcal{X}_k^i = \mathbf{f}(\mathcal{X}_{k-1}^i)$, $i = 1, \ldots, s$, the UT approximations to the predicted mean and covariance matrix are

$$\hat{\mathbf{x}}_{k|k-1} = \sum_{i=1}^{s} w^i \mathcal{X}_k^i, \tag{2.44}$$

$$\mathbf{P}_{k|k-1} = \mathbf{Q}_k + \sum_{i=1}^{s} w^i (\mathcal{X}_k^i - \hat{\mathbf{x}}_{k|k-1})(\mathcal{X}_k^i - \hat{\mathbf{x}}_{k|k-1})^T. \tag{2.45}$$

Predicted measurement statistics

The predicted statistics for the measurement $\mathbf{y}_k$ are moments of the transformation

$$\mathbf{y}_k = \mathbf{h}(\mathbf{x}_k) + \mathbf{w}_k, \tag{2.46}$$

where the statistics of $\mathbf{x}_k$ given $\mathbf{y}^{k-1}$ are available from the prediction step. We have

$$E(\mathbf{y}_k|\mathbf{y}^{k-1}) = E(\mathbf{h}(\mathbf{x}_k)|\mathbf{y}^{k-1}), \tag{2.47}$$

$$\text{cov}(\mathbf{y}_k|\mathbf{y}^{k-1}) = \text{cov}(\mathbf{h}(\mathbf{x}_k)|\mathbf{y}^{k-1}) + \text{cov}(\mathbf{w}_k), \tag{2.48}$$

$$\text{cov}(\mathbf{x}_k, \mathbf{y}_k|\mathbf{y}^{k-1}) = \text{cov}(\mathbf{x}_k, \mathbf{h}(\mathbf{x}_k)|\mathbf{y}^{k-1}). \tag{2.49}$$

Let $\mathcal{X}_k^1, \ldots, \mathcal{X}_k^s$ and $w^1, \ldots, w^s$ denote the sigma points and weights, respectively, selected to match the predicted mean and covariance matrix. The transformed sigma points are $\mathcal{Y}_k^i = \mathbf{h}(\mathcal{X}_k^i)$, $i = 1, \ldots, s$. The UT approximations to the moments (2.47)–(2.49) are

$$\hat{\mathbf{y}}_{k|k-1} = \sum_{i=1}^{s} w^i \mathcal{Y}_k^i, \tag{2.50}$$

$$\mathbf{S}_k = \mathbf{R}_k + \sum_{i=1}^{s} w^i (\mathcal{Y}_k^i - \hat{\mathbf{y}}_{k|k-1})(\mathcal{Y}_k^i - \hat{\mathbf{y}}_{k|k-1})^T, \tag{2.51}$$

$$\mathbf{\Psi}_k = \sum_{i=1}^{s} w^i (\mathcal{X}_k^i - \hat{\mathbf{x}}_{k|k-1})(\mathcal{Y}_k^i - \hat{\mathbf{y}}_{k|k-1})^T. \tag{2.52}$$

Under the Gaussian approximation, the moment approximations (2.44), (2.45) and (2.50)–(2.52) provide a complete description of an approximate predicted joint distribution of the state and measurement. It remains to substitute (2.39) and (2.40) into (2.38) and re-arrange the resulting equation. The necessary manipulations are well known and can be found, for instance, in Example 3.2 of Anderson and Moore (1979). The resulting expression for the posterior PDF is

$$p(\mathbf{x}_k|\mathbf{y}^k) = N(\mathbf{x}_k; \hat{\mathbf{x}}_{k|k}, \mathbf{P}_{k|k}), \tag{2.53}$$

where

$$\hat{\mathbf{x}}_{k|k} = \hat{\mathbf{x}}_{k|k-1} + \mathbf{\Psi}_k \mathbf{S}_k^{-1}(\mathbf{y}_k - \hat{\mathbf{y}}_{k|k-1}), \tag{2.54}$$

$$\mathbf{P}_{k|k} = \mathbf{P}_{k|k-1} - \mathbf{\Psi}_k \mathbf{S}_k^{-1} \mathbf{\Psi}_k^T. \tag{2.55}$$

A summary of the UKF is given by Algorithm 3. As with the EKF, the UKF is strongly reminiscent of the KF with the main difference being the dependence of the posterior covariance matrix on the observed measurements. In the UKF this arises because the sigma points used in the moment approximations are determined based on state estimates which are measurement-dependent. The UKF is of the same order of computational expense as the EKF but generally performs

---

Algorithm 3 Unscented Kalman filter

1: Determine sigma points $\mathcal{X}_{k-1}^1, \ldots, \mathcal{X}_{k-1}^s$ and weights $w^1, \ldots, w^s$ to match a mean $\hat{\mathbf{x}}_{k-1|k-1}$ and covariance matrix $\mathbf{P}_{k-1|k-1}$.

2: Compute the transformed sigma points $\mathcal{X}_k^i = \mathbf{f}(\mathcal{X}_{k-1}^i), i = 1, \ldots, s$.

3: Compute the predicted state statistics:

$$\hat{\mathbf{x}}_{k|k-1} = \sum_{i=1}^s w^i \mathcal{X}_k^i,$$

$$\mathbf{P}_{k|k-1} = \mathbf{Q}_k + \sum_{i=1}^s w^i (\mathcal{X}_k^i - \hat{\mathbf{x}}_{k|k-1})(\mathcal{X}_k^i - \hat{\mathbf{x}}_{k|k-1})^T.$$

4: Determine sigma points $\mathcal{X}_k^1, \ldots, \mathcal{X}_k^s$ and weights $w^1, \ldots, w^s$ to match mean $\hat{\mathbf{x}}_{k|k-1}$ and covariance matrix $\mathbf{P}_{k|k-1}$.

5: Compute the transformed sigma points $\mathcal{Y}_k^i = \mathbf{h}(\mathcal{X}_k^i), i = 1, \ldots, s$.

6: Compute the predicted measurement statistics:

$$\hat{\mathbf{y}}_{k|k-1} = \sum_{i=1}^s w^i \mathcal{Y}_k^i,$$

$$\mathbf{S}_k = \mathbf{R}_k + \sum_{i=1}^s w^i (\mathcal{Y}_k^i - \hat{\mathbf{y}}_{k|k-1})(\mathcal{Y}_k^i - \hat{\mathbf{y}}_{k|k-1})^T,$$

$$\boldsymbol{\Psi}_k = \sum_{i=1}^s w^i (\mathcal{X}_k^i - \hat{\mathbf{x}}_{k|k-1})(\mathcal{Y}_k^i - \hat{\mathbf{y}}_{k|k-1})^T.$$

7: Compute the posterior mean and covariance matrix:

$$\hat{\mathbf{x}}_{k|k} = \hat{\mathbf{x}}_{k|k-1} + \boldsymbol{\Psi}_k \mathbf{S}_k^{-1}(\mathbf{y}_k - \hat{\mathbf{y}}_{k|k-1}),$$

$$\mathbf{P}_{k|k} = \mathbf{P}_{k|k-1} - \boldsymbol{\Psi}_k \mathbf{S}_k^{-1} \boldsymbol{\Psi}_k^T.$$

---

much better. This can be attributed to increased accuracy of the moment approximations in the UKF. In fact, the UKF achieves the same level of accuracy as the second-order EKF (Jazwinski, 1970), which requires the derivation of Jacobian *and* Hessian matrices.

Before concluding, we note that the assumption of additive noise in the dynamic and measurement equations has been made only for the sake of convenience. The UT is equally applicable for non-additive noise. For instance, if the dynamic equation is $\mathbf{x}_k = \mathbf{g}(\mathbf{x}_{k-1}, \mathbf{v}_k)$, then the quantity undergoing a non-linear transformation is the augmented variable $\mathbf{z}_k = [\mathbf{x}_{k-1}^T, \mathbf{v}_k^T]^T$, which has the statistics

$E(z_k|y^{k-1}) = [\hat{x}_{k-1|k-1}^T, 0_{n,1}^T]^T$ and $\text{cov}(z_k|y^{k-1}) = \text{diag}(P_{k-1|k-1}, Q_k)$. The UT can be applied to the random variable z_k transformed through the function g to approximate the predicted state statistics. Since z_k is of dimension $n_x + n_v$ a larger number of sigma points will be required than for the case where the dynamic noise is additive. Similar comments hold for a measurement equation which is non-linear in the measurement noise.

2.5 The point mass filter

In cases where exact Bayesian solution is unavailable the EKF and UKF make the approximations necessary to arrive at closed-form expressions for the posterior PDF. In this sense the EKF and UKF may be thought of as analytic approximations to the optimal Bayesian filter. Although the EKF and UKF are computationally efficient their accuracy is limited by the validity of the approximations required for analyticity. If sufficient computational resources are available increased accuracy can be obtained by attempting a numerical approximation to the posterior PDF. One such approximation is the point mass filter (PMF) in which a discrete approximation to the posterior PDF is computed. None of the assumptions **A1–A3** are required in the PMF.

At time $k - 1$ a region of the state space is partitioned into n equi-volume hypercubes. Let x_{k-1}^i denote the centre of the ith cube for $i = 1, \ldots, n$. Each hyper-cube is associated with a weight $w_{k-1|k-1}^i$, $i = 1, \ldots, n$ such that $\sum_{i=1}^n w_{k-1|k-1}^i = 1$. The collection of hyper-cubes and weights form a discrete approximation to the posterior PDF at time $k - 1$,

$$p(x_{k-1}|y^{k-1}) \approx \sum_{i=1}^n w_{k-1|k-1}^i \delta(x_{k-1} - x_{k-1}^i). \tag{2.56}$$

Loosely speaking, $w_{k-1|k-1}^i \approx p(x_{k-1}^i|y_{k-1})$. A similar finite-dimensional approximation to the posterior PDF at time k is sought.

2.5.1 Transition and prediction densities

Assume that the object dynamics are subject to additive noise so that

$$x_k = f(x_{k-1}) + v_k, \tag{2.57}$$

and the transition density is $p(x_k|x_{k-1}) = p_{v_k}(x_k - f(x_{k-1}))$. The prediction density can be found by substituting the transition density into the CKE (2.4),

$$p(x_k|y^{k-1}) = \int p_{v_k}(x_k - f(x_{k-1})) p(x_{k-1}|y^{k-1}) \, dx_{k-1}. \tag{2.58}$$

Substituting the PMF approximation (2.56) to $p(\mathbf{x}_{k-1}|\mathbf{y}^{k-1})$ into (2.58) gives

$$p(\mathbf{x}_k|\mathbf{y}^{k-1}) \approx \int p_{\mathbf{v}_k}(\mathbf{x}_k - \mathbf{f}(\mathbf{x}_{k-1})) \sum_{i=1}^{n} w_{k-1|k-1}^i \delta(\mathbf{x}_{k-1} - \mathbf{x}_{k-1}^i) \, d\mathbf{x}_{k-1}$$

$$= \sum_{i=1}^{n} w_{k-1|k-1}^i p_{\mathbf{v}_k}(\mathbf{x}_k - \mathbf{f}(\mathbf{x}_{k-1}^i)).$$

A finite-dimensional representation of the prediction density is obtained by partitioning a region of the state space into n hyper-cubes of equal volume. It is not necessary to use the same number of points at each time although this will be done here for notational convenience. Let $\mathbf{x}_k^i$ denote the centre of the ith hyper-cube for $i = 1, \ldots, n$. Then, the PMF approximation to the prediction density is

$$p(\mathbf{x}_k|\mathbf{y}^{k-1}) \approx \sum_{i=1}^{n} w_{k|k-1}^i \delta(\mathbf{x}_k - \mathbf{x}_k^i), \qquad (2.59)$$

where, for $i = 1, \ldots, n$,

$$w_{k|k-1}^i = \sum_{j=1}^{n} w_{k-1|k-1}^j p_{\mathbf{v}_k}(\mathbf{x}_k^i - \mathbf{f}(\mathbf{x}_{k-1}^j)). \qquad (2.60)$$

2.5.2 The likelihood function and normalization factor

Assume that the measurement noise is additive so that

$$\mathbf{y}_k = \mathbf{h}(\mathbf{x}_k) + \mathbf{w}_k, \qquad (2.61)$$

and the likelhiood of $\mathbf{x}_k$ is $p(\mathbf{y}_k|\mathbf{x}_k) = p_{\mathbf{w}_k}(\mathbf{y}_k - \mathbf{h}(\mathbf{x}_k))$. The normalizing factor $p(\mathbf{y}_k|\mathbf{y}^{k-1})$ can be expanded as

$$p(\mathbf{y}_k|\mathbf{y}^{k-1}) = \int p_{\mathbf{w}_k}(\mathbf{y}_k - \mathbf{h}(\mathbf{x}_k)) p(\mathbf{x}_k|\mathbf{y}^{k-1}) \, d\mathbf{x}_k. \qquad (2.62)$$

Substituting (2.59) into (2.62) gives the PMF approximation to the normalizing factor,

$$p(\mathbf{y}_k|\mathbf{y}^{k-1}) \approx \int p_{\mathbf{w}_k}(\mathbf{y}_k - \mathbf{h}(\mathbf{x}_k)) \sum_{i=1}^{n} w_{k|k-1}^i \delta(\mathbf{x}_k - \mathbf{x}_k^i) \, d\mathbf{x}_k$$

$$= \sum_{i=1}^{n} w_{k|k-1}^i p_{\mathbf{w}_k}(\mathbf{y}_k - \mathbf{h}(\mathbf{x}_k^i)). \qquad (2.63)$$

2.5.3 Conditional density

The PMF approximation to the posterior PDF at time k can be found by substituting (2.59), (2.63) and the likelihood into (2.9),

$$p(\mathbf{x}_k | \mathbf{y}^k) \approx \frac{p_{\mathbf{w}_k}(\mathbf{y}_k - \mathbf{h}(\mathbf{x}_k)) \sum_{i=1}^{n} w_{k|k-1}^i \delta(\mathbf{x}_k - \mathbf{x}_k^i)}{\sum_{i=1}^{n} w_{k|k-1}^i p_{\mathbf{w}_k}(\mathbf{y}_k - \mathbf{h}(\mathbf{x}_k^i))}$$

$$= \sum_{i=1}^{n} w_{k|k}^i \delta(\mathbf{x}_k - \mathbf{x}_k^i), \tag{2.64}$$

where

$$w_{k|k}^i = w_{k|k-1}^i p_{\mathbf{w}_k}(\mathbf{y}_k - \mathbf{h}(\mathbf{x}_k^i)) \Big/ \sum_{j=1}^{n} w_{k|k-1}^j p_{\mathbf{w}_k}(\mathbf{y}_k - \mathbf{h}(\mathbf{x}_k^j)). \tag{2.65}$$

Thus the weights used in the posterior PDF approximation are obtained by multiplying the weights of the prediciton PDF by the likelihood of the corresponding state value and then normalizing. This is a quantitative illustration of correction using the current measurement. A point estimate of the state can be easily obtained as

$$\hat{\mathbf{x}}_{k|k} = \int \mathbf{x}_k p(\mathbf{x}_k | \mathbf{y}^k) \, d\mathbf{x}_k \approx \sum_{i=1}^{n} w_{k|k}^i \mathbf{x}_k^i. \tag{2.66}$$

2.5.4 The point mass filter equations

A recursion of the PMF, beginning with the posterior PDF approximation (2.56) at time $k-1$, is given by Algorithm 4. An issue of some importance which will not be examined here is the selection of the grid points over which the posterior PDF is approximated. The grid of points must be selected so that all regions of interest are included and regions which are not of interest are excluded. The difficulty is that the grid must be selected without knowing precisely which regions are of interest. One way of addressing this issue is to use prior statistics to predict the region of interest (Bucy and Senne, 1971). An adaptive way to choose grid positions using Chebychev's inequality theorem was explored in the recursive Bayesian context in Challa (1998).

The main argument against the PMF is computational complexity. There are two aspects of the algorithm which contribute to this. First, straightforward computation of the prediction PDF weights via step 2 of Algorithm 4 requires n^2 operations. This computational burden can be reduced using the method of ellipsoid

---

Algorithm 4 Point mass filter

1: Select grid points $\mathbf{x}_k^1, \ldots, \mathbf{x}_k^n$.
2: Compute the weights for prediction density, for $i = 1, \ldots, n$:

$$w_{k|k-1}^i = \sum_{j=1}^{n} w_{k-1|k-1}^j p_{\mathbf{v}_k}(\mathbf{x}_k^i - \mathbf{f}(\mathbf{x}_{k-1}^j)).$$

3: Compute the weights for posterior PDF, for $i = 1, \ldots, n$:

$$w_{k|k}^i = w_{k|k-1}^i p_{\mathbf{w}_k}(\mathbf{y}_k - \mathbf{h}(\mathbf{x}_k^i)) \bigg/ \sum_{j=1}^{n} w_{k|k-1}^j p_{\mathbf{w}_k}(\mathbf{y}_k - \mathbf{h}(\mathbf{x}_k^j)).$$

4: Compute a state estimate:

$$\hat{\mathbf{x}}_{k|k} = \sum_{i=1}^{n} w_{k|k}^i \mathbf{x}_k^i.$$

---

mode tracking (Bucy and Senne, 1971), which still results in $O(n^2)$ computations, or by using the fast Fourier transform (Kramer and Sorenson, 1988), which results in $O(n \log(n))$ operations. The second problem is that the relative approximation error is $O(n^{-1/n_x})$ so that the number of grid points required for a certain level of accuracy increases exponentially with the dimension of the state. Accurate approximation with reasonable computational expense becomes impossible even for moderate state dimensions. This is usually referred to as the curse of dimensionality.

2.6 The particle filter

Like the PMF, particle filters (PFs) seek a discrete approximation to the posterior PDF. However in a PF the discrete support is chosen stochastically rather than deterministically. This is advantageous in several ways. Implementation is much simpler since it is no longer necessary to devise rules for determining the grid points. The sampling procedure will automatically move sample points to the region of the state space of interest. Computational complexity is decreased because the error convergence rate is independent of the dimension of the state.

PFs are most commonly formulated as sequential importance sampling (SIS) methods. SIS involves drawing samples from an importance density such that samples of the trajectory $\mathbf{x}_{0:k}$ are obtained by appending samples for time k to samples from the trajectory up to time $k - 1$. Usually it is not necessary to retain the whole state trajectory since sampling at the current time will often depend only on the sample at the previous time. Given samples $\mathbf{x}_{k-1}^1, \ldots, \mathbf{x}_{k-1}^n$ and weights $w_{k-1}^1, \ldots, w_{k-1}^n$ which represent the posterior PDF at time $k - 1$, weights

---

Algorithm 5 Basic particle filter

1: **for** $i = 1, \ldots, n$ **do**
2:　　Draw samples $(\mathbf{x}_k^i, t^i) \sim q$.
3:　　Compute the weight update factor:

$$e_k^i = \frac{p(\mathbf{y}_k | \mathbf{x}_k, \mathbf{y}_{1:k-1}) p(\mathbf{x}_k | \mathbf{x}_{k-1}^{t^i}, \mathbf{y}_{1:k-1})}{q(\mathbf{x}_k^i, t^i)}.$$

4: **end for**
5: Compute the updated weights:

$$w_k^i = w_{k-1}^i e_k^i \left/ \sum_{j=1}^n w_{k-1}^j e_k^j \right., \quad i = 1, \ldots, n.$$

---

and samples representing the posterior PDF at time k are produced as shown in Algorithm 5.

The design of a PF for a particular tracking problem involves selection and derivation of the importance density q. To achieve good performance with a reasonable sample size the importance density q must be selected appropriately. Several different versions of the importance density q for the problem of single-object tracking will be considered in this section. It will be assumed throughout that the dynamic and measurement equations include additive noise:

$$\mathbf{x}_k = \mathbf{f}(\mathbf{x}_{k-1}) + \mathbf{v}_k, \tag{2.67}$$

$$\mathbf{y}_k = \mathbf{h}(\mathbf{x}_k) + \mathbf{w}_k. \tag{2.68}$$

2.6.1 The particle filter for single-object tracking

The importance density for the bootstrap filter (BF) can be found as

$$q(\mathbf{x}_k, t) = w_{k-1}^t p_{\mathbf{v}_k}(\mathbf{x}_k - \mathbf{f}(\mathbf{x}_{k-1}^t)). \tag{2.69}$$

The use of the importance density (2.69) is summarized in Algorithm 6. The advantages of the BF are simplicity of implementation and generality of application. The BF can be applied to any system from which the transition density can be sampled and the likelihood can be computed. The main deficiency of the BF is that samples of the mixture index and object state are drawn without considering the current measurement. The current measurement is used only to assess the quality of the samples. This is inefficient, in the sense that a large number of samples is required for accurate approximation, especially if the measurements are precise compared to the object dynamics.

Algorithm 6 Bootstrap filter for single-object tracking

1: **for** $i = 1, \ldots, n$ **do**
2:     Draw a mixture index t^i such that $\Pr(t^i = l) = w^l_{k-1}$.
3:     Draw $\mathbf{v}^i_k \sim p_{\mathbf{v}_k}$ and compute the sample object state $\mathbf{x}^i_k = \mathbf{f}(\mathbf{x}^{t^i}_{k-1}) + \mathbf{v}^i_k$.
4:     Compute the weight update $e^i_k = p_{\mathbf{w}_k}(\mathbf{y}_k - \mathbf{h}(\mathbf{x}^i_k))$.
5: **end for**
6: Compute the updated weights:

$$w^i_k = w^i_{k-1} e^i_k \bigg/ \sum_{j=1}^{n} w^j_{k-1} e^j_k , \quad i = 1, \ldots, n.$$

7: Compute a state estimate:

$$\hat{\mathbf{x}}_{k|k} = \sum_{i=1}^{n} w^i_k \mathbf{x}^i_k.$$

2.6.2 The OID-PF for single-object tracking

The optimal importance density (OID) is

$$q(\mathbf{x}_k, t) = \psi^t_k \, p(\mathbf{x}_k | \mathbf{x}^t_{k-1}, \mathbf{y}_{1:k}), \tag{2.70}$$

where

$$\psi^t_k = w^t_{k-1} p(\mathbf{y}_k | \mathbf{x}^t_{k-1}, \mathbf{y}_{1:k-1}) \bigg/ \sum_{i=1}^{n} w^i_{k-1} p(\mathbf{y}_k | \mathbf{x}^i_{k-1}, \mathbf{y}_{1:k-1}) . \tag{2.71}$$

The sampling density for the object state can be expanded, using Bayes' rule, as

$$p(\mathbf{x}_k | \mathbf{x}_{k-1}, \mathbf{y}_{1:k}) = \frac{p(\mathbf{y}_k | \mathbf{x}_k, \mathbf{y}_{1:k-1}) p(\mathbf{x}_k | \mathbf{x}_{k-1}, \mathbf{y}_{1:k-1})}{p(\mathbf{y}_k | \mathbf{x}_{k-1}, \mathbf{y}_{1:k-1})}$$

$$= \frac{p_{\mathbf{w}_k}(\mathbf{y}_k - \mathbf{h}(\mathbf{x}_k)) p_{\mathbf{v}_k}(\mathbf{x}_k - \mathbf{f}(\mathbf{x}_{k-1}))}{\int p_{\mathbf{w}_k}(\mathbf{y}_k - \mathbf{h}(\boldsymbol{\xi}_k)) p_{\mathbf{v}_k}(\boldsymbol{\xi}_k - f(\mathbf{x}_{k-1})) \, d\boldsymbol{\xi}_k}.$$

No analytical solution exists for the OID in general. It is necessary to assume **A2**, i.e., the process and measurement noises are Gaussian, and that the measurement equation is linear in the object state, i.e., $\mathbf{h}(\mathbf{x}_k) = \mathbf{H}\mathbf{x}_k$. Then, the OID can be written as

$$q(\mathbf{x}_k | \mathbf{y}_k, \mathbf{x}_{k-1}) = \frac{N(\mathbf{y}_k; \mathbf{H}_k \mathbf{x}_k, \mathbf{R}_k) N(\mathbf{x}_k; \mathbf{f}_k(\mathbf{x}_{k-1}), \mathbf{Q}_k)}{\int N(\mathbf{y}_k; \mathbf{H}\boldsymbol{\xi}_k, \mathbf{R}_k) N(\boldsymbol{\xi}_k; \mathbf{f}_k(\mathbf{x}_{k-1}), \mathbf{Q}_k) \, d\boldsymbol{\xi}_k}. \tag{2.72}$$

By Theorem 2.1,

$$N(\mathbf{y}_k; \mathbf{H}\mathbf{x}_k, \mathbf{R}_k)N(\mathbf{x}_k; \mathbf{f}(\mathbf{x}_{k-1}), \mathbf{Q}_k) = N(\mathbf{y}_k; \boldsymbol{\gamma}_k, \mathbf{S}_k)N(\mathbf{x}_k; \boldsymbol{\mu}_k, \boldsymbol{\Sigma}_k), \qquad (2.73)$$

where

$$\boldsymbol{\gamma}_k = \mathbf{H}\mathbf{f}(\mathbf{x}_{k-1}),$$
$$\mathbf{S}_k = \mathbf{H}\mathbf{Q}_k\mathbf{H}^T + \mathbf{R}_k,$$
$$\boldsymbol{\mu}_k = \mathbf{f}(\mathbf{x}_{k-1}) + \mathbf{K}_k(\mathbf{y}_k - \boldsymbol{\gamma}_k),$$
$$\boldsymbol{\Sigma}_k = \mathbf{Q}_k - \mathbf{K}_k\mathbf{H}\mathbf{Q}_k,$$

with $\mathbf{K}_k = \mathbf{Q}_k\mathbf{H}^T\mathbf{S}_k^{-1}$. Substituting (2.73) into (2.72) gives

$$p(\mathbf{x}_k|\mathbf{y}_k, \mathbf{x}_{k-1}) = N(\mathbf{x}_k; \boldsymbol{\mu}_k, \boldsymbol{\Sigma}_k). \qquad (2.74)$$

The weights use the normalizing factor in (2.72), which can be found using (2.73) as

$$p(\mathbf{y}_k|\mathbf{x}_{k-1}) = \int N(\mathbf{y}_k; \mathbf{H}\mathbf{x}_k, \mathbf{R}_k)N(\mathbf{x}_k; \mathbf{f}(\mathbf{x}_{k-1}), \mathbf{Q}_k)\, d\mathbf{x}_k = N(\mathbf{y}_k; \boldsymbol{\gamma}_k, \mathbf{S}_k). \qquad (2.75)$$

A recursion of the OID-PF for single-object tracking is given by Algorithm 7. The sampling density for the object state is the same as that given by the Kalman filter with perfect information at the previous time. For a given sample size, the OID-PF provides a more accurate approximation than the BF. Intuitively, this is because it uses the current measurement in the sampling of mixture indices and object states. However, the conditions required for derivation of the sampling densities, Gaussian process and measurement noises and a linear measurement equation are too strict for many object tracking applications. The principle of measurement-directed sampling can still be employed through the use of auxiliary PFs (APFs).

2.6.3 Auxiliary bootstrap filter for single-object tracking

The auxiliary bootstrap filter (ABF) uses the importance density

$$q(\mathbf{x}_k, t) = \xi_k^t p_{\mathbf{v}_k}(\mathbf{x}_k - \mathbf{f}(\mathbf{x}_{k-1}^t)), \qquad (2.76)$$

where

$$\xi_k^t = w_{k-1}^t p_{\mathbf{w}_k}(\mathbf{y}_k - \mathbf{h}(\boldsymbol{\mu}_k^t)) \left/ \sum_{s=1}^{n} w_{k-1}^s p_{\mathbf{w}_{\rho_k^s}}(\mathbf{y}_k - \mathbf{h}_{\rho_k^s}(\boldsymbol{\mu}_k^s)) \right., \qquad (2.77)$$

with $\boldsymbol{\mu}_k^t = \mathbf{f}(\mathbf{x}_{k-1}^t) + \mathbf{v}_k^t$, $\mathbf{v}_k^t \sim p_{\mathbf{v}_k}$. A recursion of the ABF is given by Algorithm 8. The ABF has the same generality as the BF but will perform better for a given

---

Algorithm 7 Optimal importance density particle filter for single-object tracking

---

1: Compute: $\mathbf{S}_k = \mathbf{H}\mathbf{Q}_k\mathbf{H}^T + \mathbf{R}_k$, $\mathbf{K}_k = \mathbf{Q}_k\mathbf{H}^T\mathbf{S}_k^{-1}$ and $\Sigma_k = \mathbf{Q}_k - \mathbf{K}_k\mathbf{H}\mathbf{Q}_k$.

2: **for** $i = 1, \ldots, n$ **do**

3:     Compute:

$$\boldsymbol{\gamma}_k^i = \mathbf{H}\mathbf{f}(\mathbf{x}_{k-1}^i),$$
$$\boldsymbol{\mu}_k^i = \mathbf{f}(\mathbf{x}_{k-1}^i) + \mathbf{K}_k(\mathbf{y}_k - \boldsymbol{\gamma}_k^i).$$

4:     Compute the first-stage weight update $a_k^i = N(\mathbf{y}_k; \boldsymbol{\gamma}_k^i, \mathbf{S}_k)$.

5: **end for**

6: Compute the first-stage weights:

$$\psi_k^t = w_{k-1}^t a_k^t \Big/ \sum_{i=1}^{n} w_{k-1}^i a_k^i, \quad t = 1, \ldots, n.$$

7: **for** $i = 1, \ldots, n$ **do**

8:     Draw a mixture index t^i such that $\Pr(t^i = l) = \psi_k^l$.

9:     Draw the sample object state $\mathbf{x}_k^i \sim N(\boldsymbol{\mu}_k^{t^i}, \Sigma_k)$.

10: **end for**

11: Compute the weights $w_k^i = 1/n, i = 1, \ldots, n$.

12: Compute a state estimate:

$$\hat{\mathbf{x}}_{k|k} = \sum_{i=1}^{n} w_k^i \mathbf{x}_k^i.$$

---

sample size as it takes into account the current measurement when drawing mixture indices. This provides a reasonable compromise between generality and ease of implementation on the one hand and approximation accuracy on the other.

2.6.4 Extended Kalman auxiliary particle filter for single-object tracking

A major disadvantage of the OID-PF is that it cannot be used when the measurement equation is non-linear in the object state. The extended Kalman APF (EK-APF) removes the need for linearity by adopting a linearized approximation to the measurement equation. Gaussian assumptions on the process and measurement noises are retained. The importance density is

$$q(\mathbf{x}_k, t) = \xi_k^t \hat{p}(\mathbf{x}_k | \mathbf{x}_{k-1}^t, \mathbf{y}_{1:k}), \tag{2.78}$$

Algorithm 8 Auxiliary bootstrap filter for single-object tracking

---

1: **for** $i = 1, \ldots, n$ **do**

2:     Draw $\tilde{\mathbf{v}}_k^i \sim p_{\mathbf{v}_k}$ and compute $\boldsymbol{\mu}_k^i = \mathbf{f}(\mathbf{x}_{k-1}^i) + \tilde{\mathbf{v}}_k^i$.

3:     Compute the first-stage weight update $a_k^i = p_{\mathbf{w}_k}(\mathbf{y}_k - \mathbf{h}(\boldsymbol{\mu}_k^i))$.

4: **end for**

5: Compute the first-stage weights:

$$\xi_k^t = w_{k-1}^t a_k^t \left/ \sum_{i=1}^n w_{k-1}^i a_k^i \right. , \quad t = 1, \ldots, n.$$

6: **for** $i = 1, \ldots, n$ **do**

7:     Draw a mixture index t^i such that $\Pr(t^i = l) = \xi_k^l$.

8:     Draw $\mathbf{v}_k^i \sim p_{\mathbf{v}_k}$ and compute the sample object state $\mathbf{x}_k^i \sim \mathbf{f}(\mathbf{x}_{k-1}^{t^i}) + \mathbf{v}_k^i$.

9:     Compute the un-normalized weight:

$$\tilde{w}_k^i = \frac{p_{\mathbf{w}_k}(\mathbf{y}_k - \mathbf{h}(\mathbf{x}_k^i))}{p_{\mathbf{w}_k}(\mathbf{y}_k - \mathbf{h}(\boldsymbol{\mu}_k^{t^i}))}.$$

10: **end for**

11: Normalize the weights:

$$w_k^i = \tilde{w}_k^i \left/ \sum_{j=1}^n \tilde{w}_k^j \right. , \quad i = 1, \ldots, n.$$

12: Compute a state estimate:

$$\hat{\mathbf{x}}_{k|k} = \sum_{i=1}^n w_k^i \mathbf{x}_k^i.$$

---

where

$$\xi_k^t = w_{k-1}^t \hat{p}(\mathbf{y}_k | \mathbf{x}_{k-1}^t, \mathbf{y}_{1:k-1}) \left/ \sum_{i=1}^n w_{k-1}^i \hat{p}(\mathbf{y}_k | \mathbf{x}_{k-1}^i, \mathbf{y}_{1:k-1}) \right. . \qquad (2.79)$$

The sampling density and update factor for the first-stage weights are

$$\hat{p}(\mathbf{x}_k | \mathbf{x}_{k-1}, \mathbf{y}_{1:k}) = \frac{p_{\mathbf{w}_k}(\mathbf{y}_k - \mathbf{h}(\mathbf{f}(\mathbf{x}_{k-1})) - \mathbf{H}(\mathbf{x}_k - \mathbf{f}(\mathbf{x}_{k-1}))) p_{\mathbf{v}_k}(\mathbf{x}_k - \mathbf{f}(\mathbf{x}_{k-1}))}{\hat{p}(\mathbf{y}_k | \mathbf{x}_{k-1}, \mathbf{y}_{1:k-1})},$$

$$\qquad (2.80)$$

$$\hat{p}(\mathbf{y}_k | \mathbf{x}_{k-1}, \mathbf{y}_{1:k-1}) = \int p_{\mathbf{w}_k}(\mathbf{y}_k - \mathbf{h}(\mathbf{f}(\mathbf{x}_{k-1}))$$

$$\qquad - \mathbf{H}(\boldsymbol{\xi}_k - \mathbf{f}(\mathbf{x}_{k-1}))) p_{\mathbf{v}_k}(\boldsymbol{\xi}_k - f(\mathbf{x}_{k-1})) d\boldsymbol{\xi}_k, \qquad (2.81)$$

Algorithm 9 Extended Kalman auxiliary particle filter for single-object tracking

---

1: **for** $i = 1, \ldots, n$ **do**
2:     Compute the Jacobian $\mathbf{H}_k^i = \nabla_{\mathbf{x}^T} \mathbf{h}(\mathbf{x})|_{\mathbf{x}=\mathbf{f}(\mathbf{x}_{k-1}^i)}$.
3:     Compute:

$$
\begin{aligned}
\mathbf{x}_{k|k-1}^i &= \mathbf{f}(\mathbf{x}_{k-1}^i), & \boldsymbol{\gamma}_k^i &= \mathbf{h}(\mathbf{x}_{k|k-1}^i), \\
\mathbf{S}_k^i &= \mathbf{H}_k^i \mathbf{Q}_k (\mathbf{H}_k^i)^T + \mathbf{R}_k, & \mathbf{K}_k^i &= \mathbf{Q}_k (\mathbf{H}_k^i)^T (\mathbf{S}_k^i)^{-1}, \\
\boldsymbol{\mu}_k^i &= \mathbf{f}(\mathbf{x}_{k-1}^i) + \mathbf{K}_k^i (\mathbf{y}_k - \boldsymbol{\gamma}_k^i), & \boldsymbol{\Sigma}_k^i &= \mathbf{Q}_k - \mathbf{K}_k \mathbf{H}_k^i \mathbf{Q}_k.
\end{aligned}
$$

4:     Compute the first-stage weight update $a_k^i = N(\mathbf{y}_k; \boldsymbol{\gamma}_k^i, \mathbf{S}_k^i)$.
5: **end for**
6: Compute the first-stage weights:

$$
\psi_k^t = w_{k-1}^t a_k^t \bigg/ \sum_{i=1}^n w_{k-1}^i a_k^i, \quad t = 1, \ldots, n.
$$

7: **for** $i = 1, \ldots, n$ **do**
8:     Draw a mixture index t^i such that $\Pr(t^i = l) = \psi_k^l$.
9:     Draw the sample object state $\mathbf{x}_k^i \sim N(\boldsymbol{\mu}_k^{t^i}, \boldsymbol{\Sigma}_k^{t^i})$.
10:    Compute the un-normalized weight:

$$
\tilde{w}_k^i = \frac{p_{\mathbf{w}_k}(\mathbf{y}_k - \mathbf{h}(\mathbf{x}_k^i))}{p_{\mathbf{w}_k}(\mathbf{y}_k - \boldsymbol{\gamma}_k^{t^i} - \mathbf{H}_k^{t^i}(\mathbf{x}_k^i - \mathbf{x}_{k|k-1}^{t^i}))}.
$$

11: **end for**
12: Normalize the weights:

$$
w_k^i = \tilde{w}_k^i \bigg/ \sum_{j=1}^n \tilde{w}_k^j, \quad i = 1, \ldots, n.
$$

13: Compute a state estimate:

$$
\hat{\mathbf{x}}_{k|k} = \sum_{i=1}^n w_k^i \mathbf{x}_k^i.
$$

---

where $\mathbf{H} = \nabla_{\mathbf{x}^T} \mathbf{h}(\mathbf{x})|_{\mathbf{x}=\mathbf{f}(\mathbf{x}_{k-1})}$. The quantities (2.80) and (2.81) can be derived as for the OID-PF to give the procedure of Algorithm 9. The weight calculation for the EK-APF, in step 10, is interesting. It is here that the linearization used in the sampling procedure is accounted for. It is computationally expensive to linearize about each predicted sample $\mathbf{x}^i_{k|k-1}$. An alternative is to linearize about the mean predicted sample. This is much less computationally expensive because the same innovation covariance matrix, gain matrix and posterior covariance matrix is then used for all samples. Although the EK-APF does not have the generality of the ABF, since Gaussian process and measurement noises are assumed, it does not require the restrictive assumption of a linear measurement equation.

It should be noted that all of the PFs described in this section are capable of optimal performance as the sample size $n \to \infty$. Performance differences between the various algorithms arise when they are used for finite sample sizes.

2.7 Performance bounds

We have seen that the optimal Bayesian filter can be realized only under restrictive conditions, for example, if the dynamic and measurement equations are linear and Gaussian. In many cases of practical interest these conditions are not met and it is necessary to resort to a sub-optimal approximation. A useful way of assessing these sub-optimal estimators is to compare their performance with the best possible performance in some sense. Indicators of best possible performance, referred to as performance bounds, usually take the form of a lower bound on the mean square error (MSE). Examples of performance bounds include the Cramér–Rao bound (CRB), Barankin bound (BB) and Weiss–Weinstein bound (WWB). Here we will focus on the CRB due to the ease with which it can be computed. Other bounds are often tighter than the CRB and can be computed under less restrictive conditions but are considerably more difficult to compute.

The CRB places a lower bound on the variance of any unbiased estimator of a determinstic parameter. Specifically, consider a determinstic vector parameter $\boldsymbol{\theta} = [\theta_1, \ldots, \theta_q]^T$ and a collection of measurements $\mathbf{y} = [y_1, \ldots, y_n]^T$ dependent upon $\boldsymbol{\theta}$. Then, the covariance of any estimator $\hat{\boldsymbol{\theta}}(\mathbf{y})$ satisfying $\mathsf{E}[\hat{\boldsymbol{\theta}}(\mathbf{y})] = \boldsymbol{\theta}$ satisfies

$$\mathsf{cov}(\hat{\boldsymbol{\theta}}(\mathbf{y})) \geq \mathbf{J}^{-1}, \tag{2.82}$$

where $\mathbf{J}$ is the Fisher information matrix,

$$\mathbf{J} = -\mathsf{E}[\nabla_{\boldsymbol{\theta}} \nabla_{\boldsymbol{\theta}}^T \log p(\mathbf{y}|\boldsymbol{\theta})], \tag{2.83}$$

with $\nabla_{\boldsymbol{\theta}} = [\partial/\partial\theta_1, \ldots, \partial/\partial\theta_q]^T$. Equation (2.82) holds provided that the derivatives and expectations in (2.83) exist.

Since the CRB applies to deterministic parameters it cannot be directly applied to the filtering problem where we are interested in estimating random parameters. A counterpart to the CRB for use with random parameters, known as the posterior CRB (PCRB), was formulated in Van Trees (1968). The PCRB is defined as follows. Consider a random vector parameter $\boldsymbol{\theta} = [\theta_1, \ldots, \theta_q]^T$ and observations $\mathbf{y} = [y_1, \ldots, y_n]^T$. The mean square error of the estimator $\hat{\boldsymbol{\theta}}(\mathbf{y})$ satisfies

$$\mathsf{mse}(\hat{\boldsymbol{\theta}}(\mathbf{y})) = \mathsf{E}[\hat{\boldsymbol{\theta}}(\mathbf{y}) - \boldsymbol{\theta}][\hat{\boldsymbol{\theta}}(\mathbf{y}) - \boldsymbol{\theta}]^T \geq \mathbf{J}^{-1}, \tag{2.84}$$

where

$$\mathbf{J} = -\mathsf{E}[\nabla_{\boldsymbol{\theta}} \nabla_{\boldsymbol{\theta}}^T \log p(\mathbf{y}, \boldsymbol{\theta})]. \tag{2.85}$$

Note that the expectation in (2.85) is taken over both the measurements $\mathbf{y}$ and the parameter $\boldsymbol{\theta}$. In addition to requiring that the derivatives and expectations in (2.85) exist, (2.84) requires that, for $i = 1, \ldots, q$,

$$\lim_{\theta_i \to \infty} \mathbf{b}(\boldsymbol{\theta}) p(\boldsymbol{\theta}) = \mathbf{0}, \qquad \lim_{\theta_i \to -\infty} \mathbf{b}(\boldsymbol{\theta}) p(\boldsymbol{\theta}) = \mathbf{0}, \tag{2.86}$$

where $\mathbf{b}(\boldsymbol{\theta})$ is the conditional bias,

$$\mathbf{b}(\boldsymbol{\theta}) = \int [\hat{\boldsymbol{\theta}}(\mathbf{y}) - \boldsymbol{\theta}] p(\mathbf{y}|\boldsymbol{\theta}) \, d\mathbf{y}.$$

The conditions (2.86) may be viewed as analogous to the unbiased condition in the CRB for deterministic parameters.

To obtain a bound for the filtering problem at time k we consider a vector parameter composed of object states up to the current time, $\boldsymbol{\theta} = \mathbf{x}^k = [\mathbf{x}_0^T, \ldots, \mathbf{x}_k^T]^T$, and the observation history $\mathbf{y} = \mathbf{y}^k = [\mathbf{y}_1^T, \ldots, \mathbf{y}_k^T]^T$. Note that, although at present we are interested only in bounding the MSE of estimators of $\mathbf{x}_k$, it is necessary to include all past states in the vector parameter. Let $\mathbf{J}_k(\boldsymbol{\theta})$ be such that $\mathsf{mse}(\hat{\boldsymbol{\theta}}(\mathbf{y}^k)) \geq \mathbf{J}_k(\boldsymbol{\theta})^{-1}$, i.e., the inverse of $\mathbf{J}_k(\boldsymbol{\theta})$ lower bounds the MSE of estimators of $\boldsymbol{\theta}$ based on observations up to time k. According to this notation we wish to compute $\mathbf{J}_k(\mathbf{x}_k)$ for $k = 1, 2, \ldots$. For $\boldsymbol{\theta} = \mathbf{x}^k$, we have

$$\mathbf{J}_k(\mathbf{x}^k) = \begin{bmatrix} \mathbf{A}_k & \mathbf{B}_k \\ \mathbf{B}_k^T & \mathbf{C}_k \end{bmatrix}, \tag{2.87}$$

where

$$\mathbf{A}_k = -\mathsf{E}[\nabla_{\mathbf{x}^{k-1}} \nabla_{\mathbf{x}^{k-1}}^T \log p(\mathbf{y}^k, \mathbf{x}^k)],$$

$$\mathbf{B}_k = -\mathsf{E}[\nabla_{\mathbf{x}^{k-1}} \nabla_{\mathbf{x}_k}^T \log p(\mathbf{y}^k, \mathbf{x}^k)],$$

$$\mathbf{C}_k = -\mathsf{E}[\nabla_{\mathbf{x}_k} \nabla_{\mathbf{x}_k}^T \log p(\mathbf{y}^k, \mathbf{x}^k)].$$

The desired quantity can then be found as

$$\mathbf{J}_k(\mathbf{x}_k) = \mathbf{C}_k - \mathbf{B}_k^T \mathbf{A}_k^{-1} \mathbf{B}_k. \tag{2.88}$$

Note that direct implementation of (2.88) as a method of computing the filtering bound requires inversion of a $n_x(k-1) \times n_x(k-1)$ matrix at time k for a state of dimension n_x. This operation will become computationally expensive for even a modest amount of data. A better approach is to exploit the particular form of the joint density to obtain a filtering bound which can be computed recursively. This recursive method of computing the PCRB for the filtering problem was first proposed by Tichavský *et al.* (1998) and has proved to be extremely useful.

To derive this recursion, the logarithm of the joint density is expanded as

$$\log p(\mathbf{y}^k, \mathbf{x}^k) = \log p(\mathbf{x}_0) + \sum_{t=1}^{k} \left[\log p(\mathbf{y}_t|\mathbf{x}_t) + \log p(\mathbf{x}_t|\mathbf{x}_{t-1}) \right]. \qquad (2.89)$$

Recall that the transition density $p(\mathbf{x}_t|\mathbf{x}_{t-1})$ and likelihood $p(\mathbf{y}_t|\mathbf{x}_t)$, required in (2.89), are determined by the object dynamics (2.1) and the measurement equation (2.2). Using (2.89) it can be seen that $\mathbf{J}(\mathbf{x}^k)$ has the form

$$\mathbf{J}_k(\mathbf{x}^k) = \begin{bmatrix} \mathbf{J}_0(\mathbf{x}_0) + \mathbf{U}_1 & \mathbf{V}_1 & \mathbf{0} & \cdots & & \mathbf{0} \\ \mathbf{V}_1^T & \mathbf{W}_1 + \mathbf{U}_2 & \mathbf{V}_2 & & \ddots & \vdots \\ \mathbf{0} & \ddots & \ddots & & \ddots & \mathbf{0} \\ \vdots & & \ddots & & \mathbf{V}_{k-1}^T & \mathbf{W}_{k-1} + \mathbf{U}_k & \mathbf{V}_k \\ \mathbf{0} & \cdots & & \mathbf{0} & \mathbf{V}_k^T & \mathbf{W}_k \end{bmatrix}, \qquad (2.90)$$

where, for $t = 1, \ldots, k$,

$$\mathbf{U}_t = -\mathsf{E}\left[\nabla_{\mathbf{x}_{t-1}} \nabla_{\mathbf{x}_{t-1}}^T \log p(\mathbf{x}_t|\mathbf{x}_{t-1}) \right], \qquad (2.91)$$

$$\mathbf{V}_t = -\mathsf{E}\left[\nabla_{\mathbf{x}_{t-1}} \nabla_{\mathbf{x}_t}^T \log p(\mathbf{x}_t|\mathbf{x}_{t-1}) \right], \qquad (2.92)$$

$$\mathbf{W}_t = -\mathsf{E}\left[\nabla_{\mathbf{x}_t} \nabla_{\mathbf{x}_t}^T \log p(\mathbf{x}_t|\mathbf{x}_{t-1}) \right] - \mathsf{E}\left[\nabla_{\mathbf{x}_t} \nabla_{\mathbf{x}_t}^T \log p(\mathbf{y}_t|\mathbf{x}_t) \right]. \qquad (2.93)$$

Equation (2.90) can be written in the form (2.87) with

$$\mathbf{A}_k = \begin{bmatrix} \mathbf{A}_{k-1} & \mathbf{B}_{k-1} \\ \mathbf{B}_{k-1}^T & \mathbf{C}_{k-1} + \mathbf{U}_k \end{bmatrix}, \qquad (2.94)$$

$$\mathbf{B}_k = \begin{bmatrix} \mathbf{0} \\ \mathbf{V}_k \end{bmatrix}, \qquad (2.95)$$

$$\mathbf{C}_k = \mathbf{W}_k. \qquad (2.96)$$

Substituting (2.94)–(2.96) into (2.88) gives

$$\mathbf{J}_k(\mathbf{x}_k) = \mathbf{W}_k - \mathbf{V}_k^T (\mathbf{C}_{k-1} + \mathbf{U}_k - \mathbf{B}_{k-1}^T \mathbf{A}_{k-1}^{-1} \mathbf{B}_{k-1})^{-1} \mathbf{V}_k.$$

Since $\mathbf{J}_{k-1}(\mathbf{x}_{k-1}) = \mathbf{C}_{k-1} - \mathbf{B}_{k-1}^T \mathbf{A}_{k-1}^{-1} \mathbf{B}_{k-1}$ we obtain the recursive equation

$$\mathbf{J}_k(\mathbf{x}_k) = \mathbf{W}_k - \mathbf{V}_k^T [\mathbf{J}_{k-1}(\mathbf{x}_{k-1}) + \mathbf{U}_k]^{-1} \mathbf{V}_k, \qquad (2.97)$$

for $k = 1, 2, \dots$. Computation of the PCRB then amounts to computing the quantities in (2.91)–(2.93) and substituting into (2.97) for each time step. This procedure is demonstrated in the following simple example.

Example 2.3: PCRB for linear/Gaussian systems
Assume that we have linear/Gaussian dynamic and measurement equations:

$$\mathbf{x}_k = \mathbf{F}_k \mathbf{x}_{k-1} + \mathbf{v}_k,$$

$$\mathbf{y}_k = \mathbf{H}_k \mathbf{x}_k + \mathbf{w}_k,$$

where $\mathrm{cov}(\mathbf{v}_k, \mathbf{v}_l) = \mathbf{Q}_k \delta_{k-l}$, $\mathrm{cov}(\mathbf{v}_k, \mathbf{w}_l) = \mathbf{0}$ and $\mathrm{cov}(\mathbf{w}_k, \mathbf{w}_k) = \mathbf{R}_k \delta_{k-l}$. The required quantities can be found as

$$\mathbf{U}_k = -\mathbb{E}\left[\nabla_{\mathbf{x}_{t-1}} \nabla_{\mathbf{x}_{t-1}}^T (\mathbf{x}_k - \mathbf{F}_k \mathbf{x}_{k-1})^T \mathbf{Q}_k^{-1} (\mathbf{x}_k - \mathbf{F}_k \mathbf{x}_{k-1}) \right]$$

$$= \mathbf{F}_k^T \mathbf{Q}_k^{-1} \mathbf{F}_k, \qquad (2.98)$$

$$\mathbf{V}_k = -\mathbf{F}_k^T \mathbf{Q}_k^{-1}, \qquad (2.99)$$

$$\mathbf{W}_k = \mathbf{H}_k^T \mathbf{R}_k^{-1} \mathbf{H}_k + \mathbf{Q}_k^{-1}. \qquad (2.100)$$

Substituting (2.98)–(2.100) into (2.97) gives

$$\mathbf{J}_k(\mathbf{x}_k) = \mathbf{H}_k^T \mathbf{R}_k^{-1} \mathbf{H}_k + \mathbf{Q}_k^{-1} - \mathbf{Q}_k^{-1} \mathbf{F}_k [\mathbf{J}_{k-1}(\mathbf{x}_{k-1}) + \mathbf{F}_k^T \mathbf{Q}_k^{-1} \mathbf{F}_k]^{-1} \mathbf{F}_k^T \mathbf{Q}_k^{-1}. \qquad (2.101)$$

It can be shown that the recursion (2.101) is the same as that obtained by the Kalman filter for the inverse of the covariance matrix. Thus the PCRB and the Kalman filter estimator covariance matrix are equal for a linear/Gaussian system.

An important aspect of computing the PCRB which does not arise in the linear/ Gaussian example is the evaluation of the expectations in (2.91)–(2.93). In most cases these expectations cannot be computed in closed-form and it is necessary to use an approximation. The most common approximation is Monte Carlo integration, which involves simulating a large number of object trajectories and replacing integrals by summations. In later sections we will encounter examples which require such an approach.

A limitation of the above PCRB recursion is that it applies only to the case where the transition density is non-singular. In an additive noise dynamic model, for example, this means that the noise covariance matrix must be non-singular. This limitation is ovecome in the recursion proposed by Bergman (2001).

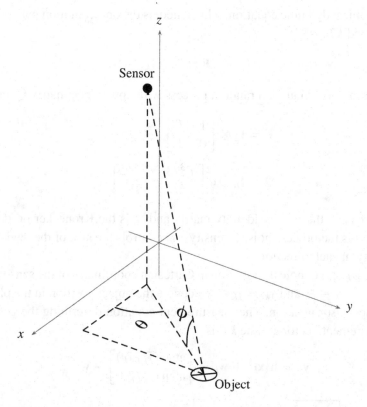

Figure 2.2 Depiction of the example tracking scenario. The angle θ is the azimuth and ϕ is the elevation.

2.8 Illustrative example

Angle tracking

The behavior of the algorithms described in this chapter will now be illustrated in a typical tracking scenario. In the scenario considered here an airborne sensor observes a object moving along the ground. The sensor measures the azimuth and elevation of the object as shown in Figure 2.2. The measurement equation is non-linear and it will be necessary to approximate the optimal Bayesian solution. We will consider the EKF, UKF and two PFs.

It is assumed that measurements are obtained at equi-spaced intervals of T seconds. We need an equation which describes the object motion beween each of these sampling instants. This requires the use of some prior knowledge regarding the likely form of object motion. Here it is assumed that the object moves with a velocity subject to small random perturbations. For this object model the object state at time kT is $\mathbf{x}_k = [x_k, \dot{x}_k, y_k, \dot{y}_k]^T$ where (x_k, y_k) is the object position in Cartesian coordinates and the dot notation denotes differentiation with respect to

time. The object dynamic equation, which models the object motion between times
$(k-1)T$ and kT, is

$$\mathbf{x}_k = \mathbf{F}\mathbf{x}_{k-1} + \mathbf{v}_k,$$

where $\mathbf{v}_k$ is a white Gaussian random process with covariance matrix $\mathbf{Q}$ and

$$\mathbf{F} = \mathbf{I}_2 \otimes \begin{bmatrix} 1 & T \\ 0 & 1 \end{bmatrix},$$

$$\mathbf{Q} = \mathbf{I}_2 \otimes q \begin{bmatrix} T^3/3 & T^2/2 \\ T^2/2 & T \end{bmatrix}.$$

The matrix $\mathbf{I}_m$ is the $m \times m$ identity matrix and $\otimes$ is the Kronecker product. The
parameter q is the process noise intensity and controls the size of the deviations of
the velocity in each direction.

Let (ξ_k, ψ_k, ζ_k) denote the position in Cartesian coordinates of the sensor at time
kT and $\bar{x}_k = x_k - \xi_k$ and $\bar{y}_k = y_k - \psi_k$ denote the object position in the plane relative to the sensor position. The measurement equation describing the generation
of a measurement vector at time kT is

$$\mathbf{y}_k = \mathbf{h}(\mathbf{x}_k) + \mathbf{w}_k = \begin{bmatrix} \text{atan}\,(\bar{y}_k/\bar{x}_k) \\ \text{atan}\,(\zeta_k/\rho_k) \end{bmatrix} + \mathbf{w}_k,$$

where $\rho_k = \sqrt{\bar{x}_k^2 + \bar{y}_k^2}$ and $\mathbf{w}_k$ is a white Gaussian random process, independent
of $\mathbf{v}_k$, with covariance matrix $\mathbf{R}$.

Computation of the PCRB requires the matrices $\mathbf{U}_k$, $\mathbf{V}_k$ and $\mathbf{W}_k$ of (2.91)–
(2.93). The matrices $\mathbf{U}_k$ and $\mathbf{V}_k$, which depend only on the state transition equation, can be found as in the linear/Gaussian example. Thus $\mathbf{U}_k = \mathbf{F}^T\mathbf{Q}^{-1}\mathbf{F}$ and
$\mathbf{V}_k = -\mathbf{F}^T\mathbf{Q}^{-1}$. The matrix $\mathbf{W}_k$, which depends on the non-linear measurement
equation, can be found as (Tichavský *et al.*, 1998)

$$\mathbf{W}_k = \mathsf{E}[\mathbf{H}(\mathbf{x}_k)^T\mathbf{R}^{-1}\mathbf{H}(\mathbf{x}_k)] + \mathbf{Q}^{-1}, \tag{2.102}$$

where the expectation is over the object state $\mathbf{x}_k$ and $\mathbf{H}(\mathbf{x}_k) = [\nabla_\mathbf{x}\mathbf{h}(\mathbf{x})^T|_{\mathbf{x}=\mathbf{x}_k}]^T$.
Evaluation of $\mathbf{W}_k$ requires the Jacobian $\mathbf{H}(\mathbf{x}_k)$, which can be found as

$$\mathbf{H}(\mathbf{x}_k) = \begin{bmatrix} -\bar{y}_k/\rho_k^2 & 0 & \bar{x}_k/\rho_k^2 & 0 \\ -\zeta_k\bar{x}_k/[\rho_k(\rho_k^2+\zeta_k^2)] & 0 & -\zeta_k\bar{y}_k/[\rho_k(\rho_k^2+\zeta_k^2)] & 0 \end{bmatrix}. \tag{2.103}$$

Although the distribution of $\mathbf{x}_k$ can be found using the state transition equation, the
expectation in (2.102) cannot be evaluated due to the form of the Jacobian (2.103).
We can instead resort to a Monte Carlo approximation. Let $\mathbf{x}_k^1, \dots \mathbf{x}_k^n$ denote a
collection of samples from the distribution of the state vector at time kT. Then the

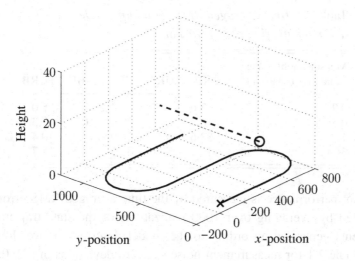

Figure 2.3 Simulation scenario for performance analysis of single-object tracking algorithms.

matrix $\mathbf{W}_k$ can be approximated as

$$\mathbf{W}_k \approx \mathbf{Q}^{-1} + 1/n \sum_{i=1}^{n} \mathbf{H}(\mathbf{x}_k^i)^T \mathbf{R}^{-1} \mathbf{H}(\mathbf{x}_k^i).$$

The samples can be generated recursively by simulating the state transition equation, i.e., begin by drawing samples from the prior distribution, $\mathbf{x}_0^i \sim \pi_0$ for $i = 1, \ldots, n$. Then draw $\mathbf{x}_1^i = \mathbf{F}_0 \mathbf{x}_0^i + \mathbf{w}_0^i$, $i = 1, \ldots, n$ at time T, $\mathbf{x}_2^i = \mathbf{F}_0 \mathbf{x}_1^i + \mathbf{w}_1^i$, $i = 1, \ldots, n$ at time $2T$ and so on. The variables $\mathbf{w}_k^i$ are independent Gaussian random variables with covariance matrix $\mathbf{Q}$.

The performances of several filtering approximations will be examined. Two PF implementations, the BF and the ABF, will be considered in addition to the EKF and the UKF. The PFs are implemented with 10 000 particles. The particular scenario used in the numerical simulations is shown in Figure 2.3. The object trajectory is the solid line with a cross indicating the object position at time t_1. The dashed line is the sensor trajectory with a circle indicating the starting point. The turns executed by the sensor are coordinated turns in which the sensor speed remains constant. The process noise intensity is $q = 1/200$. The initial object state is distributed as $N(\mathbf{x}_0, \mathbf{P}_0)$ where $\mathbf{x}_0 = [350, 0, 0, 6]^T$ and $\mathbf{P}_0 = \text{diag}(225, 25, 225, 25)$. The object is observed for 150 s with measurements taken at equi-spaced intervals of $T = 3$s. The measurement noise covariance matrix is $\mathbf{R} = \kappa^2 \mathbf{I}_2$, where the standard deviation κ will be varied to examine the effect of measurement accuracy.

Table 2.1 *Time-averaged RMS errors for single,*
non-maneuvering object tracking.

Measurement noise standard deviation (°)	EKF	UKF	BF	ABF	PCRB
0.2	7.6	7.7	7.8	7.7	5.0
0.5	10.7	10.7	22.6	10.8	7.3
1	33.6	14.9	25.6	15.5	10.4
3	112.1	52.7	33.5	28.5	18.7

Algorithm performance is measured by the time-averaged RMS error. This is approximated by averaging over 1000 realizations of the state trajectories and measurements generated according to the model described above. Results are shown in Table 2.1 for measurement noise standard deviations of 0.2, 0.5, 1 and 3 degrees. The analytic approximations perform well when the measurements are precise but are poor even for moderate amounts of measurement noise. Of the analytic approximations, the UKF significantly outperforms the EKF. This can be attributed to the increased accuracy of the UT moment approximations compared to linearized moment approximations. As expected the PFs are capable of performing better than the analytic approximations although they do not approach the PCRB for the sample sizes used here. The importance of PF design is clearly evident. The BF, which blindly samples both particle indices and object states, produces divergent estimates even with a relatively large sample size of 10 000. The ABF, which is only a slight modification of the BF to enable measurement-directed sampling of particle indices, is far superior. The relatively poor performance of the ABF compared to the UKF and EKF for $\kappa = 0.2$ degrees is due to the fact that, when the measurements become precise, blind sampling of the object state will result in an insufficient number of samples being drawn in the desired part of the state space. This can be remedied by measurement-directed sampling of the state, as in the EK-APF (Algorithm 9).

2.9 Summary

The problem considered in this chapter is that of tracking a single, non-maneuvering object. Even for this, the simplest of all object tracking problems, the optimal Bayesian solution is available only under restrictive conditions. In most cases of practical interest these conditions are not met thus approximations are required. The various approximations which have been proposed can be broadly categorized into analytic approximations and numerical approximations. Analytic approximations are computationally efficient but are limited as to how accurately they can approach optimality. Techniques belonging to this class include the EKF and the UKF. Numerical approximations are relatively computationally expensive but are capable of nearly optimal performance given sufficient computational resources. The PMF and PFs are examples of numerical approximations.

No attempt has been made here to study all aspects of the single, non-maneuvering object tracking problem. Interesting topics which have not been covered include various filtering approximations, such as variational Bayes approximations and methods which approximate the posterior PDF by a mixture (Alspach and Sorenson, 1971; Ito and Xiong, 2000), generalized edge worth series and Gauss hermite quadrature (Challa, 2000) and the development of exact finite-dimensional non-linear filters (Daum, 1986). These areas of study have been omitted to focus on the most popular filtering approximations and show how they can be derived as approximations to the optimal Bayesian solution. In the context of object tracking, the main interest in these algorithms is that they form the basis of algorithms designed for more complicated problems involving maneuvers, clutter and multiple targets. These problems are considered in the next chapters.

3

Maneuvering object tracking

Maneuvering objects are those objects whose dynamical behavior changes over time. An object that suddenly turns or accelerates displays a maneuvering behavior with regard to its tracked position. While the definition of a maneuvering object extends beyond the tracking of position and speed, historically it is in this context that maneuvering object tracking theory developed. This chapter presents a unified derivation of some of the most common maneuvering object tracking algorithms in the Chapman–Kolmogorov–Bayesian framework.

3.1 Modeling for maneuvering object tracking

In general, maneuvering object tracking refers to the problem of state estimation where the system model undergoes abrupt changes. The standard Kalman filter with a single motion model is limited in performance for such problems because it does not effectively respond to the changes in the dynamics as the object maneuvers. A large number of approaches to the maneuvering object tracking problem have been developed including process noise adaptation (Singer *et al.*, 1974; Moose, 1975; Gholson and Moose, 1977; Ricker and Williams, 1978; Moose *et al.*, 1979; Farina and Studer, 1985), input estimation (Chan *et al.*, 1979), variable dimension filtering (Bar-Shalom and Birmiwal, 1982) and multiple models (MM) (Ackerson and Fu, 1970; Mori *et al.*, 1986; Blom and Bar-Shalom, 1988; Bar-Shalom and Li, 1993), etc. These apparently diverse approaches may be grouped into two broad categories:

1. single model with state augmentation;
2. multiple models with Markovian jumps.

The first category requires maneuver detection and compensation procedures and generally attempts to reduce the filter bias that arises due to a change in the object maneuver mode. In process noise adaptation approaches, object maneuvering is treated as an additional system process noise and the filter switches to an appropriate noise level upon maneuver detection. However, the improvement is limited since the object maneuver process, in general, is not noise. Input estimation and variable dimension filtering are based on the assumption that the object acceleration is constant or slowly varying. Therefore, if the object acceleration changes rapidly, these approaches do not give satisfactory tracking performance.

The second category is based on a stochastic hybrid system with multiple models for the object. Multiple model (MM) estimation assumes that the object dynamic mode at any time matches one of a finite set of predetermined models (modes) and the real object maneuver mode can jump randomly from one to another. If knowledge of the object motion pattern is known and can be described by a small number of models, the MM approach yields good tracking performance when compared to variable dimension filtering and input estimation approaches (Farooq *et al.*, 1992). Implementable MM algorithms include the generalized pseudo-Bayesian (GPB) algorithm (Ackerson and Fu, 1970; Jaffer and Gupta, 1971) and the interacting multiple model (IMM) algorithm (Blom and Bar-Shalom, 1988; Blair *et al.*, 1991; Munir and Atherton, 1994). Many modified or extended versions can be found in the literature, such as the interacting multiple bias model algorithm (Blair and Watson, 1992), the second-order IMM algorithm (Blair *et al.*, 1993), etc. Among MM algorithms, the IMM algorithm is one of the most computationally efficient approaches (Blom and Bar-Shalom, 1988; Bar-Shalom and Li, 1993; Bar-Shalom *et al.*, 2005). Many researchers have successfully applied the IMM algorithm to various maneuvering object tracking applications (Tugnait, 1982; Blom and Bar-Shalom, 1988; Farooq *et al.*, 1992; Bar-Shalom and Li, 1993).

A limitation of all MM algorithms is the fact that they have a fixed structure, i.e., they use a fixed number of models. If the dynamic behavior of the underlying object can be approximated by a small set of models, any MM estimator can provide reasonable tracking performance. In addition, the MM approach assumes that one of the set of models matches the system true mode which is in effect at any given instant. But in reality, knowledge of such a model set is rarely known. Consequently, modeling the possible maneuvering object patterns such that the true mode is always contained in the models chosen may result in a very large model set. As pointed out in Li and Bar-Shalom (1996), apart from an increase in computational load, the inclusion of more models in the model set does not necessarily improve the tracking performance and to the contrary has the potential to degrade the performance. An attractive approach to overcome this problem is the variable structure multiple model (VSMM) algorithm.

The VSMM approach was first considered in Li and Bar-Shalom (1992) and the fundamental framework was laid down in Li (1994) and Li and Bar-Shalom

(1996). VSMM is an extension of MM estimation and appears to be more general in the sense that it includes the standard fixed structure MM (FSMM) approach as a special case. The main model set adaptation schemes are based on digraph switching (DS), adaptive grids (AD) or recursive adaptive model sets (RAMS) (Li and Bar-Shalom, 1992; Li, 2000). Ideally, VSMM should operate with as small a model set as possible at any particular time to limit computation while achieving the accuracy of a large model set. Existing algorithms include the model-group switching (MSG) algorithm (Li *et al.*, 1999), the likely model set algorithm (LMS) (Li and Zhang, 2000) and the minimum sub-model set switching (MSMSS) algorithm (Wang *et al.*, 2003). The MSMSS algorithm is based on the observation that the dynamics (acceleration) of a maneuvering object is a continuous (vector-valued) variable and represents a particular point in the model space at any time. Hence, only a small number of models that cover that point at a given time are necessary candidates for the current "model set."

3.1.1 Single model via state augmentation

When object acceleration is included as a component of the state vector in linear object models, the resulting state space model is usually called the constant acceleration (CA) model, otherwise it is known as the constant velocity (CV) model (Bar-Shalom and Fortmann, 1988; Bar-Shalom and Li, 1993). In other words, the state vector of the CV model is denoted as

$$x = [\text{position, velocity}]',$$

and the state vector of the CA model is denoted as

$$\mathbf{x} = [\text{position, velocity, acceleration}]' = [x, u]',$$

where u stands for the object acceleration (vector). The recursion for the conditional probability density function of the state is

$$p(\mathbf{x}_k|\mathbf{y}^k) = \frac{1}{c} p(\mathbf{y}_k|\mathbf{x}_k, \mathbf{y}^{k-1}) p(\mathbf{x}_k|\mathbf{y}^{k-1})$$

$$= \frac{1}{c} p(\mathbf{y}_k|\mathbf{x}_k) \int p(\mathbf{x}_k|\mathbf{x}_{k-1}) p(\mathbf{x}_{k-1}|\mathbf{y}^{k-1}) d\mathbf{x}_{k-1}, \qquad (3.1)$$

where $\mathbf{y}^k = (\mathbf{y}_1, \mathbf{y}_2, \dots, \mathbf{y}_k)$ is the set of measurements up to time k. The first term of (3.1) is the likelihood function, c is a normalization constant and the integration term in (3.1) is the well-known Chapman–Kolmogrov equation. The recursion shows that this conditional pdf is completely determined by the likelihood function $p(\mathbf{y}_k|\mathbf{x}_k)$, the transition density $p(x_k|x_{k-1})$ and the prior (past history of the conditional pdf) summarized by the term $p(x_{k-1}|\mathbf{y}^{k-1})$. Under linear Gaussian assumptions, all these terms are Gaussian and equation (3.1) results in the standard Kalman filter.

3.1.2 Multiple-model-based approaches

Equation (3.1) can also be written as

$$p(\mathbf{x}_k, \mathbf{u}_k | \mathbf{y}^k)$$
$$= \frac{1}{c} p(\mathbf{y}_k | \mathbf{x}_k, \mathbf{u}_k) \int_{\mathbf{x}_k} \int_{u_k} p(\mathbf{x}_k, \mathbf{u}_k | \mathbf{x}_{k-1}, \mathbf{u}_{k-1}) p(\mathbf{x}_{k-1}, \mathbf{u}_{k-1} | \mathbf{y}^{k-1}) d\mathbf{x}_{k-1} d\mathbf{u}_{k-1},$$
(3.2)

by replacing $(\mathbf{x}_k)$ by $(\mathbf{x}_k, \mathbf{u}_k)$, i.e., writing $(\mathbf{x}_k)$ as the augmentation of $\mathbf{x}_k$ and u_k. It turns out that by discretizing $\mathbf{u}_k$ and focusing only on a few parts of its state space, one can show that it results in the standard MM approach.

In the MM approach, the object acceleration $\mathbf{u}_k$ is assumed to belong to a discrete model set, i.e., $\mathbf{u}_k \in \{s_1, s_2, \ldots, s_N\}$ and the transition within this set of models is governed by a Markov chain. Thus, (3.2) can be rewritten as

$$p(\mathbf{x}_k, \mathbf{u}_k | \mathbf{y}^k) = p(\mathbf{x}_k, \mathbf{u}_k = s_i | \mathbf{y}^k) \quad \forall\, i = 1, 2, \ldots, N. \tag{3.3}$$

Applying Bayes' theorem to each of $p(\mathbf{x}_k, \mathbf{u}_k = s_i | \mathbf{y}^k)$, we have

$$p(\mathbf{x}_k, \mathbf{u}_k = s_i | \mathbf{y}^k)$$
$$= \frac{1}{c} p(\mathbf{y}_k | \mathbf{x}_k, \mathbf{u}_k = s_i) \int \sum_{s_j \in S} p(\mathbf{x}_k, \mathbf{u}_k = s_i | \mathbf{x}_{k-1}, \mathbf{u}_{k-1} = s_j)$$
$$\times p(\mathbf{x}_{k-1}, \mathbf{u}_{k-1} = s_j | \mathbf{y}^{k-1}) d\mathbf{x}_{k-1}$$
$$= \int \sum_{s_j \in S} p(\mathbf{x}_k | \mathbf{x}_{k-1}, \mathbf{u}_{k-1} = s_j) p(\mathbf{u}_k = s_i | \mathbf{u}_{k-1} = s_j)$$
$$\times p(\mathbf{x}_{k-1}, \mathbf{u}_{k-1} = s_j | \mathbf{y}^{k-1}) d\mathbf{x}_{k-1}, \tag{3.4}$$

for $i = 1, 2, \ldots, N$, where the integration is over both the position and velocity spaces while the summation is over the discrete (mode) set S; $p(\mathbf{u}_k = s_i | \mathbf{u}_{k-1} = s_j)$ is the (i, j)th entries of the mode transition probability matrix, which is usually assumed to be independent of the system state x. This is in contrast to the integral in (3.2) when $\mathbf{u}_k$ was assumed continuous. The following observations can be made:

- Equation (3.4) is the optimal solution for recursively determining the conditional PDF of the hybrid state in MM estimation.

- In view of the fact that the overall conditional pdf of the base state is given by

$$p(\mathbf{x}_k | \mathbf{y}^k) = \sum_{s_i \in S} p(\mathbf{x}_k, \mathbf{u}_k = s_i | \mathbf{y}^k)$$

$$= \sum_{s_i \in S} p(\mathbf{x}_k | \mathbf{u}_k = s_i, \mathbf{y}^k) P(\mathbf{u}_k = s_i | \mathbf{y}^k), \qquad (3.5)$$

  the conditional pdf of the base state in the MM approach is a Gaussian mixture rather than a Gaussian, where $p(u_k = s_i | \mathbf{y}^k)$ is the conditional probability based on the hypothesis that object true mode is s_i at time k.

- Applying a Gaussian approximation to the conditional pdf of the base state at each sampling time k yields the sub-optimal GPB1 algorithm. Moreover, if the conditional pdf of the base state is approximated by a Gaussian after every $n - 1$ sampling intervals, equations (3.4) and (3.5) yield the GPB-n algorithm.

- In the limiting case where the number N of modes in the set S goes to infinity, i.e., $N \to \infty$ and the transition probability of the modes is modeled by a Gaussian kernel, the sum operation in (3.4) becomes an integral and equation (3.4) reduces to equation (3.1) (CA model case), which is the worst case due to the higher order (CA) model involved (Bar-Shalom and Fortmann, 1988). It turns out that if the set of object true modes is unknown, increasing the number of acceleration models for the MM algorithm essentially reduces to increasing N towards infinity. Since the CA model is known to give poor performance, an MM filter with a large number of models will also lead to poor performance.

- The sub-optimal VSMM approaches reduce the computational load by choosing the transition probability matrix at each time such that the summations in both (3.4) and (3.5) consider only those models with model probabilities exceeding a threshold.

Fixed and variable multiple models account for most of the approaches used for maneuvering object tracking. Below we introduce the optimal Bayesian filter for single maneuvering object tracking and derive the MM algorithms GPB1, GPB2, IMM, VS-IMM and MSMSS. The primary focus of this chapter concerns object models which can jump between models according to a Markov law, usually referred to as jump-Markov systems (JMS).

3.2 The optimal Bayesian filter

3.2.1 Process, measurement and noise models

In jump-Markov systems, the object dynamics are assumed to belong to the set of models defined by

$$\mathbf{x}_k = \mathbf{f}_{r_k}(\mathbf{x}_{k-1}, \mathbf{u}_k) + \mathbf{v}_{r_k} \quad r_k \in \{1, 2, \ldots, d\}, \qquad (3.6)$$

where the process noise $\mathbf{v}_{r_k}$ is assumed additive and r_k is assumed to be a random variable satisfying a homogeneous discrete-time Markov chain with state space $\{1, \ldots, d\}$ and transition probability matrix Γ, where

$$\Gamma_{ji} = \Pr(r_k = i | r_{k-1} = j),$$

with initial conditions $\Pr(r_0 = i) = \pi_0(i)$. The measurements, in a fairly general sense, are assumed to be model dependent and related to the true object state through

$$\mathbf{y}_k = \mathbf{h}_{r_k}(\mathbf{x}_k) + \mathbf{w}_{r_k}. \tag{3.7}$$

If the model in effect at time k, r_k is known a priori, then the set of equations (3.6) and (3.7) reduce to (2.1) and (2.2) respectively and one of the filters derived in earlier chapters i.e., KF or EKF, can be used to solve for the conditional densities. However, the model in effect is not known a priori. Moreover, since the model jumping process is assumed to be a random process governed by an underlying Markov chain, the true model can never be known. However, it can be estimated by treating it as a joint random variable with the object state and forming a hybrid state as $(\mathbf{x}_k, r_k)$. The measurements $\mathbf{y}^k = (\mathbf{y}_1, \mathbf{y}_2, \ldots, \mathbf{y}_k)$ have information on this continuous-discrete random variable and the optimal Bayesian filter involves deriving recursions for the joint probability density function $p(\mathbf{x}_k, r_k | \mathbf{y}^k)$. Since r_k is a discrete random variable, taking values in a discrete set $\{1, 2, \ldots, d\}$, the joint density can be decomposed into d components as follows:

$$p(\mathbf{x}_k, r_k = i | \mathbf{y}^k) \quad i = \{1, 2, \ldots, d\}.$$

In object tracking, including the case of maneuvering object tracking, the density of interest is the conditional density $p(\mathbf{x}_k | \mathbf{y}^k)$. The joint density recursions form an intermediate step in determining this conditional density. Using the total probability theorem, this conditional density can be obtained by summing up the individual components of the joint density as follows:

$$p(\mathbf{x}_k | \mathbf{y}^k) = \sum_{i=1}^{d} p(\mathbf{x}_k, r_k = i | \mathbf{y}^k).$$

Using the conditional probability lemma, the joint density on the right-hand side of the above equation can be broken up into two components,

$$p(\mathbf{x}_k, r_k = i | \mathbf{y}^k) = p(\mathbf{x}_k | r_k = i, \mathbf{y}^k) p(r_k = i | \mathbf{y}^k). \tag{3.8}$$

Using $\mu_{k|k}(i) = p(r_k = i|\mathbf{y}^k)$ and the above decomposition, the conditional density equation can be written as

$$p(\mathbf{x}_k|\mathbf{y}^k) = \sum_{i=1}^{d} p(\mathbf{x}_k|r_k = i, \mathbf{y}^k)\mu_{k|k}(i). \qquad (3.9)$$

The first component is the conditional density of the object state conditioned on both the model used and the measurements. The second component is the conditional model probability of the ith model conditioned on all the measurements taken up to and including time k. In this approach, Bayes' recursions are derived for both components.

3.2.2 The conditional density and the conditional model probability

The optimal Bayesian recursion for the first component can be derived by first expanding the set of measurements $\mathbf{y}^k$ into $\{\mathbf{y}_k, \mathbf{y}^{k-1}\}$ and then invoking Bayes' theorem:

$$\begin{aligned} p(\mathbf{x}_k|r_k = i, \mathbf{y}^k) &= p(\mathbf{x}_k|r_k = i, \mathbf{y}_k, \mathbf{y}^{k-1}) \\ &= \frac{p(\mathbf{y}_k|\mathbf{x}_k, r_k = i, \mathbf{y}^{k-1})}{p(\mathbf{y}_k|r_k = i, \mathbf{y}^{k-1})} p(\mathbf{x}_k|r_k = i, \mathbf{y}^{k-1}), \qquad (3.10) \end{aligned}$$

where:

- $p(\mathbf{x}_k|r_k = i, \mathbf{y}^{k-1})$ is the prediction density;
- $p(\mathbf{y}_k|\mathbf{x}_k, r_k = i, \mathbf{y}^{k-1})$ is the likelihood function; and
- $p(\mathbf{y}_k|r_k = i, \mathbf{y}^{k-1})$ is the normalization factor.

Similarly, the second component of the joint density in (3.8), the conditional model probability $\mu_{k|k}(i) = p(r_k = i|\mathbf{y}^k)$ can also be recursively calculated. Once again, expanding $\mathbf{y}^k$ into $\{\mathbf{y}_k, \mathbf{y}^{k-1}\}$ and invoking Bayes' theorem,

$$\mu_{k|k}(i) = p(r_k = i|\mathbf{y}_k, \mathbf{y}^{k-1}) = \frac{p(\mathbf{y}_k|r_k = i, \mathbf{y}^{k-1})p(r_k = i|\mathbf{y}^{k-1})}{p(\mathbf{y}_k|\mathbf{y}^{k-1})}.$$

Letting $p(r_k = i|\mathbf{y}^{k-1}) = \mu_{k|k-1}(i)$, the conditional probability recursion can be rewritten as

$$\mu_{k|k}(i) = \frac{p(\mathbf{y}_k|r_k = i, \mathbf{y}^{k-1})\mu_{k|k-1}(i)}{p(\mathbf{y}_k|\mathbf{y}^{k-1})}, \qquad (3.11)$$

where:

- $\mu_{k|k-1}(i)$ is the predicted model probability;
- $p(\mathbf{y}_k|r_k = i, \mathbf{y}^{k-1})$ is the likelihood function; and
- $p(\mathbf{y}_k|\mathbf{y}^{k-1})$ is the normalization factor.

3.2.3 Optimal estimation

The prediction density and the predicted model probability

Introducing the prior object state $\mathbf{x}_{k-1}$, the prediction density can be expanded into

$$p(\mathbf{x}_k|r_k = i, \mathbf{y}^{k-1}) = \int_{\mathbf{x}_{k-1}} p(\mathbf{x}_k, \mathbf{x}_{k-1}|r_k = i, \mathbf{y}^{k-1})d\mathbf{x}_{k-1}.$$

Invoking the conditional density lemma on the joint density inside the integrand, the prediction density can be decomposed into

$$p(\mathbf{x}_k|r_k = i, \mathbf{y}^{k-1}) = \int_{\mathbf{x}_{k-1}} p(\mathbf{x}_k|\mathbf{x}_{k-1}, r_k = i, \mathbf{y}^{k-1})p(\mathbf{x}_{k-1}|r_k = i, \mathbf{y}^{k-1})d\mathbf{x}_{k-1}.$$

$$(3.12)$$

The first integrand can be identified as the transition density that can be derived from the object dynamical equations assumed in (3.6):

$$p(\mathbf{x}_k|\mathbf{x}_{k-1}, r_k = i, \mathbf{y}^{k-1}) = p_{\mathbf{v}_i}(\mathbf{x}_k - \mathbf{f}_{\mathbf{r_k}=\mathbf{i}}(\mathbf{x}_{k-1}, u_{\mathbf{k}})),$$

where $\mathbf{v}_i = v_{r_k=i}$ and $_\mathbf{i}(\cdot) = \mathbf{f}_{\mathbf{r_k}=\mathbf{i}}(\cdot)$ are used in the above equation.

The second term of (3.12) can be rewritten according to the total probability theorem using all the models at time $k - 1$:

$$p(\mathbf{x}_{k-1}|r_k = i, \mathbf{y}^{k-1})$$

$$= \sum_{j=1}^{d} p(\mathbf{x}_{k-1}|r_{k-1} = j, r_k = i, \mathbf{y}^{k-1})p(r_{k-1} = j|r_k = i, \mathbf{y}^{k-1})$$

$$= \sum_{j=1}^{d} p(\mathbf{x}_{k-1}|r_{k-1} = j, \mathbf{y}^{k-1})p(r_{k-1} = j|r_k = i, \mathbf{y}^{k-1})$$

$$= \sum_{j=1}^{d} p(\mathbf{x}_{k-1}|r_{k-1} = j, \mathbf{y}^{k-1})\mu_{k-1, j|i}.$$

$$(3.13)$$

Equation (3.13) is called the "mixing" step. The mixing probabilities $\mu_{k-1,j|i} = p(r_{k-1} = j|r_k = i, \mathbf{y}^{k-1})$ are calculated as

$$
\begin{aligned}
\mu_{k-1,j|i} &= p(r_{k-1} = j|r_k = i, \mathbf{y}^{k-1}) \\
&= \frac{p(r_k = i|r_{k-1} = j\mathbf{y}^{k-1})p(r_{k-1} = j|\mathbf{y}^{k-1})}{p(r_k = i|\mathbf{y}^{k-1})} \\
&= \frac{1}{c_{2j}}\Gamma_{ji}\mu_{k-1|k-1}(j),
\end{aligned}
$$

where $c_{2j} = \sum_j \Gamma_{ji}\mu_{k-1|k-1}(j)$ and $p(r_{k-1} = j|\mathbf{y}^{k-1}) = \mu_{k-1|k-1}(j)$.
The mixing step of (3.13) is calculated as

$$
p(\mathbf{x}_{k-1}|r_k = i, \mathbf{y}^{k-1}) = \sum_{j=1}^{d} \mu_{k-1,j|i}\, p(\mathbf{x}_{k-1}|r_{k-1} = j, \mathbf{y}^{k-1}). \tag{3.14}
$$

Using (3.14), the predicted pdf of (3.12) can be calculated as

$$
\begin{aligned}
p(\mathbf{x}_k|r_k &= i, \mathbf{y}^{k-1}) \\
&= \int_{\mathbf{x}_{k-1}} p(\mathbf{x}_k|\mathbf{x}_{k-1}, r_k = i, \mathbf{y}^{k-1}) \sum_{j=1}^{d} \mu_{k-1,j|i}\, p(\mathbf{x}_{k-1}|r_{k-1} = j, \mathbf{y}^{k-1})d\mathbf{x}_{k-1} \\
&= \sum_{j=1}^{d} \mu_{k-1,j|i} \int_{\mathbf{x}_{k-1}} p(\mathbf{x}_k|\mathbf{x}_{k-1}, r_k = i, \mathbf{y}^{k-1})p(\mathbf{x}_{k-1}|r_{k-1} = j, \mathbf{y}^{k-1})d\mathbf{x}_{k-1}.
\end{aligned}
$$

$$\tag{3.15}$$

The predicted probability of the object maneuver is given by

$$
\mu_{k|k-1}(i) = \sum_{j=1}^{d} \Gamma_{ji}\, \mu_{k-1|k-1}(j). \tag{3.16}
$$

The recursive relations for the maneuver probability update can be further simplified to

$$
\mu_{k|k}(i) = \frac{p(\mathbf{y}_k|r_k = i, \mathbf{y}^{k-1}) \sum_{j=1}^{d} \Gamma_{ji}\, \mu_{k-1|k-1}(j)}{p(\mathbf{y}_k|\mathbf{y}^{k-1})}. \tag{3.17}
$$

The likelihood functions

In both (3.10) and (3.17), the likelihoods are yet to be determined. The likelihood functions are always derived from the measurement equation (3.7) and given by

$$
\begin{aligned}
p(\mathbf{y}_k|\mathbf{x}_k, r_k = i, \mathbf{y}^{k-1}) &= p(\mathbf{y}_k|\mathbf{x}_k, r_k = i) \\
&= p_{w_{r_k=i}}(\mathbf{y}_k - h_{r_k=i}(\mathbf{x}_k)) \\
&= p_{w_i}(\mathbf{y}_k - h_i(\mathbf{x}_k)), \quad (3.18)
\end{aligned}
$$

for the conditional density considered in (3.10). The likelihood function in the maneuver probability update equation (3.17), can also be evaluated in a similar way by expanding it into

$$
\begin{aligned}
p(\mathbf{y}_k|r_k = i, \mathbf{y}^{k-1}) &= \int_{\mathbf{x}_k} p(\mathbf{y}_k, \mathbf{x}_k|r_k = i, \mathbf{y}^{k-1}) d\mathbf{x}_k \\
&= \int_{\mathbf{x}_k} p(\mathbf{y}_k|\mathbf{x}_k, r_k = i, \mathbf{y}^{k-1}) p(\mathbf{x}_k|r_k = i, \mathbf{y}^{k-1}) d\mathbf{x}_k, \quad (3.19)
\end{aligned}
$$

where the first term in the integrand is given in (3.18) and the second term is given in (3.12).

The normalization factors

In (3.10) the normalization factor is $p(\mathbf{y}_k|r_k = i, \mathbf{y}^{k-1})$ and is given in (3.19) for each $i \in \{1, \ldots, d\}$. In (3.17), the normalization factor is $p(\mathbf{y}_k|\mathbf{y}^{k-1})$, which can be evaluated by expanding

$$
\begin{aligned}
p(\mathbf{y}_k|\mathbf{y}^{k-1}) &= \sum_{i=1}^{d} p(\mathbf{y}_k|r_k = i, \mathbf{y}^{k-1}) \, p(r_k = i|\mathbf{y}^{k-1}) \\
&= \sum_{i=1}^{d} p(\mathbf{y}_k|r_k = i, \mathbf{y}^{k-1}) \, \mu_{k|k-1}(i) \\
&= \sum_{i=1}^{d} p(\mathbf{y}_k|r_k = i, \mathbf{y}^{k-1}) \sum_{j=1}^{d} \Gamma_{ji} \, \mu_{k-1|k-1}(j), \quad (3.20)
\end{aligned}
$$

using (3.16).

The optimal estimates

Substituting (3.10) and (3.17) in (3.9), the conditional density of the object state $p(\mathbf{x}_k|\mathbf{y}^k)$ can be determined. From this, the minimum variance object state estimate and the associated covariance can be obtained using

$$\hat{\mathbf{x}}_{k|k} = \int_{\mathbf{x}_k} \mathbf{x}_k p(\mathbf{x}_k|\mathbf{y}^k)d\mathbf{x}_k,$$

$$P_{k|k} = \int_{\mathbf{x}_k} [\mathbf{x}_k - \hat{x}_{k|k}][\mathbf{x}_k - \hat{\mathbf{x}}_{k|k}]^T p(\mathbf{x}_k|\mathbf{y}^k)d\mathbf{x}_k.$$

Using (3.9), the mean and covariance equations can be further simplified to

$$\hat{\mathbf{x}}_{k|k} = \int_{\mathbf{x}_k} \mathbf{x}_k \sum_{i=1}^{d} p(\mathbf{x}_k|r_k = i, \mathbf{y}^k)\mu_{k|k}(i)d\mathbf{x}_k,$$

$$P_{k|k} = \int_{\mathbf{x}_k} [\mathbf{x}_k - \hat{\mathbf{x}}_k][\mathbf{x}_k - \hat{\mathbf{x}}_k]^T \sum_{i=1}^{d} p(\mathbf{x}_k|r_k = i, \mathbf{y}^k)\mu_{k|k}(i)d\mathbf{x}_k.$$

Rearranging the summation and integrals, the conditional mean and covariance are given by

$$\hat{\mathbf{x}}_{k|k} = \sum_{i=1}^{d} \left(\int_{\mathbf{x}_k} \mathbf{x}_k p(\mathbf{x}_k|r_k = i, \mathbf{y}^k)d\mathbf{x}_k \right) \mu_{k|k}(i), \tag{3.21}$$

$$P_{k|k} = \sum_{i=1}^{d} \left(\int_{\mathbf{x}_k} [\mathbf{x}_k - \hat{\mathbf{x}}_k][\mathbf{x}_k - \hat{\mathbf{x}}_k]^T p(\mathbf{x}_k|r_k = i, \mathbf{y}^k)d\mathbf{x}_k \right) \mu_{k|k}(i). \tag{3.22}$$

The popular multiple-model tracking algorithms, such as GPB1, GPB2 and IMM, differ only in the way they approximate the optimal Bayesian recursions. These filters are considered next.

3.3 Generalized pseudo-Bayesian filters

3.3.1 Generalized pseudo-Bayesian filter of order 1

The system equations assumed for general maneuvering object tracking problems are given in Section 3.2.1. In the GPB1 filter (Ackerson and Fu, 1970; Bar-Shalom *et al.*, 2001), additional assumptions are imposed on the object dynamics. First, it is assumed that $\mathbf{f}_{r_k}(\cdot)$ is a linear function of the object state. Second, the additive noise is assumed to be a zero-mean Gaussian with covariance $\mathbf{Q}_{r_k}$. Third, the functional form of the prior object state pdf at time 0 is assumed to be Gaussian with mean $\hat{\mathbf{x}}_0$ and covariance $\mathbf{P}_0$. These assumptions yield object dynamics equations of the form $\mathbf{x}_k = \mathbf{F}\mathbf{x}_{k-1} + \mathbf{u}_k + \mathbf{v}_k$, where $\mathbf{u}_k$ is an unknown system input. Input estimation treats the unknown input to the system as a deterministic quantity and estimates the unknown input using least squares estimation once a object maneuver is detected. The idea first appeared in Chan *et al.* (1979) and was further enhanced in Bar-Shalom and Fortmann (1988) and Bar-Shalom and Li (1993). In the GPB1 filter, the trio $\{\mathbf{F}, \mathbf{u}_k, \mathbf{v}_k\}$ are functions of a random variable r_k, which is modeled as a homogeneous discrete-time Markov chain with state space $\{1, \ldots, d\}$ and transition probability matrix Γ, where

$$\Gamma_{ji} = \Pr(r_k = i | r_{k-1} = j),$$

with initial conditions $\Pr(r_0 = i) = \pi_0(i)$. Thus the system models used in the GPB1 estimation approach can be represented using

$$\mathbf{x}_k = \mathbf{F}_{r_k} \mathbf{x}_{k-1} + \mathbf{u}_{r_k} + \mathbf{v}_{r_k}. \qquad (3.23)$$

Measurements are assumed to be linear functions of the object state and independent of the modal state r_k. Under these assumptions, (3.7) reduces to

$$\mathbf{y}_k = \mathbf{H}\mathbf{x}_k + \mathbf{w}_k. \qquad (3.24)$$

The conditional density recursion, as derived in (3.10), is given by

$$p(\mathbf{x}_k | r_k = i, \mathbf{y}^k) = \frac{p(\mathbf{y}_k | \mathbf{x}_k, r_k = i, \mathbf{y}^{k-1}) p(\mathbf{x}_k | r_k = i, \mathbf{y}^{k-1})}{p(\mathbf{y}_k | r_k = i, \mathbf{y}^{k-1})},$$

and the conditional model probability recursion, obtained in (3.17), is given by

$$\mu_{k|k}(i) = \frac{p(\mathbf{y}_k | r_k = i, \mathbf{y}^{k-1}) \sum_{j=1}^{d} \Gamma_{ji} \, \mu_{k-1|k-1}(j)}{p(\mathbf{y}_k | \mathbf{y}^{k-1})}.$$

Using Gaussian assumptions, the components of the above recursions are derived in this section.

The transition density

The transition density, for dynamical models defined in (3.6), is given by

$$p(\mathbf{x}_k | \mathbf{x}_{k-1}, r_k = i, \mathbf{y}^{k-1}) = p_{\mathbf{v}_i}(\mathbf{x}_k - \mathbf{F}_i \mathbf{x}_{k-1} - \mathbf{u}_i). \qquad (3.25)$$

Since $\mathbf{v}_i$ is modeled as a zero mean white Gaussian noise with covariance $\mathbf{Q}_i$, the transition density can be simplified to

$$p(\mathbf{x}_k | \mathbf{x}_{k-1}, r_k = i, \mathbf{y}^{k-1}) = N(\mathbf{x}_k; \mathbf{F}_i \mathbf{x}_{k-1} + \mathbf{u}_i, \mathbf{Q}_i). \qquad (3.26)$$

The prediction density and predicted model probability

Substituting (3.25) in (3.12), an expanded version of the predicted density is derived:

$$p(\mathbf{x}_k | r_k = i, \mathbf{y}^{k-1}) = \int_{\mathbf{x}_{k-1}} p(\mathbf{x}_k | \mathbf{x}_{k-1}, r_k = i, \mathbf{y}^{k-1}) p(\mathbf{x}_{k-1} | \mathbf{y}^{k-1}) d\mathbf{x}_{k-1}.$$

GPB1 approximation   The last term of the integrand, the object state prior density, is approximated as

$$p(\mathbf{x}_{k-1}|\mathbf{y}^{k-1}) \approx N(\mathbf{x}_{k-1}; \hat{\mathbf{x}}_{k-1|k-1}, \mathbf{P}_{k-1|k-1}),$$

and the first term in the integrand, the transition density, is given in (3.26). Thus, one has

$$p(\mathbf{x}_k|r_k = i, \mathbf{y}^{k-1})$$

$$= \int_{\mathbf{x}_{k-1}} N(\mathbf{x}_k; \mathbf{F}_i\mathbf{x}_{k-1} + \mathbf{u}_i, \mathbf{Q}_i) N(\mathbf{x}_{k-1}; \hat{\mathbf{x}}_{k-1|k-1}, \mathbf{P}_{k-1|k-1}) d\mathbf{x}_{k-1}.$$

The integral is reduced to a normal density (see Appendix C)

$$p(\mathbf{x}_k|r_k = i, \mathbf{y}^{k-1}) = N(\mathbf{x}_k; \hat{\mathbf{x}}^i_{k|k-1}, \mathbf{P}^i_{k|k-1}), \tag{3.27}$$

where the mean and covariance are given by

$$\hat{\mathbf{x}}^i_{k|k-1} = \mathbf{F}_i\hat{\mathbf{x}}_{k-1|k-1} + \mathbf{u}_i, \tag{3.28}$$

$$\mathbf{P}^i_{k|k-1} = \mathbf{F}_i\mathbf{P}_{k-1|k-1}\mathbf{F}_i^T + \mathbf{Q}_i. \tag{3.29}$$

The predictor of the second component is given by

$$p(r_k = i|\mathbf{y}^{k-1}) = \sum_{j=1}^{d} p(r_k = i|r_{k-1} = j, \mathbf{y}^{k-1}) p(r_{k-1} = j|\mathbf{y}^{k-1})$$

$$= \sum_{j=1}^{d} \Gamma_{ji} \, p(r_{k-1} = j|\mathbf{y}^{k-1}). \tag{3.30}$$

There is no need for any approximation here and it is evaluated straightforwardly. Equations (3.28), (3.29) and (3.30) form the predictor equations of the GPB1 filter.

The likelihood functions

There are two likelihood functions that need to be evaluated – one corresponding to the object's kinematic state derived in (3.18) and the other corresponding to the object's modal state derived in (3.19). Both are derived from the measurement equation in (3.7), where it is assumed that the measurements are not model dependent and reduce to (3.24), where $\mathbf{w}_k = \mathbf{y}_k - \mathbf{H}\mathbf{x}_k$. Note the conditional independence property of the measurements. Moreover, $p_{\mathbf{w}_k}(\cdot)$ is a Gaussian density, leading the likelihood function in (3.18) to be

$$p(\mathbf{y}_k|\mathbf{x}_k, r_k = i, \mathbf{y}^{k-1}) = p(\mathbf{y}_k|\mathbf{x}_k) = N(\mathbf{y}_k; \mathbf{H}\mathbf{x}_k, \mathbf{R}_k).$$

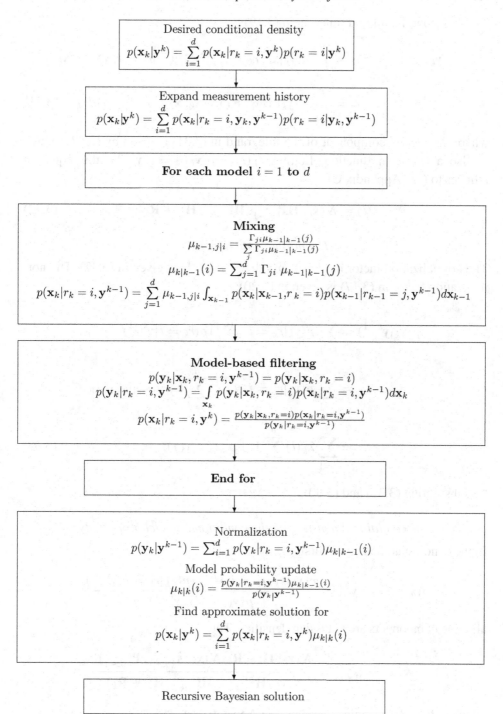

Summary of the recursive Bayesian framework for maneuvering object tracking.

The second likelihood function in (3.19) is

$$p(\mathbf{y}_k|r_k = i, \mathbf{y}^{k-1}) = \int_{\mathbf{x}_k} p(\mathbf{y}_k|\mathbf{x}_k, r_k = i, \mathbf{y}^{k-1})p(\mathbf{x}_k|r_k = i, \mathbf{y}^{k-1})d\mathbf{x}_k$$

$$= \int_{\mathbf{x}_k} N(\mathbf{y}_k; \mathbf{H}x_k, \mathbf{R}_k)N(\mathbf{x}_k; \hat{\mathbf{x}}^i_{k|k-1}, \mathbf{P}^i_{k|k-1})d\mathbf{x}_k, \quad (3.31)$$

where the second component of the integrand in (3.31) is given by (3.27), which is also a Gaussian function. Letting $\lambda_k(i) = p(\mathbf{y}_k|r_k = i, \mathbf{y}^{k-1})$, this equation reduces to (see Appendix C)

$$\lambda_k(i) = N(\mathbf{y}_k; \mathbf{H}\hat{\mathbf{x}}^i_{k|k-1}, \mathbf{H}\mathbf{P}^i_{k|k-1}\mathbf{H}^T + \mathbf{R}_k). \quad (3.32)$$

The normalization factors

The normalization factor in (3.10), $p(\mathbf{y}_k|r_k = i, \mathbf{y}^{k-1})$ is given in (3.32). The normalization factor in (3.17) is given in (3.20):

$$p(\mathbf{y}_k|\mathbf{y}^{k-1}) = \sum_{i=1}^{d} p(\mathbf{y}_k|r_k = i, \mathbf{y}^{k-1})p(r_k = i|\mathbf{y}^{k-1})$$

$$= \sum_{i=1}^{d} p(\mathbf{y}_k|r_k = i, \mathbf{y}^{k-1}) \sum_{j=1}^{d} \Gamma_{ji} \, \mu_{k-1|k-1}(j)$$

$$= \sum_{i=1}^{d} \lambda_k(i) \sum_{j=1}^{d} \Gamma_{ji} \, \mu_{k-1|k-1}(j), \quad (3.33)$$

by substituting (3.32) and (3.30).

The conditional density and the conditional model probability

In the conditional density recursion,

$$p(\mathbf{x}_k|r_k = i, \mathbf{y}^k) = \frac{p(\mathbf{y}_k|\mathbf{x}_k, r_k = i, \mathbf{y}^{k-1})p(\mathbf{x}_k|r_k = i, \mathbf{y}^{k-1})}{p(\mathbf{y}_k|r_k = i, \mathbf{y}^{k-1})},$$

all three components are Gaussian densities. Thus

$$p(\mathbf{x}_k|r_k = i, \mathbf{y}^k) = \frac{N(\mathbf{y}_k; \mathbf{H}x_k, \mathbf{R}_k)N(\mathbf{x}_k; \hat{\mathbf{x}}^i_{k|k-1}, \mathbf{P}^i_{k|k-1})}{N(\mathbf{y}_k; \mathbf{H}\hat{\mathbf{x}}^i_{k|k-1}, \mathbf{H}\mathbf{P}^i_{k|k-1}\mathbf{H}^T + \mathbf{R}_k)},$$

which reduces to a single Gaussian (see Appendix C),

$$p(\mathbf{x}_k|r_k = i, \mathbf{y}^k) = N(\mathbf{x}_k; \hat{\mathbf{x}}^i_{k|k}, \mathbf{P}^i_{k|k}),$$

where the mean and covariance are given by

$$\hat{\mathbf{x}}_{k|k}^i = \hat{\mathbf{x}}_{k|k-1}^i + \mathbf{P}_{k|k-1}^i \mathbf{H}^T (\mathbf{H}\mathbf{P}_{k|k-1}^i \mathbf{H}^T + \mathbf{R}_k)^{-1}(\mathbf{y}_k - \mathbf{H}\hat{\mathbf{x}}_{k|k-1}^i), \quad (3.34)$$

$$\mathbf{P}_{k|k}^i = \mathbf{P}_{k|k-1}^i - \mathbf{P}_{k|k-1}^i \mathbf{H}^T (\mathbf{H}\mathbf{P}_{k|k-1}^i \mathbf{H}^T + \mathbf{R}_k)^{-1}\mathbf{H}\mathbf{P}_{k|k-1}^i. \quad (3.35)$$

The conditional model probability $p(r_k = i|\mathbf{y}^k)$ is derived in (3.17) as

$$\mu_{k|k}(i) = \frac{p(\mathbf{y}_k|r_k = i, \mathbf{y}^{k-1}) \sum_{j=1}^d \Gamma_{ji} \, \mu_{k-1|k-1}(j)}{p(\mathbf{y}_k|\mathbf{y}^{k-1})}.$$

Substituting (3.32) and (3.33) in the above equation the recursion simplifies to

$$\mu_{k|k}(i) = \frac{\lambda_k(i) \sum_{j=1}^d \Gamma_{ji} \, \mu_{k-1|k-1}(j)}{\sum_{l=1}^d \lambda_k(l) \sum_{m=1}^d \Gamma_{ml} \, \mu_{k-1|k-1}(m)}. \quad (3.36)$$

The GPB1 estimates

The conditional density is a Gaussian mixture, given by

$$p(\mathbf{x}_k|\mathbf{y}^k) = \sum_{i=1}^d p(\mathbf{x}_k|r_k = i, \mathbf{y}^k)\mu_{k|k}(i),$$

where $p(\mathbf{x}_k|r_k = i, \mathbf{y}^k)$ is Gaussian for all i. This Gaussian mixture is approximated by a single Gaussian:

$$p(\mathbf{x}_k|\mathbf{y}^k) \approx N(\mathbf{x}_k; \hat{\mathbf{x}}_{k|k}, \mathbf{P}_{k|k}),$$

where the conditional mean $\hat{\mathbf{x}}_{k|k}$ and covariance $\mathbf{P}_{k|k}$ are given by substituting (3.34), (3.35) and (3.36) in (3.21) and (3.22) respectively:

$$\hat{\mathbf{x}}_{k|k} = \sum_{i=1}^d \hat{\mathbf{x}}_{k|k}^i \mu_{k|k}(i),$$

$$\mathbf{P}_{k|k} = \sum_{i=1}^d \mu_{k|k}(i)\{\mathbf{P}_{k|k}^i + [\hat{\mathbf{x}}_{k|k}^i - \hat{\mathbf{x}}_{k|k}][\hat{\mathbf{x}}_{k|k}^i - \hat{\mathbf{x}}_{k|k}]^T\}.$$

The GPB1 estimate operation is effectively a Gaussian mix operation, denoted by

$$[\hat{\mathbf{x}}_{k|k}, \mathbf{P}_{k|k}] = \text{GMix}\left[\left\{\hat{\mathbf{x}}_{k|k}^i, \mathbf{P}_{k|k}^i, \mu_{k|k}(i)\right\}_i\right],$$

and returns the mean and covariance of a probability distribution which is defined as a mixture of Gaussian distributions, where each Gaussian distribution i is defined by its mean $\hat{\mathbf{x}}_{k|k}^i$, covariance $\mathbf{P}_{k|k}^i$ and relative weight $\mu_{k|k}(i) \in [0, 1]$ and

$\sum_i \mu_{k|k}(i) = 1$:

$$\hat{\mathbf{x}}_{k|k} = \sum_i \mu_{k|k}(i)\hat{\mathbf{x}}^i_{k|k},$$

$$\mathbf{P}_{k|k} = \sum_i \mu_{k|k}(i)\left(\mathbf{P}^i_{k|k} + \hat{\mathbf{x}}^i_{k|k}(\hat{\mathbf{x}}^i_{k|k})^T\right) - \hat{\mathbf{x}}_{k|k}\hat{\mathbf{x}}^T_{k|k}.$$

3.3.2 Generalized pseudo-Bayesian filter of order 2

Object dynamics, sensor models and noise models

The GPB2 filter (Bar-Shalom *et al.*, 2001) proceeds as the GPB1 filter but takes into account the dynamical model at the previous stage. The derivations are extended to consideration of r_{k-1} in addition to r_k. The GPB2 computations are

---

Algorithm 10 GPB1 Algorithm recursion equations at time k

1: **for** each model i **do**

$$\hat{\mathbf{x}}^i_{k|k-1} = \mathbf{F}_i\hat{\mathbf{x}}_{k-1|k-1} + u_i,$$

$$\mathbf{P}^i_{k|k-1} = \mathbf{F}_i\mathbf{P}_{k-1|k-1}\mathbf{F}^T_i + \mathbf{Q}_i,$$

$$\hat{\mathbf{x}}^i_{k|k} = \hat{\mathbf{x}}^i_{k|k-1} + \mathbf{P}^i_{k|k-1}\mathbf{H}^T(\mathbf{HP}^i_{k|k-1}\mathbf{H}^T + \mathbf{R}_k)^{-1}(\mathbf{y}_k - \mathbf{H}\hat{\mathbf{x}}^i_{k|k-1}),$$

$$\mathbf{P}^i_{k|k} = \mathbf{P}^i_{k|k-1} - \mathbf{P}^i_{k|k-1}\mathbf{H}^T(\mathbf{HP}^i_{k|k-1}\mathbf{H}^T + \mathbf{R}_k)^{-1}\mathbf{HP}^i_{k|k-1},$$

$$\lambda_k(i) = N(\mathbf{y}_k; \mathbf{H}\hat{\mathbf{x}}^i_{k|k-1}, \mathbf{HP}^i_{k|k-1}\mathbf{H}^T + \mathbf{R}_k).$$

2: **end for**

3: **for** each model i **do** {model probability update}

$$\mu_{k|k}(i) = \frac{\lambda_k(i)\sum_{j=1}^d \Gamma_{ji}\,\mu_{k-1|k-1}(j)}{\sum_{l=1}^d \lambda_k(l)\sum_{m=1}^d \Gamma_{ml}\,\mu_{k-1|k-1}(m)}.$$

4: **end for**

5: Combined conditional state and covariance:

$$\hat{\mathbf{x}}_{k|k} = \sum_{i=1}^d \hat{\mathbf{x}}^i_{k|k}\mu_{k|k}(i),$$

$$\mathbf{P}_{k|k} = \sum_{i=1}^d \mu_{k|k}(i)\{\mathbf{P}^i_{k|k} + [\hat{\mathbf{x}}^i_{k|k} - \hat{\mathbf{x}}_{k|k}][\hat{\mathbf{x}}^i_{k|k} - \hat{\mathbf{x}}_{k|k}]^T\}.$$

---

more extensive than those of GPB1 for a gain in performance. As in the case of the GPB1 filter, it is assumed that $\mathbf{f}_{r_k}(\cdot)$ is a linear function of the object state, the additive noise is assumed to be a zero mean Gaussian with covariance $\mathbf{Q}_{r_k}$. These assumptions yield the system state equations used in the GPB1 estimation approach,

$$\mathbf{x}_k = \mathbf{F}_{r_k}\mathbf{x}_{k-1} + \mathbf{u}_{r_k} + \mathbf{v}_{r_k},$$

and the measurements are also assumed to be a linear function of object states given by

$$\mathbf{y}_k = \mathbf{H}x_k + \mathbf{w}_k,$$

where $\mathbf{w}_k$ is assumed to be a zero mean white Gaussian noise with covariance $\mathbf{R}_k$. The density of interest in GPB2, as in other tracking filters, is $p(\mathbf{x}_k|\mathbf{y}^k)$. As shown in (3.9), this function can be decomposed into

$$p(\mathbf{x}_k|\mathbf{y}^k) = \sum_{i=1}^{d} p(\mathbf{x}_k|r_k = i, \mathbf{y}^k)\mu_{k|k}(i).$$

In the GPB2 approach a further decomposition is initiated by invoking the total probability theorem, yielding

$$p(\mathbf{x}_k|\mathbf{y}^k) = \sum_{j=1}^{d}\sum_{i=1}^{d} p(\mathbf{x}_k, r_k = i, r_{k-1} = j|\mathbf{y}^k).$$

Invoking the conditional probability lemma, the right-hand side can be further expanded into

$$p(\mathbf{x}_k|\mathbf{y}^k) = \sum_{j=1}^{d}\sum_{i=1}^{d} p(\mathbf{x}_k|r_k=i, r_{k-1}=j, \mathbf{y}^k)p(r_{k-1}=j|r_k = i, \mathbf{y}^k)p(r_k = i|\mathbf{y}^k).$$

Invoking Bayes' theorem, and noting that $\mathbf{y}^k = (\mathbf{y}^{k-1}, \mathbf{y}_k)$, $p(\mathbf{x}_k|r_k = i, r_{k-1} = j, \mathbf{y}^k)$ equals

$$\frac{p(\mathbf{y}_k|\mathbf{x}_k, r_k = i, r_{k-1} = j, \mathbf{y}^{k-1})}{p(\mathbf{y}_k|r_k = i, r_{k-1} = j, \mathbf{y}^{k-1})} p(\mathbf{x}_k|r_k = i, r_{k-1} = j, \mathbf{y}^{k-1}), \qquad (3.37)$$

where $p(\mathbf{x}_k|r_k = i, r_{k-1} = j, \mathbf{y}^{k-1})$ can be identified as a prediction density.

The optimal recursion for the second component, the so-called merging probability, $\mu_{k-1|k}^{ji} = p(r_{k-1} = j | r_k = i, \mathbf{y}^k)$, is given by

$$
\begin{aligned}
\mu_{k-1|k}^{ji} &= p(r_{k-1} = j | r_k = i, \mathbf{y}_k, \mathbf{y}^{k-1}) \\
&= \frac{p(\mathbf{y}_k, r_k = i | r_{k-1} = j, \mathbf{y}^{k-1}) p(r_{k-1} = j | \mathbf{y}^{k-1})}{p(\mathbf{y}_k, r_k = i | \mathbf{y}^{k-1})} \\
&= \frac{p(\mathbf{y}_k | r_k = i, r_{k-1} = j, \mathbf{y}^{k-1}) p(r_k = i | r_{k-1} = j, \mathbf{y}^{k-1}) \, \mu_{k-1|k-1}(j)}{p(\mathbf{y}_k, r_k = i | \mathbf{y}^{k-1})} \\
&= \frac{p(\mathbf{y}_k | r_k = i, r_{k-1} = j, \mathbf{y}^{k-1}) \, \Gamma_{ji} \, \mu_{k-1|k-1}(j)}{p(\mathbf{y}_k, r_k = i | \mathbf{y}^{k-1})}.
\end{aligned}
\tag{3.38}
$$

Finally, the optimal Bayesian recursion for the third component, the conditional model probability $\mu_{k|k}(i) = p(r_k = i | \mathbf{y}^k)$, is given by

$$
\begin{aligned}
\mu_{k|k}(i) \\
&= p(r_k = i | \mathbf{y}^k) = p(r_k = i | \mathbf{y}_k, \mathbf{y}^{k-1}) \\
&= \sum_{j=1}^{d} p(r_k = i, r_{k-1} = j | \mathbf{y}_k, \mathbf{y}^{k-1}) \\
&= \frac{\sum_{j=1}^{d} p(\mathbf{y}_k | r_k = i, r_{k-1} = j, \mathbf{y}^{k-1}) p(r_k = i | r_{k-1} = j, \mathbf{y}^{k-1}) p(r_{k-1} = j | \mathbf{y}^{k-1})}{p(\mathbf{y}_k | \mathbf{y}^{k-1})} \\
&= \frac{\sum_{j=1}^{d} p(\mathbf{y}_k | r_k = i, r_{k-1} = j, \mathbf{y}^{k-1}) \, \Gamma_{ji} \, \mu_{k-1|k-1}(j)}{p(\mathbf{y}_k | \mathbf{y}^{k-1})}.
\end{aligned}
\tag{3.39}
$$

The recursive solution for the predictor equations of the GPB2 filter is obtained by solving for the likelihood functions, the normalization factors and other predictive components in (3.37), (3.38) and (3.39).

The prediction density and predicted model probability

The prediction density in (3.37) is calculated as follows:

$$
p(\mathbf{x}_k | r_k = i, r_{k-1} = j, \mathbf{y}^{k-1}) = \int_{\mathbf{x}_{k-1}} p(\mathbf{x}_k, \mathbf{x}_{k-1} | r_k = i, r_{k-1} = j, \mathbf{y}^{k-1}) d\mathbf{x}_{k-1}.
$$

Invoking the conditional density lemma, $p(\mathbf{x}_k | r_k = i, r_{k-1} = j, \mathbf{y}^{k-1})$ becomes

$$
\int_{\mathbf{x}_{k-1}} p(\mathbf{x}_k | \mathbf{x}_{k-1}, r_k = i, r_{k-1} = j, \mathbf{y}^{k-1}) p(\mathbf{x}_{k-1} | r_k = i, r_{k-1} = j, \mathbf{y}^{k-1}) d\mathbf{x}_{k-1}.
$$

Given $\mathbf{x}_{k-1}$ and r_k, the transition density $p(\mathbf{x}_k|\mathbf{x}_{k-1}, r_k = i, r_{k-1} = j, \mathbf{y}^{k-1})$ is independent of r_{k-1} and $\mathbf{y}^{k-1}$. Thus

$$p(\mathbf{x}_k|\mathbf{x}_{k-1}, r_k = i, r_{k-1} = j, \mathbf{y}^{k-1}) = p(\mathbf{x}_k|\mathbf{x}_{k-1}, r_k = i).$$

In addition, since $\mathbf{x}_{k-1}$ is independent of the future maneuvers r_k,

$$p(\mathbf{x}_{k-1}|r_k = i, r_{k-1} = j, \mathbf{y}^{k-1}) = p(\mathbf{x}_{k-1}|r_{k-1} = j, \mathbf{y}^{k-1}).$$

The prediction density simplifies to

$$p(\mathbf{x}_k|r_k = i, r_{k-1} = j, \mathbf{y}^{k-1}) = \int_{\mathbf{x}_{k-1}} p(\mathbf{x}_k|\mathbf{x}_{k-1}, r_k = i) p(\mathbf{x}_{k-1}|r_{k-1} = j, \mathbf{y}^{k-1}) d\mathbf{x}_{k-1}.$$

Since the system equations are the same as in the GPB1 filter, the transition density is also the same

$$p(\mathbf{x}_k|\mathbf{x}_{k-1}, r_k = i) = p_{\mathbf{v}_i}(\mathbf{x}_k - \mathbf{F}_i\mathbf{x}_{k-1} - \mathbf{u}_i)$$
$$= N(\mathbf{x}_k; \mathbf{F}_i\mathbf{x}_{k-1} + \mathbf{u}_i, \mathbf{Q}_i).$$

GPB2 approximation Now, approximating $p(\mathbf{x}_{k-1}|r_{k-1} = j, \mathbf{y}^{k-1})$ by a Gaussian

$$p(\mathbf{x}_{k-1}|r_{k-1} = j, \mathbf{y}^{k-1}) \approx N(\mathbf{x}_{k-1}; \hat{\mathbf{x}}_{k-1|k-1}^j, \mathbf{P}_{k-1|k-1}^j),$$

we obtain

$$p(\mathbf{x}_k|r_k = i, r_{k-1} = j, \mathbf{y}^{k-1})$$
$$= \int_{\mathbf{x}_{k-1}} N(\mathbf{x}_k; \mathbf{F}_i\mathbf{x}_{k-1} + \mathbf{u}_i, \mathbf{Q}_i) N(\mathbf{x}_{k-1}; \hat{\mathbf{x}}_{k-1|k-1}^j, \mathbf{P}_{k-1|k-1}^j) d\mathbf{x}_{k-1}$$
$$= N(\mathbf{x}_k; \mathbf{F}_i\hat{\mathbf{x}}_{k-1|k-1}^j + \mathbf{u}_i, \mathbf{F}_i\mathbf{P}_{k-1|k-1}^j\mathbf{F}_i^T + \mathbf{Q}_i)$$
$$= N(\mathbf{x}_k; \hat{\mathbf{x}}_{k|k-1}^{ij}, \mathbf{P}_{k|k-1}^{ij}).$$

The predicted model probability is given by

$$p(r_k = i|\mathbf{y}^{k-1}) = \sum_{j=1}^{d} p(r_k = i|r_{k-1} = j, \mathbf{y}^{k-1}) p(r_{k-1} = j|\mathbf{y}^{k-1})$$
$$= \sum_{j=1}^{d} \Gamma_{ji} \, p(r_{k-1} = j|\mathbf{y}^{k-1})$$
$$= \sum_{j=1}^{d} \Gamma_{ji} \, \mu_{k-1|k-1}(j).$$

The likelihood functions

The likelihood functions in the GPB2 approach are $p(\mathbf{y}_k|\mathbf{x}_k, r_k = i, r_{k-1} = j, \mathbf{y}^{k-1})$ in (3.37), and $p(\mathbf{y}_k|r_k = i, r_{k-1} = j, \mathbf{y}^{k-1})$ in (3.38) and (3.39). These likelihoods are evaluated using the measurements equation. Given $\mathbf{x}_k$, $\mathbf{y}_k$ is independent of the rest of the terms in the conditioning and the likelihood $p(\mathbf{y}_k|\mathbf{x}_k, r_k = i, r_{k-1} = j, \mathbf{y}^{k-1})$ reduces to $p(\mathbf{y}_k|\mathbf{x}_k)$. From the measurements equation $p(\mathbf{y}_k|\mathbf{x}_k) = p_{\mathbf{w}_k}(\mathbf{y}_k - \mathbf{H}\mathbf{x}_k) = N(\mathbf{y}_k; \mathbf{H}\mathbf{x}_k, \mathbf{R}_k)$. The second likelihood function, conditioned on both the models $r_{k-1} = j$ and $r_k = i$, symbolically represented by $\lambda_k(ij)$, is evaluated using

$$
\begin{aligned}
\lambda_k(ij) &= p(\mathbf{y}_k|r_k = i, r_{k-1} = j, \mathbf{y}^{k-1}) \\
&= \int_{\mathbf{x}_k} p(\mathbf{y}_k, \mathbf{x}_k|r_k = i, r_{k-1} = j, \mathbf{y}^{k-1}) d\mathbf{x}_k \\
&= \int_{\mathbf{x}_k} p(\mathbf{y}_k|\mathbf{x}_k, r_k = i, r_{k-1} = j, \mathbf{y}^{k-1}) p(\mathbf{x}_k|r_k = i, r_{k-1} = j, \mathbf{y}^{k-1}) d\mathbf{x}_k \\
&= \int_{\mathbf{x}_k} N(\mathbf{y}_k; \mathbf{H}\mathbf{x}_k, \mathbf{R}_k) N(\mathbf{x}_k; \hat{\mathbf{x}}_{k|k-1}^{ij}, \mathbf{P}_{k|k-1}^{ij}) d\mathbf{x}_k.
\end{aligned}
$$

The above integral simplifies to a single Gaussian $\lambda_k(ij) = N(\mathbf{y}_k; \mathbf{H}\hat{\mathbf{x}}_{k|k-1}^{ij}, \mathbf{H}\mathbf{P}_{k|k-1}^{ij}\mathbf{H}^T + \mathbf{R}_k)$.

The normalization factors

The first normalization factor in (3.37) can be identified as $\lambda_k(ij)$. The second normalization factor in (3.38) is

$$
\begin{aligned}
p(\mathbf{y}_k, r_k = i|\mathbf{y}^{k-1}) &\\
&= \sum_{j=1}^{d} p(\mathbf{y}_k, r_k = i, r_{k-1} = j|\mathbf{y}^{k-1}) \\
&= \sum_{j=1}^{d} p(\mathbf{y}_k|r_k = i, r_{k-1} = j, \mathbf{y}^{k-1}) p(r_k = i|r_{k-1} = j, \mathbf{y}^{k-1}) p(r_{k-1} = j|\mathbf{y}^{k-1}) \\
&= \sum_{j=1}^{d} \lambda_k(ij) \, \Gamma_{ji} \, \mu_{k-1|k-1}(j).
\end{aligned}
$$

The third and final normalization factor in (3.39) is

$$p(\mathbf{y}_k|\mathbf{y}^{k-1}) = \sum_{i=1}^{d} p(\mathbf{y}_k, r_k = i|\mathbf{y}^{k-1})$$

$$= \sum_{i=1}^{d}\sum_{j=1}^{d} p(\mathbf{y}_k, r_k = i, r_{k-1} = j|\mathbf{y}^{k-1})$$

$$= \sum_{i=1}^{d}\sum_{j=1}^{d} p(\mathbf{y}_k|r_k = i, r_{k-1} = j, \mathbf{y}^{k-1})$$

$$\times p(r_k = i|r_{k-1} = j, \mathbf{y}^{k-1}) p(r_{k-1} = j|\mathbf{y}^{k-1})$$

$$= \sum_{i=1}^{d}\sum_{j=1}^{d} \lambda_k(ij)\, \Gamma_{ji}\, \mu_{k-1|k-1}(j).$$

The conditional density and conditional model probability

Substituting terms in (3.37), the conditional density is

$$p(\mathbf{x}_k|r_k = i, r_{k-1} = j, \mathbf{y}^k) = \frac{N(\mathbf{y}_k; \mathbf{H}\mathbf{x}_k, \mathbf{R}_k) N(\mathbf{x}_k; \hat{\mathbf{x}}_{k|k-1}^{ij}, \mathbf{P}_{k|k-1}^{ij})}{N(\mathbf{y}_k; \mathbf{H}\hat{\mathbf{x}}_{k|k-1}^{ij}, \mathbf{H}\mathbf{P}_{k|k-1}^{ij}\mathbf{H}^T + \mathbf{R}_k)}$$

$$= N(\mathbf{x}_k; \hat{\mathbf{x}}_{k|k}^{ij}, \mathbf{P}_{k|k}^{ij}), \tag{3.40}$$

where the conditional mean and covariance are given by (see Appendix A)

$$\hat{\mathbf{x}}_{k|k}^{ij} = \hat{\mathbf{x}}_{k|k-1}^{ij} + \mathbf{P}_{k|k-1}^{ij}\mathbf{H}^T(\mathbf{H}\mathbf{P}_{k|k-1}^{ij}\mathbf{H}^T + \mathbf{R}_k)^{-1}(\mathbf{y}_k - \mathbf{H}\hat{\mathbf{x}}_{k|k-1}^{ij}), \tag{3.41}$$

$$\mathbf{P}_{k|k}^{ij} = \mathbf{P}_{k|k-1}^{ij} - \mathbf{P}_{k|k-1}^{ij}\mathbf{H}^T(\mathbf{H}\mathbf{P}_{k|k-1}^{ij}\mathbf{H}^T + \mathbf{R}_k)^{-1}\mathbf{H}\mathbf{P}_{k|k-1}^{ij}. \tag{3.42}$$

Substituting in (3.38), the final expression for the merging probability is obtained as

$$\mu_{k-1|k}^{ji} = \frac{\lambda_k(ij)\, \Gamma_{ji}\, \mu_{k-1|k-1}(j)}{\sum_{m=1}^{d} \lambda_k(im)\, \Gamma_{mi}\, \mu_{k-1|k-1}(m)}.$$

Finally, the model probability update is

$$\mu_{k|k}(i) = \frac{\sum_{j=1}^{d} \lambda_k(ij)\, \Gamma_{ji}\, \mu_{k-1|k-1}(j)}{\sum_{l=1}^{d}\sum_{m=1}^{d} \lambda_k(lm)\, \Gamma_{ml}\, \mu_{k-1|k-1}(m)}.$$

The GPB2 estimates

The conditional density is a Gaussian mixture, given by

$$p(\mathbf{x}_k|\mathbf{y}^k) = \sum_{i=1}^{d}\sum_{j=1}^{d} p(\mathbf{x}_k|r_k = i, r_{k-1} = j, \mathbf{y}^k)\, \mu_{k-1|k}^{ji}\, \mu_{k|k}(i), \quad (3.43)$$

where $p(\mathbf{x}_k|r_k = i, r_{k-1} = j, \mathbf{y}^k)$ is given in (3.40), (3.41) and (3.42) for all (i, j). The internal summation can be identified as

$$p(\mathbf{x}_k|r_k = i, \mathbf{y}^k) = \sum_{j=1}^{d} p(\mathbf{x}_k|r_k = i, r_{k-1} = j, \mathbf{y}^k)\, \mu_{k-1|k}^{ji}.$$

This Gaussian mixture is approximated by a single Gaussian with the same mean and covariance:

$$\left[\hat{\mathbf{x}}_{k|k}^i, \mathbf{P}_{k|k}^i\right] = \mathrm{GMix}\left[\left\{\hat{\mathbf{x}}_{k|k}^{ij}, \mathbf{P}_{k|k}^{ij}, \mu_{k-1|k}^{ji}\right\}_j\right]. \quad (3.44)$$

Substituting (3.44) in (3.43) yields a further simplification:

$$p(\mathbf{x}_k|\mathbf{y}^k) = \sum_{i=1}^{d} p(\mathbf{x}_k|r_k = i, \mathbf{y}^k)\mu_{k|k}(i).$$

This mixture is again approximated by a single Gaussian with the state estimate and covariance given by

$$\left[\hat{\mathbf{x}}_{k|k}, \mathbf{P}_{k|k}\right] = \mathrm{GMix}\left[\left\{\hat{\mathbf{x}}_{k|k}^i, \mathbf{P}_{k|k}^i, \mu_{k|k}(i)\right\}_i\right].$$

3.4 Interacting multiple model filter

The interacting multiple model (IMM) estimator (Blom, 1984a, 1984b; Blom and Bar-Shalom, 1988) has enjoyed remarkable success since its introduction (Bar-Shalom and Li, 1995; Mazor *et al.*, 1998; Sworder and Boyd, 1999; Bar-Shalom *et al.*, 2001). It performs well on systems characterized by multiple modes of behavior, including maneuvering object tracking (Bar-Shalom, 1990; Blair *et al.*, 1991, 1993; Bar-Shalom and Li, 1993, 1995; Blair and Watson, 1994), automatic track formation (Bar-Shalom *et al.*, 1989, 1990), air traffic control, object classification and data association, and others (Bar-Shalom, 1978; Bar-Shalom and Li, 1993, 1995; Cutaia and O'Sullivan, 1995; Li and Zhang, 2000; Wang *et al.*, 2003). Its relationship to the optimal estimates is presented in Bar-Shalom *et al.* (2005). The computational requirements of IMM are essentially those of GPB1 (Blom and Bar-Shalom, 1988). However, IMM performs almost as well as the GPB2

Algorithm 11 GPB2 Algorithm recursion equations at time k

1: **for** each model i **do**
2:     **for** each model j **do**

- $\hat{\mathbf{x}}_{k|k-1}^{ij} = \mathbf{F}_i \hat{\mathbf{x}}_{k-1|k-1}^{j} + \mathbf{u}_i,$

- $\mathbf{P}_{k|k-1}^{ij} = \mathbf{F}_i \mathbf{P}_{k-1|k-1}^{j} \mathbf{F}_i^T + \mathbf{Q}_i,$

- $\hat{\mathbf{x}}_{k|k}^{ij} = \hat{\mathbf{x}}_{k|k-1}^{ij} + \mathbf{P}_{k|k-1}^{ij} \mathbf{H}^T (\mathbf{H}\mathbf{P}_{k|k-1}^{ij}\mathbf{H}^T + \mathbf{R}_k)^{-1}(\mathbf{y}_k - \mathbf{H}\hat{\mathbf{x}}_{k|k-1}^{ij}),$

- $\mathbf{P}_{k|k}^{ij} = \mathbf{P}_{k|k-1}^{ij} - \mathbf{P}_{k|k-1}^{ij} \mathbf{H}^T (\mathbf{H}\mathbf{P}_{k|k-1}^{ij}\mathbf{H}^T + \mathbf{R}_k)^{-1}\mathbf{H}\mathbf{P}_{k|k-1}^{ij},$

- $\lambda_k(ij) = N(\mathbf{y}_k; \mathbf{H}\hat{\mathbf{x}}_{k|k-1}^{ij}, \mathbf{H}\mathbf{P}_{k|k-1}^{ij}\mathbf{H}^T + \mathbf{R}_k),$

- $\mu_{k-1|k}^{ji} = \{\lambda_k(ij)\,\Gamma_{ji}\,\mu_{k-1|k-1}(j)\}\{\sum_{m=1}^{d} \lambda_k(im)\,\Gamma_{mi}\,\mu_{k-1|k-1}(m)\}^{-1}.$

3:     **end for**

- $\mu_{k|k}(i) = \{\sum_{j=1}^{d} \lambda_k(ij)\,\Gamma_{ji}\mu_{k-1|k-1}(j)\}$
  $\times \{\sum_{l=1}^{d}\sum_{m=1}^{d} \lambda_k(lm)\,\Gamma_{ml}\,\mu_{k-1|k-1}(m)\}^{-1},$

- $\hat{\mathbf{x}}_{k|k}^{i} = \sum_{j=1}^{d} \hat{\mathbf{x}}_{k|k}^{ij}\,\mu_{k-1|k}^{ji},$

- $\mathbf{P}_{k|k}^{i} = \sum_{j=1}^{d} \mu_{k-1|k}^{ji}\{\mathbf{P}_{k|k}^{ij} + [\hat{\mathbf{x}}_{k|k}^{ij} - \hat{\mathbf{x}}_{k|k}^{i}][\hat{\mathbf{x}}_{k|k}^{ij} - \hat{\mathbf{x}}_{k|k}^{i}]^T\}.$

4: **end for**

- $\hat{\mathbf{x}}_{k|k} = \sum_{i=1}^{d} \hat{\mathbf{x}}_{k|k}^{i}\,\mu_{k|k}(i),$

- $\mathbf{P}_{k|k} = \sum_{i=1}^{d} \mu_{k|k}(i)\{\mathbf{P}_{k|k}^{i} + [\hat{\mathbf{x}}_{k|k}^{i} - \hat{\mathbf{x}}_{k|k}][\hat{\mathbf{x}}_{k|k}^{i} - \hat{\mathbf{x}}_{k|k}]^T\}.$

---

filter, which takes into account d^2 model states, where d is the number of models and the object dynamics are assumed to belong to the set of models defined by

$$\mathbf{x}_k = f_{r_k}(\mathbf{x}_{k-1}, \mathbf{u}_k) + \mathbf{v}_{r_k} \quad r_k \in \{1, 2, \ldots, d\}.$$

In the above equation, r_k is assumed to be a random variable satisfying a homogeneous discrete-time Markov chain with state space $\{1, \ldots, d\}$. The IMM estimator, which is a d modes algorithm for the estimation of the object state, proceeds in a similar way as the two generalized pseudo-Bayesian filters but makes different approximations. GPB1 approximates the prior object state $p(\mathbf{x}_{k-1}|\mathbf{y}^{k-1})$ by a Gaussian distribution. This results in a mixture of Gaussians for the posterior distribution of the object state $p(\mathbf{x}_k|\mathbf{y}^k)$ at the end of the cycle. This mixture is in turn approximated by a single Gaussian distribution, which is the approximation first mentioned, as the posterior distribution $p(\mathbf{x}_k|\mathbf{y}^k)$ is the prior distribution at the next stage. GPB2 approximates the conditional prior distribution $p(\mathbf{x}_{k-1}|r_{k-1} = j, \mathbf{y}^{k-1})$ by a Gaussian distribution. This results in the conditional posterior distribution $p(\mathbf{x}_k|r_k = i, \mathbf{y}^k)$ and the posterior distribution $p(\mathbf{x}_k|\mathbf{y}^k)$ to be Gaussian mixtures. They are both approximated by single Gaussian distributions.

IMM also approximates $p(\mathbf{x}_{k-1}|r_{k-1} = j, \mathbf{y}^{k-1})$ by a Gaussian distribution, as in GPB2. However, it further approximates to

$$\sum_{j=1}^{d} p(\mathbf{x}_k|r_{k-1} = j, \mathbf{y}^{k-1}) p(r_{k-1} = j|r_k = i, \mathbf{y}^{k-1}),$$

by a single Gaussian distribution. This step is what makes IMM retain the computational simplicity of the GPB1 filter while having almost the same performance as the GPB2.

3.4.1 The IMM filter equations

The posterior probability density of the object state $p(\mathbf{x}_k|\mathbf{y}^k)$ can be obtained by summing up the individual components of the joint density as follows:

$$p(\mathbf{x}_k|\mathbf{y}^k) = \sum_{i=1}^{d} p(\mathbf{x}_k, r_k = i|\mathbf{y}^k).$$

Using the conditional probability lemma, the joint density in the right-hand side of the above equation can be broken up into two components,

$$p(\mathbf{x}_k, r_k = i|\mathbf{y}^k) = p(\mathbf{x}_k|r_k = i, \mathbf{y}^k) p(r_k = i|\mathbf{y}^k).$$

Letting $\mu_{k|k}(i) = p(r_k = i|\mathbf{y}^k)$ and the above decomposition, the posterior density equation can be rewritten as

$$p(\mathbf{x}_k|\mathbf{y}^k) = \sum_{i=1}^{d} p(\mathbf{x}_k|r_k = i, \mathbf{y}^k) \mu_{k|k}(i). \tag{3.45}$$

The optimal Bayesian recursion for the first component can be derived by first expanding the set of measurements $\mathbf{y}^k$ into $\{\mathbf{y}_k, \mathbf{y}^{k-1}\}$ and then invoking Bayes' theorem:

$$p(\mathbf{x}_k|r_k = i, \mathbf{y}^k) = p(\mathbf{x}_k|r_k = i, \mathbf{y}_k, \mathbf{y}^{k-1})$$

$$= \frac{p(\mathbf{y}_k|\mathbf{x}_k, r_k = i, \mathbf{y}^{k-1})}{p(\mathbf{y}_k|r_k = i, \mathbf{y}^{k-1})} p(\mathbf{x}_k|r_k = i, \mathbf{y}^{k-1}),$$

where $p(\mathbf{x}_k|r_k = i, \mathbf{y}^{k-1})$ is the predicted density, $p(\mathbf{y}_k|\mathbf{x}_k, r_k = i, \mathbf{y}^{k-1})$ is the likelihood function and $p(\mathbf{y}_k|r_k = i, \mathbf{y}^{k-1})$ is the normalization factor. The second component of the joint density in (3.45), the posterior model probability $\mu_{k|k}(i) = p(r_k = i|\mathbf{y}^k)$, can also be recursively calculated. Once again, expanding

$\mathbf{y}^k$ into $\{\mathbf{y}_k, \mathbf{y}^{k-1}\}$ and invoking Bayes' theorem,

$$\mu_{k|k}(i) = p(r_k = i|\mathbf{y}_k, \mathbf{y}^{k-1}) = \frac{p(\mathbf{y}_k|r_k = i, \mathbf{y}^{k-1})p(r_k = i|\mathbf{y}^{k-1})}{p(\mathbf{y}_k|\mathbf{y}^{k-1})}.$$

Letting $p(r_k = i|\mathbf{y}^{k-1}) = \mu_{k|k-1}(i)$, the probability recursion can be rewritten as

$$\mu_{k|k}(i) = \frac{p(\mathbf{y}_k|r_k = i, \mathbf{y}^{k-1})\mu_{k|k-1}(i)}{p(\mathbf{y}_k|\mathbf{y}^{k-1})},$$

where $\mu_{k|k-1}(i)$ is the predicted model probability, $p(\mathbf{y}_k|r_k = i, \mathbf{y}^{k-1})$ is the likelihood function and $p(\mathbf{y}_k|\mathbf{y}^{k-1})$ is the normalization factor.

The prediction density and predicted model probability

So far, the derivation is the same as that of GPB1. Like GPB2, IMM also conditions on r_{k-1}, but on a different term. While GPB2 expands $p(\mathbf{x}_k|r_k = i, \mathbf{y}^k)$, IMM expands the term $p(\mathbf{x}_k|r_k = i, \mathbf{y}^{k-1})$. Using the laws of probability,

$$p(\mathbf{x}_k|r_k = i, \mathbf{y}^{k-1}) = \sum_{j=1}^{d} p(\mathbf{x}_k, r_{k-1} = j|r_k = i, \mathbf{y}^{k-1})$$

$$= \sum_{j=1}^{d} p(\mathbf{x}_k|r_{k-1} = j, r_k = i, \mathbf{y}^{k-1})p(r_{k-1} = j|r_k = i, \mathbf{y}^{k-1})$$

$$= \sum_{j=1}^{d} p(\mathbf{x}_k|r_{k-1} = j, r_k = i, \mathbf{y}^{k-1})\mu_{k-1|k}^{j|i},$$

where $\mu_{k-1|k}^{j|i} = p(r_{k-1} = j|r_k = i, \mathbf{y}^{k-1})$ are called the mixing probabilities (Bar-Shalom *et al.*, 2005). The mixing probabilities relate to the transition probabilities as follows:

$$\mu_{k-1|k}^{j|i} = p(r_{k-1} = j|r_k = i, \mathbf{y}^{k-1})$$

$$= \frac{p(r_k = i|r_{k-1} = j, \mathbf{y}^{k-1})p(r_{k-1} = j|\mathbf{y}^{k-1})}{p(r_k = i|\mathbf{y}^{k-1})}$$

$$= \frac{p(r_k = i|r_{k-1} = j)p(r_{k-1} = j|\mathbf{y}^{k-1})}{\sum_{m=1}^{d} p(r_k = i|r_{k-1} = m, \mathbf{y}^{k-1})p(r_{k-1} = m|\mathbf{y}^{k-1})}$$

$$= \frac{p(r_k = i|r_{k-1} = j)p(r_{k-1} = j|\mathbf{y}^{k-1})}{\sum_{m=1}^{d} p(r_k = i|r_{k-1} = m)p(r_{k-1} = m|\mathbf{y}^{k-1})}$$

$$= \frac{\Gamma_{ji}\,\mu_{k-1|k-1}(j)}{\sum_{m=1}^{d} \Gamma_{mi}\,\mu_{k-1|k-1}(m)}.$$

Note that the mixing probabilities $\mu_{k-1|k}^{j|i}$ are different from the merging probabilities $\mu_{k-1|k}^{ji}$ of GPB2. They do not depend on the last measurement $\mathbf{y}_k$. Introducing the prior object state $\mathbf{x}_{k-1}$, the prediction density can be expanded into

$$
p(\mathbf{x}_k | r_k = i, \mathbf{y}^{k-1}) = \sum_{j=1}^{d} \int_{\mathbf{x}_{k-1}} p(\mathbf{x}_k, \mathbf{x}_{k-1} | r_{k-1} = j, r_k = i, \mathbf{y}^{k-1}) \mu_{k-1|k}^{j|i} d\mathbf{x}_{k-1}.
$$

Invoking the conditional density lemma on the joint density inside the integrand, the prediction density can be decomposed into

$$
p(\mathbf{x}_k | r_k = i, \mathbf{y}^{k-1})
$$

$$
= \sum_{j=1}^{d} \int_{\mathbf{x}_{k-1}} p(\mathbf{x}_k | \mathbf{x}_{k-1}, r_{k-1} = j, r_k = i, \mathbf{y}^{k-1})
$$

$$
\times p(\mathbf{x}_{k-1} | r_{k-1} = j, r_k = i, \mathbf{y}^{k-1}) \mu_{k-1|k}^{j|i} d\mathbf{x}_{k-1}
$$

$$
= \sum_{j=1}^{d} \int_{\mathbf{x}_{k-1}} p(\mathbf{x}_k | \mathbf{x}_{k-1}, r_k = i, \mathbf{y}^{k-1}) p(\mathbf{x}_{k-1} | r_{k-1} = j, \mathbf{y}^{k-1}) \mu_{k-1|k}^{j|i} d\mathbf{x}_{k-1}
$$

$$
= \int_{\mathbf{x}_{k-1}} p(\mathbf{x}_k | \mathbf{x}_{k-1}, r_k = i, \mathbf{y}^{k-1}) \sum_{j=1}^{d} p(\mathbf{x}_{k-1} | r_{k-1} = j, \mathbf{y}^{k-1}) \mu_{k-1|k}^{j|i} d\mathbf{x}_{k-1}.
$$

In the above, we have used the fact that the object state at time $k - 1$, $\mathbf{x}_{k-1}$ is not dependent on the model at time k, r_k. The first integrand can be identified as the transition density that can be derived from the object dynamical equations:

$$
p(\mathbf{x}_k | \mathbf{x}_{k-1}, r_k = i, \mathbf{y}^{k-1}) = p_{\mathbf{v}_i}(\mathbf{x}_k - \mathbf{F}_i(\mathbf{x}_{k-1}, \mathbf{u}_k)),
$$

where $\mathbf{v}_i = \mathbf{v}_{r_k=i}$ and $\mathbf{F}_i(\cdot) = \mathbf{f}_{r_k=i}(\cdot)$. Since $\mathbf{v}_i$ is modeled as a zero mean white Gaussian noise with covariance $\mathbf{Q}_i$, the transition density is

$$
p(\mathbf{x}_k | \mathbf{x}_{k-1}, r_k = i, \mathbf{y}^{k-1}) = N(\mathbf{x}_k; F_i \mathbf{x}_{k-1} + \mathbf{u}_i, \mathbf{Q}_i).
$$

The second term in the integral is approximated by a Gaussian distribution.

IMM approximations   Approximating $p(\mathbf{x}_{k-1} | r_{k-1} = j, \mathbf{y}^{k-1})$ by a Gaussian and the resulting mixture of Gaussians also by a Gaussian,

$$
\sum_{j=1}^{d} p(\mathbf{x}_{k-1} | r_{k-1} = j, \mathbf{y}^{k-1}) \mu_{k-1|k}^{j|i},
$$

we proceed in obtaining the IMM filter:

$$p(\mathbf{x}_{k-1}|r_{k-1} = j, \mathbf{y}^{k-1}) \approx N(\mathbf{x}_{k-1}; \hat{\mathbf{x}}^j_{k-1|k-1}, \mathbf{P}^j_{k-1|k-1})$$

$$\sum_{j=1}^d p(\mathbf{x}_{k-1}|r_{k-1} = j, \mathbf{y}^{k-1})\mu^{j|i}_{k-1|k} \approx N(\mathbf{x}_{k-1}; \hat{\mathbf{x}}^{0i}_{k-1|k-1}, \mathbf{P}^{0i}_{k-1|k-1}),$$

where

$$\left[\hat{\mathbf{x}}^{0i}_{k-1|k-1}, \mathbf{P}^{0i}_{k-1|k-1}\right] = \text{GMix}\left[\left\{\hat{\mathbf{x}}^j_{k-1|k-1}, \mathbf{P}^j_{k-1|k-1}, \mu^{j|i}_{k-1|k}\right\}_i\right].$$

The prediction step can now be carried out:

$$p(\mathbf{x}_k|r_k = i, \mathbf{y}^{k-1})$$

$$= \int_{\mathbf{x}_{k-1}} p(\mathbf{x}_k|\mathbf{x}_{k-1}, r_k = i, \mathbf{y}^{k-1}) \sum_{j=1}^d p(\mathbf{x}_{k-1}|r_{k-1} = j, \mathbf{y}^{k-1})\mu^{j|i}_{k-1|k}d\mathbf{x}_{k-1}$$

$$= \int_{\mathbf{x}_{k-1}} N(\mathbf{x}_k; F_i x_{k-1} + \mathbf{u}_i, \mathbf{Q}_i)N(\mathbf{x}_{k-1}; \hat{\mathbf{x}}^{0i}_{k-1|k-1}, \mathbf{P}^{0i}_{k-1|k-1})d\mathbf{x}_{k-1}$$

$$= N(\mathbf{x}_k; \hat{\mathbf{x}}^i_{k|k-1}, \mathbf{P}^i_{k|k-1}),$$

where the mean and covariance are given by

$$\hat{\mathbf{x}}^i_{k|k-1} = \mathbf{F}_i\hat{\mathbf{x}}^{0i}_{k-1|k-1} + \mathbf{u}_i,$$

$$\mathbf{P}^i_{k|k-1} = \mathbf{F}_i\mathbf{P}^{0i}_{k-1|k-1}\mathbf{F}^T_i + \mathbf{Q}_i.$$

At this point, IMM returns to a derivation similar to that of GPB1, after having conditioned on r_{k-1} as in GPB2, but with different approximations. It is this step that makes IMM almost as efficient as GPB2, while retaining the computational load of GPB1. The predicted model probability can be obtained by expanding $\mu_{k|k-1}(i)$ as follows:

$$\mu_{k|k-1}(i) = p(r_k = i|\mathbf{y}^{k-1})$$

$$= \sum_{j=1}^d p(r_k = i|r_{k-1} = j, \mathbf{y}^{k-1})p(r_{k-1} = j|\mathbf{y}^{k-1})$$

$$= \sum_{j=1}^d \Gamma_{ji}\, p(r_{k-1} = j|\mathbf{y}^{k-1})$$

$$= \sum_{j=1}^d \Gamma_{ji}\, \mu_{k-1|k-1}(j).$$

This step is knows as the IMM mixing step and is denoted by

$$\left[\left\{\mu_{k|k-1}(i), \hat{\mathbf{x}}^i_{k|k-1}, \mathbf{P}^i_{k|k-1}\right\}_i\right]$$

$$= \text{IMM}_{\text{MP}}\left[\left\{\mu_{k-1|k-1}(i), \hat{\mathbf{x}}^i_{k-1|k-1}, \mathbf{P}^i_{k-1|k-1}, \mathbf{F}_i, \mathbf{Q}_i\right\}_i, \boldsymbol{\Gamma}\right].$$

The likelihood functions

The first likelihood function $p(\mathbf{y}_k|\mathbf{x}_k, r_k = i, \mathbf{y}^{k-1})$ simplifies due to the assumptions of the measurement model. Since given $\mathbf{x}_k$, $\mathbf{y}_k$ does not depend on r_k and $\mathbf{y}^{k-1}$ (conditional independence property of measurements), the likelihood is

$$p(\mathbf{y}_k|\mathbf{x}_k, r_k = i, \mathbf{y}^{k-1}) = p(\mathbf{y}_k|\mathbf{x}_k) = N(\mathbf{y}_k; \mathbf{H}\mathbf{x}_k, \mathbf{R}_k).$$

The second likelihood function needs further simplification:

$$p(\mathbf{y}_k|r_k = i, \mathbf{y}^{k-1}) = \int_{\mathbf{x}_k} p(\mathbf{y}_k, \mathbf{x}_k|r_k = i, \mathbf{y}^{k-1})dx_k$$

$$= \int_{\mathbf{x}_k} p(\mathbf{y}_k|\mathbf{x}_k, r_k = i, \mathbf{y}^{k-1})p(\mathbf{x}_k|r_k = i, \mathbf{y}^{k-1})dx_k$$

$$= \int_{\mathbf{x}_k} N(\mathbf{y}_k; \mathbf{H}\mathbf{x}_k, \mathbf{R}_k)N(\mathbf{x}_k; \hat{\mathbf{x}}^i_{k|k-1}, \mathbf{P}^i_{k|k-1})dx_k.$$

Letting $p(\mathbf{y}_k|r_k = i, \mathbf{y}^{k-1}) = \lambda_k(i)$, it reduces to $\lambda_k(i) = N(\mathbf{y}_k; \mathbf{H}\hat{\mathbf{x}}^i_{k|k-1}, \mathbf{H}\mathbf{P}^i_{k|k-1}\mathbf{H}^T + \mathbf{R}_k)$.

The normalization factors

The normalization factor $p(\mathbf{y}_k|r_k = i, \mathbf{y}^{k-1})$ is $\lambda_k(i)$ given above. The second normalization factor is evaluated using the derivation of the predicted model $p(r_k = i|\mathbf{y}^{k-1})$:

$$p(\mathbf{y}_k|\mathbf{y}^{k-1}) = \sum_{i=1}^{d} p(\mathbf{y}_k|r_k = i, \mathbf{y}^{k-1})p(r_k = i|\mathbf{y}^{k-1})$$

$$= \sum_{i=1}^{d} \lambda_k(i) \sum_{j=1}^{d} \Gamma_{ji} \, \mu_{k-1|k-1}(j).$$

The conditional density and conditional model probability

The conditional density is

$$p(\mathbf{x}_k|r_k = i, \mathbf{y}^{k}) = \frac{p(\mathbf{y}_k|\mathbf{x}_k, r_k = i, \mathbf{y}^{k-1})p(\mathbf{x}_k|r_k = i, \mathbf{y}^{k-1})}{p(\mathbf{y}_k|r_k = i, \mathbf{y}^{k-1})}$$

$$= \frac{N(\mathbf{y}_k; \mathbf{H}\mathbf{x}_k, \mathbf{R}_k)N(\mathbf{x}_k; \hat{\mathbf{x}}^i_{k|k-1}, \mathbf{P}^i_{k|k-1})}{N(\mathbf{y}_k; \mathbf{H}\hat{\mathbf{x}}^i_{k|k-1}, \mathbf{H}\mathbf{P}^i_{k|k-1}\mathbf{H}^T + \mathbf{R}_k)},$$

which reduces to a single Gaussian $p(\mathbf{x}_k|r_k = i, \mathbf{y}^k) = N(\mathbf{x}_k; \hat{\mathbf{x}}^i_{k|k}, \mathbf{P}^i_{k|k})$, where the mean and covariance are given by

$$\hat{\mathbf{x}}^i_{k|k} = \hat{\mathbf{x}}^i_{k|k-1} + \mathbf{P}^i_{k|k-1}\mathbf{H}^T(\mathbf{H}\mathbf{P}^i_{k|k-1}\mathbf{H}^T + \mathbf{R}_k)^{-1}(\mathbf{y}_k - \mathbf{H}\hat{\mathbf{x}}^i_{k|k-1}),$$

$$\mathbf{P}^i_{k|k} = \mathbf{P}^i_{k|k-1} - \mathbf{P}^i_{k|k-1}\mathbf{H}^T(\mathbf{H}\mathbf{P}^i_{k|k-1}\mathbf{H}^T + \mathbf{R}_k)^{-1}\mathbf{H}\mathbf{P}^i_{k|k-1}.$$

The conditional model probability $p(r_k = i|\mathbf{y}^k)$ is derived as

$$\mu_{k|k}(i) = \frac{p(\mathbf{y}_k|r_k = i, \mathbf{y}^{k-1})p(r_k = i|\mathbf{y}^{k-1})}{p(\mathbf{y}_k|\mathbf{y}^{k-1})}.$$

Substituting in the above equation, the recursion simplifies to

$$\mu_{k|k}(i) = \frac{\lambda_k(i)\sum_{j=1}^d \Gamma_{ji}\,\mu_{k-1|k-1}(j)}{\sum_{l=1}^d \lambda_k(l)\sum_{m=1}^d \Gamma_{ml}\,\mu_{k-1|k-1}(m)}.$$

The IMM estimates

The posterior density is a Gaussian mixture, given by

$$p(\mathbf{x}_k|\mathbf{y}^k) = \sum_{i=1}^d p(\mathbf{x}_k|r_k = i, \mathbf{y}^k)\mu_{k|k}(i).$$

The conditional mean $\hat{\mathbf{x}}_{k|k}$ and covariance $\mathbf{P}_{k|k}$ of this Gaussian mixture are given by

$$[\hat{\mathbf{x}}_{k|k}, \mathbf{P}_{k|k}] = \mathrm{GMix}\left[\left\{\hat{\mathbf{x}}^i_{k|k}, \mathbf{P}^i_{k|k}, \mu_{k|k}(i)\right\}_i\right].$$

3.5 Particle filters for maneuvering object tracking

Particle filters (PFs) are particularly useful for approximating the multi-modal posterior density in maneuvering object tracking. Since it is necessary to estimate a maneuvering mode in addition to the usual object dynamics, samples are drawn of the hybrid state vector $[\mathbf{x}'_k, r_k]'$. Assume that an approximation to the posterior density at time $k - 1$, represented by the samples $\mathbf{x}^1_{k-1}, r^1_{k-1}, \ldots, \mathbf{x}^n_{k-1}, r^n_{k-1}$ and weights $w^1_{k-1}, \ldots, w^n_{k-1}$, is available. Given that the hybrid state evolves according to

$$p(\mathbf{x}_k, r_k = i|r_{k-1} = j, = i, \mathbf{x}_{k-1}) = \Pr(r_k|r_{k-1} = j)p(\mathbf{x}_k|r_k = i, \mathbf{x}_{k-1})$$

$$= \Gamma_{j,i}\,p_{\mathbf{v}_i}(\mathbf{x}_k - \mathbf{f}_i(\mathbf{x}_{k-1})), \qquad (3.46)$$

Algorithm 12 IMM Algorithm recursion equations at time k

1: **for** each model i **do** {model conditioned mixing}
   - Predicted model probability:

$$\mu_{k-1|k}(i) = p(r_k = i|\mathbf{y}^{k-1}) = \sum_{j=1}^{d} \Gamma_{ji} \, \mu_{k-1|k-1}(j).$$

   - Mixing probability:

$$\mu_{k-1|k}^{j|i} = p(r_{k-1} = j|r_k = i, \mathbf{y}^{k-1}) = \frac{\Gamma_{ji} \, \mu_{k-1|k-1}(j)}{\sum_{m=1}^{d} \Gamma_{mi} \mu_{k-1|k-1}(m)}.$$

   - Mixing estimate:

$$\hat{\mathbf{x}}_{k-1|k-1}^{0i} = \sum_{j=1}^{d} \hat{\mathbf{x}}_{k-1|k-1}^{j} \mu_{k-1|k}^{j|i}.$$

   - Mixing covariance:

$$\mathbf{P}_{k-1|k-1}^{0i}$$
$$= \sum_{j=1}^{d} \mu_{k-1|k}^{j|i} \{\mathbf{P}_{k-1|k-1}^{j} + [\hat{\mathbf{x}}_{k-1|k-1}^{j} - \hat{\mathbf{x}}_{k-1|k-1}^{0i}][\hat{\mathbf{x}}_{k-1|k-1}^{j} - \hat{\mathbf{x}}_{k-1|k-1}^{0i}]^{T}\}.$$

2: **end for**
3: **for** each model i **do** {model-based filtering}
   - Predicted state: $\hat{\mathbf{x}}_{k|k-1}^{i} = \mathbf{F}_i \hat{\mathbf{x}}_{k-1|k-1}^{0i} + \mathbf{u}_i.$
   - Predicted covariance: $\mathbf{P}_{k|k-1}^{i} = \mathbf{F}_i \mathbf{P}_{k-1|k-1}^{0i} \mathbf{F}_i^{T} + \mathbf{Q}_i.$
   - Updated state:
     $\hat{\mathbf{x}}_{k|k}^{i} = \hat{\mathbf{x}}_{k|k-1}^{i} + \mathbf{P}_{k|k-1}^{i} \mathbf{H}^{T} (\mathbf{H}\mathbf{P}_{k|k-1}^{i}\mathbf{H}^{T} + \mathbf{R}_k)^{-1}(\mathbf{y}_k - \mathbf{H}\hat{\mathbf{x}}_{k|k-1}^{i}).$
   - Updated covariance:
     $\mathbf{P}_{k|k}^{i} = \mathbf{P}_{k|k-1}^{i} - \mathbf{P}_{k|k-1}^{i} \mathbf{H}^{T} (\mathbf{H}\mathbf{P}_{k|k-1}^{i}\mathbf{H}^{T} + \mathbf{R}_k)^{-1}\mathbf{H}\mathbf{P}_{k|k-1}^{i}.$
   - Model likelihood: $\lambda_k(i) = N(\mathbf{y}_k; \mathbf{H}\hat{\mathbf{x}}_{k|k-1}^{i}, \mathbf{H}\mathbf{P}_{k|k-1}^{i}\mathbf{H}^{T} + \mathbf{R}_k).$
4: **end for**
   - Conditional mean: $\hat{\mathbf{x}}_{k|k} = \sum_{i=1}^{d} \hat{\mathbf{x}}_{k|k}^{i} \mu_{k|k}(i).$
   - Conditional covariance:
     $\mathbf{P}_{k|k} = \sum_{i=1}^{d} \mu_{k|k}(i)\{\mathbf{P}_{k|k}^{i} + [\hat{\mathbf{x}}_{k|k}^{i} - \hat{\mathbf{x}}_{k|k}][\hat{\mathbf{x}}_{k|k}^{i} - \hat{\mathbf{x}}_{k|k}]^{T}\}.$

---

and that measurements are generated as

$$p(\mathbf{y}_k | \mathbf{x}_k, r_k = i) = p_{\mathbf{w}_i}(\mathbf{y}_k - \mathbf{h}_i(\mathbf{x}_k)), \tag{3.47}$$

it is desired to produce a set of weighted samples representing an approximation to the posterior density at time k. This can be done using the basic particle filtering algorithm (Algorithm 5), extended to include sampling of the maneuvering mode. In this section it will be shown how these methods can be applied to the particular problem of maneuvering object tracking.

3.5.1 Bootstrap filter for maneuvering object tracking

Recall that the bootstrap filter (BF) draws samples without consideration of the current measurement and then weights these samples by their likelihood. The importance density for maneuvering object tracking can be written as

$$q(\mathbf{x}_k, r_k = i, t) = w_{k-1}^t \, \Gamma_{r_{k-1}^t, i} \, p_{\mathbf{v}_i}(\mathbf{x}_k - \mathbf{f}_i(\mathbf{x}_{k-1}^t)). \tag{3.48}$$

Samples can be drawn from the importance density (3.48) by noting that it factorizes into the marginal densities,

$$q(t) = w_{k-1}^t, \tag{3.49}$$

$$q(r_k = i | t) = \Gamma_{r_{k-1}^t, i}, \tag{3.50}$$

$$q(\mathbf{x}_k | r_k = i, t) = p_{\mathbf{v}_i}(\mathbf{x}_k - \mathbf{f}_i(\mathbf{x}_{k-1}^t)). \tag{3.51}$$

The new weights, given by the ratio of the posterior density to the importance density, are $w_k^i = C \, p(\mathbf{y}_k - \mathbf{h}_{r_k^i}(\mathbf{x}_k^i))$ with C such that the weights sum to one. A recursion of the BF for maneuvering object tracking is given in Algorithm 13. Due to the manner in which samples are drawn, the BF recursion separates nicely into a prediction step, in which samples are drawn according to the state dynamics, and a correction step, in which the quality of the samples is determined by calculating their likelihood. The correction step of the BF for maneuvering object tracking is the same as in the single non-maneuvering object tracking case. Differences arise in the prediction step since it is necessary to account for the uncertainty regarding the motion model in maneuvering object tracking.

3.5.2 Auxiliary bootstrap filter for maneuvering object tracking

The importance density for the auxiliary bootstrap filter (ABF) for maneuvering object tracking is

$$q(\mathbf{x}_k, r_k = i, t) = \xi_k^t \, \Gamma_{r_{k-1}^t, i} \, p_{\mathbf{v}_i}(\mathbf{x}_k - \mathbf{f}_i(\mathbf{x}_{k-1}^t)), \tag{3.52}$$

Algorithm 13 Bootstrap filter for maneuvering object tracking

---

1: **for** $i = 1, \ldots, n$ **do**

2:     Draw a mixture index t^i such that $\Pr(t^i = l) = w_{k-1}^l$.

3:     Draw a maneuvering mode r_k^i such that $\Pr(r_k^i = j) = \Gamma_{r_{k-1}^{t^i}, j}$.

4:     Draw $\mathbf{v}_k^i \sim p_{\mathbf{v}_{r_k^i}}$ and compute the sample object state $\mathbf{x}_k^i = \mathbf{f}_{r_k^i}(\mathbf{x}_{k-1}^{t^i}) + \mathbf{v}_k^i$.

5:     Compute the weight update $e_k^i = p_{\mathbf{w}_{r_k^i}}(\mathbf{y}_k - \mathbf{h}_{r_k^i}(\mathbf{x}_k^i))$.

6: **end for**

7: Compute the updated weights:

$$w_k^i = w_{k-1}^i e_k^i \Big/ \sum_{j=1}^{n} w_{k-1}^j e_k^j, \quad i = 1, \ldots, n.$$

8: Compute a state estimate:

$$\hat{\mathbf{x}}_{k|k} = \sum_{i=1}^{n} w_k^i \mathbf{x}_k^i, \qquad \hat{r}_k = \arg \max_{j \in \{1,\ldots,d\}} \sum_{\{t : r_k^t = j\}} w_k^t.$$

---

where

$$\xi_k^t = w_{k-1}^t p_{\mathbf{w}_{\rho_k^t}}(\mathbf{y}_k - \mathbf{h}_{\rho_k^t}(\boldsymbol{\mu}_k^t)) \Big/ \sum_{s=1}^{n} w_{k-1}^s p_{\mathbf{w}_{\rho_k^s}}(\mathbf{y}_k - \mathbf{h}_{\rho_k^s}(\boldsymbol{\mu}_k^s)), \quad (3.53)$$

with $\Pr(\rho_k^t = j) = \Gamma_{r_{k-1}^t, j}$ and $\boldsymbol{\mu}_k^t = \mathbf{f}_{\rho_k^t}(\mathbf{x}_{k-1}^t) + \mathbf{v}_k^t$, $\mathbf{v}_k^t \sim p_{\mathbf{v}_{\rho_k^t}}$. A recursion of the ABF for maneuvering object tracking is given by Algorithm 14. As with the BF, the ABF for maneuvering object tracking differs from the ABF for non-maneuvering object tracking only in the sampling step, which requires that the object kinematic state be drawn conditional on a randomly selected maneuvering mode. The weight update is the same for both non-maneuvering and maneuvering objects.

3.5.3 Extended Kalman auxiliary particle filter for maneuvering object tracking

During the derivation of the EK-APF it will be assumed that $\mathbf{v}_i \sim N(0, \mathbf{Q}_i)$ and $\mathbf{w}_i \sim N(0, \mathbf{R}_i)$. The importance density for the EK-APF for maneuvering object tracking is

$$q(\mathbf{x}_k, r_k = i, t) = \xi_k^t \hat{\Pr}(r_k = i | r_{k-1}^t = j, \mathbf{x}_{k-1}^t, \mathbf{y}_{1:k}) \, \hat{p}(\mathbf{x}_k | r_k = i, \mathbf{x}_{k-1}^t, \mathbf{y}_{1:k}),$$

$$(3.54)$$

Algorithm 14 Auxiliary bootstrap filter for maneuvering object tracking

---

1: **for** $i = 1, \ldots, n$ **do**

2:    Draw a maneuvering mode ρ_k^i such that $\Pr(\rho_k^i = j) = \Gamma_{r_{k-1}^i, j}$.

3:    Draw $\tilde{\mathbf{v}}_k^i \sim p_{\mathbf{v}_{\rho_k^i}}$ and compute $\boldsymbol{\mu}_k^i = \mathbf{f}_{\rho_k^i}(\mathbf{x}_{k-1}^i) + \tilde{\mathbf{v}}_k^i$.

4:    Compute the first-stage weight update $a_k^i = p_{\mathbf{w}_{\rho_k^i}}(\mathbf{y}_k - \mathbf{h}_{\rho_k^i}(\boldsymbol{\mu}_k^i))$.

5: **end for**

6: Compute the first-stage weights:

$$\xi_k^t = w_{k-1}^t a_k^t \bigg/ \sum_{i=1}^n w_{k-1}^i a_k^i, \quad t = 1, \ldots, n.$$

7: **for** $i = 1, \ldots, n$ **do**

8:    Draw a mixture index t^i such that $\Pr(t^i = l) = \xi_k^l$.

9:    Draw a maneuvering mode r_k^i such that $\Pr(r_k^i = j) = \Gamma_{r_{k-1}^{t^i}, j}$.

10:    Draw $\mathbf{v}_k^i \sim p_{\mathbf{v}_{r_k^i}}$ and compute the sample object state $\mathbf{x}_k^i \sim \mathbf{f}_{r_k^i}(\mathbf{x}_{k-1}^{t^i}) + \mathbf{v}_k^i$.

11:    Compute the un-normalized weight:

$$\tilde{w}_k^i = \frac{p_{\mathbf{w}_{r_k^i}}(\mathbf{y}_k - \mathbf{h}_{r_k^i}(\mathbf{x}_k^i))}{p_{\mathbf{w}_{\rho_k^{t^i}}}(\mathbf{y}_k - \mathbf{h}_{\rho_k^{t^i}}(\boldsymbol{\mu}_k^{t^i}))}.$$

12: **end for**

13: Normalize the weights:

$$w_k^i = \tilde{w}_k^i \bigg/ \sum_{j=1}^n \tilde{w}_k^j, \quad i = 1, \ldots, n.$$

14: Compute a state estimate:

$$\hat{\mathbf{x}}_{k|k} = \sum_{i=1}^n w_k^i \mathbf{x}_k^i, \qquad \hat{r}_k = \arg \max_{j \in \{1, \ldots, d\}} \sum_{\{t : r_k^t = j\}} w_k^t.$$

---

where

$$\xi_k^t = w_{k-1}^t \hat{p}(\mathbf{y}_k | \mathbf{x}_{k-1}^t, r_{k-1} = r_{k-1}^t, \mathbf{y}_{1:k-1}) \bigg/ \sum_{i=1}^n w_{k-1}^i \hat{p}$$

$$\times (\mathbf{y}_k | \mathbf{x}_{k-1}^i, r_{k-1} = r_{k-1}^i, \mathbf{y}_{1:k-1}). \tag{3.55}$$

The hat notation is used to denote probability distributions or densities computed with the linearized measurement equation approximation, $\mathbf{h}_i(\mathbf{x}) \approx \mathbf{h}(\hat{\mathbf{x}}) +$

$\mathbf{H}(\mathbf{x} - \hat{\mathbf{x}})$. The marginal densities are

$$q(t) = \xi_k^t, \tag{3.56}$$

$$q(r_k = i | t) = \hat{\Pr}(r_k = i | r_{k-1}^t = j, \mathbf{x}_{k-1}^t, \mathbf{y}_{1:k}), \tag{3.57}$$

$$q(\mathbf{x}_k | r_k = i, t) = \hat{p}(\mathbf{x}_k | r_k = i, \mathbf{x}_{k-1}^t, \mathbf{y}_{1:k}). \tag{3.58}$$

We begin by deriving the update factor for the first-stage weights. This can be expanded as

$$\hat{p}(\mathbf{y}_k | \mathbf{x}_{k-1}, r_{k-1} = j, \mathbf{y}_{1:k-1})$$

$$= \sum_{i=1}^{d} \int \hat{p}(\mathbf{y}_k, \mathbf{x}_k, r_k = i | \mathbf{x}_{k-1}, r_{k-1} = j, \mathbf{y}_{1:k-1}) \, d\mathbf{x}_k$$

$$= \sum_{i=1}^{d} \Gamma_{j,i} \int N(\mathbf{y}_k; \mathbf{h}_i(\mathbf{f}_i(\mathbf{x}_{k-1}))$$

$$\qquad - \mathbf{H}_{k,i}(\mathbf{x}_k - \mathbf{f}_i(\mathbf{x}_{k-1})), \mathbf{R}_i) N(\mathbf{x}_k; \mathbf{f}_i(\mathbf{x}_{k-1}), \mathbf{Q}_i) \, d\mathbf{x}_k, \tag{3.59}$$

where $\mathbf{H}_{k,i} = [\nabla_{\mathbf{x}} \mathbf{h}_i(\mathbf{x})^T |_{\mathbf{x}=\mathbf{f}_i(\mathbf{x}_{k-1})}]'$. The integral can be evaluated using Theorem 2.1 to give

$$\hat{p}(\mathbf{y}_k | \mathbf{x}_{k-1}, r_{k-1} = j, \mathbf{y}_{1:k-1}) = \sum_{i=1}^{d} \Gamma_{j,i} N(\mathbf{y}_k; \hat{\mathbf{y}}_{k,i}, \mathbf{S}_{k,i}), \tag{3.60}$$

where $\hat{\mathbf{y}}_{k,i} = \mathbf{h}_i(\mathbf{f}_i(\mathbf{x}_{k-1}))$ and $\mathbf{S}_{k,i} = \mathbf{H}_{k,i} \mathbf{Q}_i \mathbf{H}_{k,i}^T + \mathbf{R}_i$.

Using Bayes' theorem, the sampling distribution for the maneuvering mode can be written as

$$\hat{\Pr}(r_k = i | r_{k-1} = j, \mathbf{x}_{k-1}, \mathbf{y}_{1:k}) = \frac{\hat{p}(\mathbf{y}_k | r_k = i, \mathbf{x}_{k-1}, \mathbf{y}_{1:k-1}) \Pr(r_k = i | r_{k-1} = j)}{\hat{p}(\mathbf{y}_k | r_{k-1} = j, \mathbf{x}_{k-1}, \mathbf{y}_{1:k})}. \tag{3.61}$$

Note that the denominator is the weight update for the first-stage weights and the numerator is a summand in the weight update for the first-stage weights. Therefore,

$$\hat{\Pr}(r_k = i | r_{k-1} = j, \mathbf{x}_{k-1}, \mathbf{y}_{1:k}) = \frac{\Gamma_{j,i} N(\mathbf{y}_k; \hat{\mathbf{y}}_{k,i}, \mathbf{S}_{k,i})}{\displaystyle\sum_{a=1}^{d} \Gamma_{j,a} N(\mathbf{y}_k; \hat{\mathbf{y}}_{k,a}, \mathbf{S}_{k,a})}. \tag{3.62}$$

It is interesting to compare (3.62) with the sampling distribution used for the maneuvering mode in the BF. The sampling probability for a particular maneuvering mode in the BF is given by the transition probability. The EK-APF incorporates the current measurement by scaling the transition probability for each maneuvering mode by a quantity which is related to how much the current measurement favors the maneuvering mode.

The sampling distribution for the object state can be expanded using Bayes' rule as

$$\hat{p}(\mathbf{x}_k | r_k = i, \mathbf{x}_{k-1}, \mathbf{y}_{1:k})$$

$$= \frac{\hat{p}(\mathbf{y}_k | \mathbf{x}_k, r_k = i) p(\mathbf{x}_k | r_k = i, \mathbf{x}_{k-1})}{p(\mathbf{y}_k | r_k = i, \mathbf{x}_{k-1}, \mathbf{y}_{1:k-1})}$$

$$\propto N(\mathbf{y}_k; \hat{\mathbf{y}}_{k,i} - \mathbf{H}_{k,i}(\mathbf{x}_k - \mathbf{f}_i(\mathbf{x}_{k-1})), \mathbf{R}_i) N(\mathbf{x}_k; \mathbf{f}_i(\mathbf{x}_{k-1}), \mathbf{Q}_i). \qquad (3.63)$$

Using Theorem 2.1 gives

$$\hat{p}(\mathbf{x}_k | r_k = i, \mathbf{x}_{k-1}, \mathbf{y}_{1:k}) = N(\mathbf{x}_k; \hat{\boldsymbol{\mu}}_{k,i}, \boldsymbol{\Sigma}_{k,i}), \qquad (3.64)$$

where, with $\mathbf{K}_{k,i} = \mathbf{Q}_i \mathbf{H}_{k,i}^T \mathbf{S}_{k,i}^{-1}$,

$$\boldsymbol{\mu}_{k,i} = \mathbf{f}_i(\mathbf{x}_{k-1}) + \mathbf{K}_{k,i}(\mathbf{y}_k - \hat{\mathbf{y}}_{k,i}), \qquad (3.65)$$

$$\boldsymbol{\Sigma}_{k,i} = \mathbf{Q}_i - \mathbf{K}_{k,i} \mathbf{H}_{k,i} \mathbf{Q}_i. \qquad (3.66)$$

This completes the derivation of the marginal sampling densities for the EK-APF for maneuvering object tracking. The weights, which are given by the ratio of the posterior density to the importance density, are given by the ratio of the likelihood of the samples to the linearized likelihood of the samples. This is the same as the weight calculation for non-maneuvering object tracking using the EK-APF. This is to be expected since, in both cases, the weight update is accounting for the linearized approximation used in the sampling step. A recursion of the EK-APD for maneuvering object tracking is given by Algorithm 15.

3.6 Performance bounds

It is desirable to compare the various maneuvering object tracking algorithms with an appropriate performance bound. In Section 2.7 the posterior Cramér–Rao bound (PCRB) was introduced as a lower bound on the mean square error for tracking a non-maneuvering object. Recall that one of the conditions required for the PCRB to hold is twice differentiability of the logarithm of the joint density of the observations and the random parameter vector. In the case of a maneuvering object the discrete-valued maneuvering mode must be added to the vector of random parameters to be estimated. As a result the twice differentiability condition of the logarithm of the joint density is not met and the PCRB cannot be applied and it is necessary to consider alternatives.

Algorithm 15 Extended Kalman auxiliary particle filter for maneuvering object tracking

1: **for** $i = 1, \ldots, n$ **do**
2: **for** $j = 1, \ldots, d$ **do**
3:   Compute the Jacobian $\mathbf{H}^i_{k,j} = \nabla_{\mathbf{x}^T} \mathbf{h}_j(\mathbf{x})|_{\mathbf{x}=\mathbf{f}_j(\mathbf{x}^i_{k-1})}$.
4:   Compute:

$$\mathbf{x}^i_{k|k-1,j} = \mathbf{f}_j(\mathbf{x}^i_{k-1}), \qquad\qquad \hat{\mathbf{y}}^i_{k,j} = \mathbf{h}_j(\mathbf{x}^i_{k|k-1,j}),$$

$$\mathbf{S}^i_{k,j} = \mathbf{H}^i_{k,j} \mathbf{Q}_j (\mathbf{H}^i_{k,j})^T + \mathbf{R}_j, \qquad \mathbf{K}^i_{k,j} = \mathbf{Q}_j (\mathbf{H}^i_{k,j})^T (\mathbf{S}^i_{k,j})^{-1},$$

$$\boldsymbol{\mu}^i_{k,j} = \mathbf{x}^i_{k|k-1,j} + \mathbf{K}^i_{k,j}(\mathbf{y}_k - \hat{\mathbf{y}}^i_{k,j}), \quad \boldsymbol{\Sigma}^i_{k,j} = \mathbf{Q}_j - \mathbf{K}^i_{k,j} \mathbf{H}^i_{k,j} \mathbf{Q}_j.$$

5:   Compute the un-normalized sampling probability for the maneuvering mode:

$$\tilde{b}^i_{k,j} = \Gamma_{r^i_{k-1},j} N(\mathbf{y}_k; \hat{\mathbf{y}}^i_{k,j}, \mathbf{S}^i_{k,j}).$$

6: **end for**
7: Compute the first-stage weight update $a^i_k = \sum^d_{j=1} \tilde{b}^i_{k,j}$ and normalize the maneuvering mode sampling probabilities $b^i_{k,j} = \tilde{b}^i_{k,j}/a^i_k, j = 1, \ldots, d$.
8: **end for**
9: Compute the first-stage weights:

$$\xi^t_k = w^t_{k-1} a^t_k \bigg/ \sum^n_{i=1} w^i_{k-1} a^i_k, \quad t = 1, \ldots, n.$$

10: **for** $i = 1, \ldots, n$ **do**
11: Draw a mixture index t^i such that $\Pr(t^i = l) = \xi^l_k$.
12: Draw a maneuvering mode r^i_k such that $\Pr(r^i_k = j) = b^{r^{t^i}_{k-1}}_{k,j}$.
13: Draw the sample object state $\mathbf{x}^i_k \sim N(\boldsymbol{\mu}^{t^i}_{k,r^i_k}, \boldsymbol{\Sigma}^{t^i}_{k,r^i_k})$.
14: Compute the un-normalized weight:

$$\tilde{w}^i_k = \frac{p_{\mathbf{w}_k}(\mathbf{y}_k - \mathbf{h}(\mathbf{x}^i_k))}{p_{\mathbf{w}_k}(\mathbf{y}_k - \hat{\mathbf{y}}^{t^i}_{k,r^i_k} - \mathbf{H}^{t^i}_{k,r^i_k}(\mathbf{x}^i_k - \mathbf{x}^{t^i}_{k|k-1,r^i_k}))}.$$

15: **end for**
16: Normalize the weights:

$$w^i_k = \tilde{w}^i_k \bigg/ \sum^n_{j=1} \tilde{w}^j_k, \quad i = 1, \ldots, n.$$

17: Compute a state estimate:

$$\hat{\mathbf{x}}_{k|k} = \sum^n_{i=1} w^i_k \mathbf{x}^i_k.$$

One possibility is to consider bounds which do not require twice differentiability of the logarithm of the joint density. One example of such a bound is the Weiss–Weinstein bound (WWB), a recursive version of which was derived in Rapoport and Oshman (2004). The main problem with alternatives to the PCRB is that they are invariably more complicated to derive and compute. This is certainly the case for the WWB, which requires an optimization over a number of parameters. The recursive WWB of Rapoport and Oshman (2004) was derived for only a single collection of parameter values and thus provides an optimistic lower bound. Although the idea of using the WWB is promising, more work is required if a useful bound is to be obtained. An interesting point to note is that the recursive WWB raises the possibility of bounding the performance of estimators of the maneuvering mode in addition to the continuous part of the state vector. The bounds discussed below, although simpler, do not have this desirable property.

An alternative to the use of more complicated bounds is to apply the PCRB to a system which approximates the jump-Markov system used for maneuvering objects. The simplest of this class of techniques is to compute the bound conditional on the true maneuvering mode sequence. The resulting bound is usually far too optimistic to be useful. Another approach is to derive the PCRB for the continuous part of the state vector conditional on each possible maneuvering mode sequence and then weight each bound by the prior probability of the corresponding sequence. In practice, it is necessary to remove the least likely sequences for the sake of computational efficiency. This method tends to average the effects of a maneuver over the observation interval and so does not really capture the effect of a sudden object maneuver. Perhaps the most promising method is the recently proposed best-fitting Gaussian approach (Hernandez *et al.*, 2005).

3.7 Illustrative example

Maneuvering object tracking algorithms will be demonstrated using the same measurement system as in the illustrative example of Section 2.8. In this scenario the angular position measurements of a ground object are acquired by an airborne sensor, as depicted in Figure 2.2.

The current example differs from that of Section 2.8 in that the object will now perform maneuvers. In particular, the object motion is composed of periods of uniform motion interspersed by coordinated turn maneuvers. A multiple-model approach is used to model this situation. The object state at time kT, where T is the sampling interval, is $\mathbf{x}_k = [x_k, \dot{x}_k, y_k, \dot{y}_k, \omega_k]$, where (x_k, y_k) is the object position in Cartesian coordinates, the dot notation denotes differentiation with respect to time and ω_k is the turn rate. The manner in which the object state evolves is determined by the maneuvering mode r_k. The three possible motion models can be written in the form, for $i = 1, 2, 3$,

$$p(\mathbf{x}_k|r_k = i, \mathbf{x}_{k-1}) = N(\mathbf{x}_k; \mathbf{f}_i(\mathbf{x}_{k-1}), \mathbf{Q}_i). \tag{3.67}$$

Model 1 corresponds to uniform motion with transition function and process noise covariance matrix given by

$$\mathbf{f}_1(\mathbf{x}) = \mathbf{F}_1\mathbf{x} = \mathrm{diag}\left(\mathbf{I}_2 \otimes \begin{bmatrix} 1 & T \\ 0 & 1 \end{bmatrix}, 0\right)\mathbf{x}, \tag{3.68}$$

$$\mathbf{Q}_1 = \mathrm{diag}\left(\mathbf{I}_2 \otimes q_1 \begin{bmatrix} T^3/3 & T^2/2 \\ T^2/2 & T \end{bmatrix}, 0\right). \tag{3.69}$$

Models 2 and 3 correspond to motion under a coordinated turn. The transition function, for a state vector $\mathbf{x} = [x, \dot{x}, y, \dot{y}, \omega]$, and the process noise covariance matrix are

$$\mathbf{f}_i(\mathbf{x}) = \begin{bmatrix} 1 & \sin(\omega T)/\omega & 0 & -(1 - \cos(\omega T))/\omega & 0 \\ 0 & \cos(\omega T) & 0 & -\sin(\omega T) & 0 \\ 0 & (1 - \cos(\omega T))/\omega & 1 & \sin(\omega T)/\omega & 0 \\ 0 & \sin(\omega T) & 0 & \cos(\omega T) & 0 \\ 0 & 0 & 0 & 0 & 1 \end{bmatrix}\mathbf{x}, \tag{3.70}$$

$$\mathbf{Q}_i = \mathrm{diag}\left(\mathbf{I}_2 \otimes q_i \begin{bmatrix} T^3/3 & T^2/2 \\ T^2/2 & T \end{bmatrix}, b_i T\right). \tag{3.71}$$

Note that the transition function for coordinated turn motion is non-linear due to multiplication of trigonometric functions of the turn rate by the other elements of the object state. The use of two coordinated turn motion models enables the filter to respond quickly when the object transitions from uniform to coordinated turn motion without reducing estimation accuracy during coordinated turns. This is done by using a large process noise for model 2 and a small process noise for model 3. Model 2 will be favored at the onset of the maneuver and model 3 will be favored once the maneuver has been established.

The particular scenario used in the simulation analysis is shown in Figure 3.1. The object performs coordinated turns during the intervals (Hilton *et al.*, 1993; Blair *et al.*, 1993; Li *et al.*, 1999; Mušicki *et al.*, 2005a) with turn rates of $\pm 5°/s$. This scenario retains the geometry of the example used for non-maneuvering object tracking in Section 2.8 with the trajectories of the object and sensor in the x-y plane swapped. This permits a comparison between the estimation errors for maneuvering and non-maneuvering object tracking. Although the object trajectory is generated without process noise the filter is implemented with some process noise. The process noise intensities are $q_1 = 1/1000$ for the uniform motion model, $q_2 = 1/100$ and $b_2 = 2 \times 10^{-3}$ for motion model 2, and $q_3 = 1/1000$ and

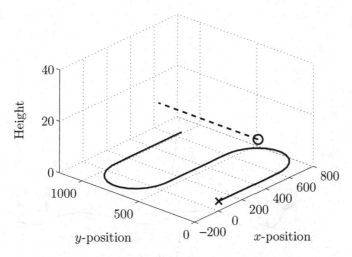

Figure 3.1 Simulation scenario for performance analysis of maneuvering object tracking algorithms.

$b_3 = 1 \times 10^{-5}$ for motion model 3. The transition matrix for the maneuvering modes is

$$\Gamma = \begin{bmatrix} 0.95 & 0.05 & 0 \\ 0.2 & 0.6 & 0.2 \\ 0 & 0.2 & 0.8 \end{bmatrix}. \tag{3.72}$$

Recall that the (j, i)th element of Γ is $P(r_k = i | r_{k-1} = j)$. Measurements are acquired at intervals of $T = 3$ s for 150 s with a standard deviation of $1°$.

The algorithms considered in the performance analysis are the IMM and the auxiliary bootstrap filter (ABF) implemented with 2000 and 5000 samples. The mode-conditioned posterior densities in the IMM are approximated using the UKF. The RMS position errors of the three algorithms, averaged over 500 realizations, are plotted against time in Figure 3.2. Also included is a plot of the PCRB for the corresponding scenario with the $x - y$ trajectories of the object and sensor swapped. It is not realistic to expect algorithm performance to achieve this bound since it applies to non-maneuvering target tracking. The purpose of including this bound is to give an idea of how object maneuvers affect tracking accuracy. It can be seen from the results of Figure 3.1 that object maneuvers greatly decrease the accuracy of all of the algorithms. The algorithm most affected by the object maneuvers is the IMM. The RMS position errors of the IMM are as high as 150 m during coordinated turn motion. The RMS position errors of the ABF do not exceed 100 m for either sample size. The difference in performance between the ABF and the IMM during uniform motion is much smaller than during coordinated turns.

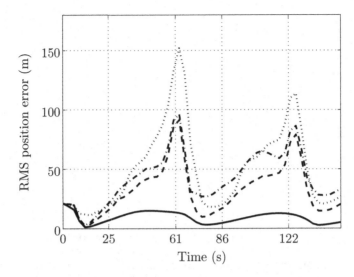

Figure 3.2 RMS position error of the IMM (dotted) and the ABF with 2000 samples (dash-dot) and 5000 samples (dashed). The solid line is the PCRB for non-maneuvering tracking. Simulation scenario for performance analysis of maneuvering object tracking algorithms.

3.8 Summary

The chapter brings together a range of maneuvering object tracking methods into the unified Chapman–Kolmogorov–Bayes formalism. The key algorithms like the GPB, IMM and the variable structure IMM along with the particle filtering approach to maneuvering object tracking are derived. The fundamental concept is that the probability density function of the state is a mixture of Gaussian and various filters are approximations to the mixture density. Reduction and approximation of mixtures of Gaussians is central to a number of object tracking problems, including the problem of tracking in clutter, and will be covered in detail in the next chapter.

4

Single-object tracking in clutter

In Chapters 2 and 3, we introduced state estimation and filtering theory and its application to idealistic object tracking problems. The fact that makes practical object tracking problems both challenging and interesting is that the sensor measurements, more often than not, contain detections from false targets. For example, in many radar and sonar applications, measurements (detections) originate not only from objects of interest, but also from thermal noise, terrain reflections, clouds, etc. Such unwanted measurements are usually termed clutter. In vision-based object tracking, where tracking can be used to count moving targets, shadows created by an afternoon sun, light reflections on snow or the movement of leaves on a tree can all generate clutter data in the images.

One of the defining characteristics of clutter or false alarms is that their number changes from one time instant to the next in a random manner and, to make matters worse, target- and clutter-originated measurements share the same measurement space and look alike. Practical tracking problems are considerably difficult since sometimes, even when there are targets in the sensor's field of view, they can go undetected or fail to appear in the set of measurements. In other words, true measurements from the target are present during each measurement scan with only a certain probability of detection. Hence, determining the state of the object using a combination of false alarms and true target returns is at the heart of all practical object tracking problems and is the subject of this chapter. In many situations the measurement origin uncertainty is a far more important impairment to tracking performance than is the noise associated with the measurements themselves and make it a more difficult problem to address than the estimation problems covered in Chapters 2 and 3.

The object tracking algorithms presented in this chapter are based on the following assumptions, unless stated otherwise:

- Object:
  - There is a single object in the surveillance area. The position of the object is a priori unknown.
  - Possible object trajectory models are assumed known, and the models are assumed to propagate as Markov chains.
  - At each scan one measurement per object may be present, described by the probability of detection, P_D, which may be different for each target. In other words, we assume point objects and they produce zero or one measurement per scan randomly.
- Clutter:
  - Clutter measurement density in the surveillance area is a priori known, otherwise we estimate it from the received measurements.
- Measurements:
  - Measurements are produced by the sensor with infinite resolution. In other words, each measurement can have only one source. When a measurement is used to update a track, it can be one of:
    * measurement (detection) of the object being tracked;
    * clutter measurement.
  - We consider here only position measurements. Additional measurement attributes, such as amplitude, Doppler speed, etc., can be included in a straightforward manner using a priori probability density functions of the measurement attributes (Lerro and Bar-Shalom, 1990, 1993; Mušicki and Evans, 2008).

These assumptions are usually not completely satisfied in practical situations. For example, sensors usually do not have infinite resolution and close measurements will be merged. In the opposite direction, with high-resolution sensors, one object may have more than one measurement per scan. However, the assumptions listed here simplify the object tracking problem considerably and are instrumental in the derivation of a majority of object tracking algorithms. As a consequence, the practical implementation of target tracking may fall short of expectations in situations where these assumptions are largely violated (Mušicki and Evans, 1995).

4.1 The optimal Bayesian filter

4.1.1 Object dynamics, sensor measurement and noise models

For object tracking in clutter, the key change is in sensor measurements. The object dynamics remain the same as for the single non-maneuvering object model,

$$\mathbf{x}_k = \mathbf{f}(\mathbf{x}_{k-1}) + \mathbf{v}_k. \tag{4.1}$$

On the other hand, sensor observations at each epoch result in a set of measurements,

$$\mathbf{y}_k = \{\mathbf{y}^k(1), \mathbf{y}^k(2), \ldots, \mathbf{y}^k(m_k)\},$$

where m_k denotes the number of received measurements at time k and $m^k = \{m_1, m_2, \ldots, m_k\}$. The posterior density of the target is not only conditioned on received measurements, $\mathbf{y}^k = (\mathbf{y}_1, \mathbf{y}_2, \ldots, \mathbf{y}_k)$, but also on the number of measurements m^k at time k. Although information on the number of measurements is embedded in the measurements, explicit representation helps in terms of clarity.

4.1.2 Conditional density

The conditional probability density of interest in the object tracking in clutter problem is $p(\mathbf{x}_k|\mathbf{y}^k, m^k)$. Using Bayes' rule, we have

$$\begin{aligned} p(\mathbf{x}_k|\mathbf{y}^k, m^k) &= p(\mathbf{x}_k|\mathbf{y}_k, m_k, \mathbf{y}^{k-1}, m^{k-1}) \\ &= \frac{p(\mathbf{y}_k, m_k|\mathbf{x}_k, \mathbf{y}^{k-1}, m^{k-1})p(\mathbf{x}_k|\mathbf{y}^{k-1}, m^{k-1})}{p(\mathbf{y}_k, m_k|\mathbf{y}^{k-1}, m^{k-1})}, \end{aligned} \tag{4.2}$$

where:

- $p(\mathbf{y}_k, m_k|\mathbf{x}_k, \mathbf{y}^{k-1}, m^{k-1})$ is the likelihood function;
- $p(\mathbf{x}_k|\mathbf{y}^{k-1}, m^{k-1})$ is the prediction density;
- $p(\mathbf{y}_k, m_k|\mathbf{y}^{k-1}, m^{k-1})$ is the normalizing factor.

4.1.3 Optimal estimation

The predicted density

The predicted density is given by the Chapman–Kolmogorov equation:

$$p(\mathbf{x}_k|\mathbf{y}^{k-1}, m^{k-1}) = \int_{\mathbf{x}_{k-1}} p(\mathbf{x}_k|\mathbf{x}_{k-1})p(\mathbf{x}_{k-1}|\mathbf{y}^{k-1}, m^{k-1})d\mathbf{x}_{k-1} \tag{4.3}$$

$$= \int_{\mathbf{x}_{k-1}} p_{\mathbf{v}_k}(\mathbf{x}_k - \mathbf{f}(\mathbf{x}_{k-1}))p(\mathbf{x}_{k-1}|\mathbf{y}^{k-1}, m^{k-1})d\mathbf{x}_{k-1}. \tag{4.4}$$

The transition density is directly derived from the object dynamics equation (4.1), and the system noise properties.

The likelihood function

The first term of (4.2), i.e., the joint likelihood, may be written as

$$p(\mathbf{y}_k, m_k|\mathbf{x}_k, \mathbf{y}^{k-1}, m^{k-1}) = p(\mathbf{y}_k(1), \mathbf{y}_k(2), \ldots, \mathbf{y}_k(m_k), m_k|\mathbf{x}_k, \mathbf{y}^{k-1}, m^{k-1}).$$

$$\tag{4.5}$$

Let:

- $\theta_k(0)$ denote the association event that none of the measurements in $\mathbf{y}_k$ is target-originated; and
- $\theta_k(i)$ denote the association event that the ith measurement in $\mathbf{y}_k$ is target-originated, and the rest of the measurements are from clutter, $i = 1, 2, \ldots, m(k)$

Therefore, the set of association events $\{\theta_k(0), \theta_k(1), \ldots, \theta_k(m)\}$ forms a mutually exclusively and exhaustive set of events at each epoch. Using the law of total probability in (4.5) we have

$$p(\mathbf{y}_k, m_k | \mathbf{x}_k, \mathbf{y}^{k-1}, m^{k-1})$$

$$= \sum_{i=0}^{m_k} p(\mathbf{y}_k(1), \mathbf{y}_k(2), \ldots, \mathbf{y}_k(m_k), m_k, \theta_k(i) | \mathbf{x}_k, \mathbf{y}^{k-1}, m^{k-1})$$

$$= \sum_{i=0}^{m_k} p(\mathbf{y}_k(1), \mathbf{y}_k(2), \ldots, \mathbf{y}_k(m_k) | \mathbf{x}_k, m_k, \theta_k(i), \mathbf{y}^{k-1}, m^{k-1})$$

$$\times p(\theta_k(i) | \mathbf{x}_k, m_k, \mathbf{y}^{k-1}, m^{k-1}) p(m_k | \mathbf{x}_k, \mathbf{y}^{k-1}, m^{k-1}). \qquad (4.6)$$

Under white measurement noise assumptions, the first term of (4.6) reduces to

$$p(\mathbf{y}_k(1), \mathbf{y}_k(2), \ldots, \mathbf{y}_k(m_k) | \mathbf{x}_k, m_k, \theta_k(i)).$$

The number of received measurements m_k is also independent of object state $\mathbf{x}_k$. Therefore, the last term in (4.6) can be written as

$$p(m_k | \mathbf{x}_k, \mathbf{y}^{k-1}, m^{k-1}) = p(m_k | \mathbf{y}^{k-1}, m^{k-1}).$$

Now consider the middle term within the summation of (4.6). Since $\theta_k(i)$ is the association event that the ith measurement is from the target and the rest are from clutter at time k, it only depends on the number of received measurements m_k. The fact that the probability is conditioned on the object state $\mathbf{x}_k$ legitimizes the association event to be specified. In other words, it would be meaningless to ask the association of the ith measurement with the object if there is no target or target state to associate with. The association event is simply an event of picking a measurement from a group of measurements and labeling it as the object-originated measurement. This association event is independent of the past measurement $\mathbf{y}^{k-1}, m^{k-1}$. Therefore, we have

$$p(\theta_k(i) | \mathbf{x}_k, m_k, \mathbf{y}^{k-1}, m^{k-1}) = p(\theta_k(i) | m_k),$$

which is identical to $\gamma_i(m_k)$ denoted in Bar-Shalom and Fortmann (1988). Thus, (4.6) can be written as

$$p(\mathbf{y}_k, m_k | \mathbf{x}_k, \mathbf{y}^{k-1}, m^{k-1})$$

$$= \sum_{i=0}^{m_k} p(\mathbf{y}_k(1), \mathbf{y}_k(2), \ldots, \mathbf{y}_k(m_k) | \mathbf{x}_k, m_k, \theta_k(i)) p(\theta_k(i) | m_k) p(m_k | \mathbf{y}^{k-1}, m^{k-1}).$$

$$(4.7)$$

The normalization factor

The normalization factor is

$$p(\mathbf{y}_k, m_k | \mathbf{y}^{k-1}, m^{k-1}) = \int_{\mathbf{x}_k} p(\mathbf{y}_k, m_k | \mathbf{x}_k, \mathbf{y}^{k-1}, m^{k-1}) p(\mathbf{x}_k | \mathbf{y}^{k-1}, m^{k-1}) d\mathbf{x}_k,$$

$$(4.8)$$

where the first term in the integrand is derived from (4.7) and the second term is derived from (4.4). The integral is then evaluated.

Substituting the prediction density, likelihood and normalization factor's simplyfying steps in (4.8) into (4.2) leads to the optimal Bayesian filter for object tracking in clutter. Approximations of this optimal recursion that have been successfully used and implemented include the nearest neighbor filter (NNF) and the probabilistic data association filter (PDA) and particle filters. These approximate filters will be introduced in the following sections.

4.2 The nearest neighbor filter

This nearest neighbour filter approximation is based on five strong assumptions:

1. The true object always exists and is always detected.
2. The measurement that is closest (in a statistical sense) to the predicted measurement is from the object.
3. All other measurements are from clutter.
4. The object motion obeys linear Gaussian statistics.
5. The measurement noise is white Gaussian.

In other words, only one measurement, denoted $\mathbf{y}_k(i)$, whose statistical distance to filter predicted measurement is the smallest among all validated measurements $\mathbf{y}_k$, is considered to be the target-originated measurement.

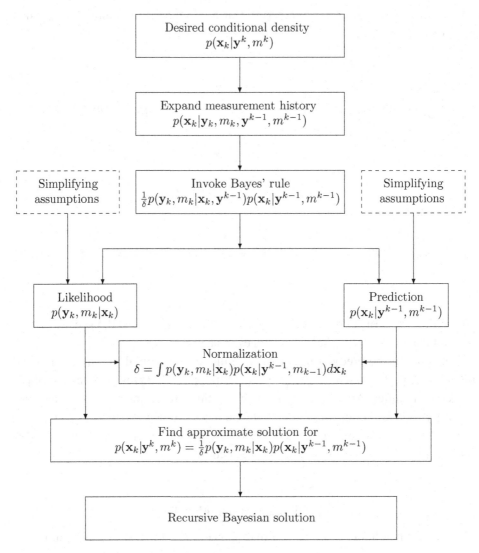

Summary of the recursive Bayesian framework for object tracking in clutter.

The target dynamics, sensor measurement and noise models

Specifically, it is done by assuming:

- the object dynamics function $\mathbf{f}(\cdot)$ is a linear function of the object state, satisfying

$$\mathbf{x}_k = \mathbf{F}\mathbf{x}_{k-1} + \mathbf{v}_k; \qquad (4.9)$$

- the sensor measurement is also a linear function of the object state, satisfying

$$\mathbf{y}_k = \mathbf{H}\mathbf{x}_k + \mathbf{w}_k; \qquad (4.10)$$

- $\mathbf{v}_k$ and $\mathbf{w}_k$ are white, uncorrelated, Gaussian noise sequences with zero mean and covariance $\mathbf{Q}_k$ and $\mathbf{R}_k$ respectively; and
- the prior conditional pdf of target state $p(\mathbf{x}_{k-1}|\mathbf{y}^{k-1})$ is a Gaussian density with mean $\hat{\mathbf{x}}_{k-1|k-1}$ and covariance $\mathbf{P}_{k-1|k-1}$.

The transition density

The target states satisfy linear dynamics given by (4.9). Hence $\mathbf{v}_k = \mathbf{x}_k - \mathbf{F}\mathbf{x}_{k-1}$ and the transition density is given by

$$p(\mathbf{x}_k|\mathbf{x}_{k-1}) = p_{\mathbf{v}_k}(\mathbf{x}_k - \mathbf{F}\mathbf{x}_{k-1}).$$

Since $p_{\mathbf{v}_k}(\cdot)$ is a Gaussian density, the transition density is given by

$$p(\mathbf{x}_k|\mathbf{x}_{k-1}) = \frac{1}{(2\pi)^{\frac{n}{2}}|\mathbf{Q}_k|^{\frac{1}{2}}} \exp\left\{-\frac{1}{2}(\mathbf{x}_k - \mathbf{F}\mathbf{x}_{k-1})^T \mathbf{Q}_k^{-1}(\mathbf{x}_k - \mathbf{F}\mathbf{x}_{k-1})\right\}.$$

This can be represented in the short form $p(\mathbf{x}_k|\mathbf{x}_{k-1}) = N(\mathbf{x}_k; \mathbf{F}\mathbf{x}_{k-1}, \mathbf{Q}_k)$.

The predicted density

The predicted density is given in (4.4) as

$$p(\mathbf{x}_k|\mathbf{y}^{k-1}, m^{k-1}) = \int_{\mathbf{x}_{k-1}} p_{\mathbf{v}_k}(\mathbf{x}_k - \mathbf{f}(\mathbf{x}_{k-1})) p(\mathbf{x}_{k-1}|\mathbf{y}^{k-1}, m^{k-1}) d\mathbf{x}_{k-1}.$$

Under linear Gaussian assumptions, the first term in the integrand, $p(\mathbf{x}_k|\mathbf{x}_{k-1}) = p_{\mathbf{v}_k}(\mathbf{x}_k - \mathbf{f}(\mathbf{x}_{k-1}))$, is $N(\mathbf{x}_k; \mathbf{F}\mathbf{x}_{k-1}, \mathbf{Q}_k)$. The second term $p(\mathbf{x}_{k-1}|\mathbf{y}^{k-1}, m^{k-1})$ is the conditional density at the previous stage (i.e., at time $k-1$), and approximated in that stage by $N(\mathbf{x}_{k-1}; \hat{\mathbf{x}}_{k-1|k-1}, \mathbf{P}_{k-1|k-1})$. Using a normal distribution theorem (see Appendix C), the predicted density can be simplified to

$$p(\mathbf{x}_k|\mathbf{y}^{k-1}, m^{k-1}) = N(\mathbf{x}_k; \hat{\mathbf{x}}_{k|k-1}, \mathbf{P}_{k|k-1}), \tag{4.11}$$

where

$$[\hat{\mathbf{x}}_{k|k-1}, \mathbf{P}_{k|k-1}] = \mathrm{KF_P}[\hat{\mathbf{x}}_{k-1|k-1}, \mathbf{P}_{k-1|k-1}, \mathbf{F}, \mathbf{Q}].$$

These equations are the famous Kalman predictor equations derived and introduced in Chapter 2.

The likelihood function

The likelihood in the NNF is approximated by choosing $\mathbf{y}_k(i)$ from the set of $\mathbf{y}_k$ by ordering the measurements based on their statistical distance to the predicted measurements. Since both the measurement noise and model process noise are Gaussian, a chi-square test (Bar-Shalom and Fortmann, 1988) determines the statistical distance.

In NNF, among all the measurements received, only one measurement is selected to associate and update the object track based on the following criterion:

$$\mathbf{y}_k(i) = \arg \min_{\mathbf{y}_k(j), \forall\, j \in \{1,\dots,m_k\}} \left[\mathbf{y}_k(j) - \mathbf{H}\hat{x}_{k|k-1}\right]^T \mathbf{S}_{k|k-1}^{-1} \left[\mathbf{y}_k(j) - \mathbf{H}\hat{x}_{k|k-1}\right],$$

(4.12)

where $\mathbf{S}_{k|k-1} = \mathbf{H}\mathbf{P}_{k|k-1}\mathbf{H}^T + \mathbf{R}_k$.

The normalization factor

The normalization factor is

$$p(\mathbf{y}_k, m_k|\mathbf{y}^{k-1}, m^{k-1}) = \int_{\mathbf{x}_k} p(\mathbf{y}_k, m_k|\mathbf{x}_k, \mathbf{y}^{k-1}, m^{k-1})\, p(\mathbf{x}_k|\mathbf{y}^{k-1}, m^{k-1})\, d\mathbf{x}_k,$$

where the first term in the integrand is $N(\mathbf{y}_k(i); \mathbf{H}\mathbf{x}_k, \mathbf{R}_k)$ and the second term is $N(\mathbf{x}_k; \hat{\mathbf{x}}_{k|k-1}, \mathbf{P}_{k|k-1})$. This results in

$$p(\mathbf{y}_k, m_k|\mathbf{y}^{k-1}, m^{k-1}) = N\left(\mathbf{y}_k(i); \mathbf{H}\hat{x}_{k|k-1}, \mathbf{S}_{k|k-1}\right).$$

(4.13)

The conditional density

Combining (4.11), (4.12) and (4.13) into (4.2), the posterior density follows as

$$p(\mathbf{x}_k|\mathbf{y}^k, m^k) = \frac{N(\mathbf{y}_k(i); \mathbf{H}\mathbf{x}_k, \mathbf{R}_k)\, N\left(\mathbf{x}_k; \hat{\mathbf{x}}_{k|k-1}, \mathbf{P}_{k|k-1}\right)}{N\left(\mathbf{y}_k(i); \mathbf{H}\hat{x}_{k|k-1}, \mathbf{S}_{k|k-1}\right)}$$

$$= N\left(\mathbf{x}_k; \hat{\mathbf{x}}_{k|k}, \mathbf{P}_{k|k}\right),$$

where

$$\left[\hat{\mathbf{x}}_{k|k}, \mathbf{P}_{k|k}\right] = \mathrm{KF_E}\left[\mathbf{y}_k(i), \hat{x}_{k|k-1}, \mathbf{P}_{k|k-1}, \mathbf{H}, \mathbf{R}_k\right].$$

4.2.1 The nearest neighbor filter equations

The use of the measurement nearest to the predicted measurement to update a track can lead to very poor performance when the density of spurious measurements or clutter is high. This algorithm does not properly account for the fact that the measurement chosen for the track update may be unrelated to the object. It also uses the philosophy of "winner takes all" when associating a measurement to an object track as any one of the measurements validated by a gating mechanism can be associated with it.

Algorithm 16 Nearest neighbor filter recursion equations at time k

---

1: Prediction:

$$\left[\hat{\mathbf{x}}_{k|k-1}, \mathbf{P}_{k|k-1}\right] = \mathrm{KF_P}\left[\hat{\mathbf{x}}_{k-1|k-1}, \mathbf{P}_{k-1|k-1}, \mathbf{F}, \mathbf{Q}\right].$$

2: Measurement selection {Section 5.5.2, (5.78)}:

$$\mathbf{y}_k(i) = \arg \min_{\mathbf{y}_k(j), \forall\, j \in \{1,\dots,m_k\}} \left[\mathbf{y}_k(j) - \mathbf{H}\hat{x}_{k|k-1}\right]^T \mathbf{S}_{k|k-1}^{-1} \left[\mathbf{y}_k(j) - \mathbf{H}\hat{x}_{k|k-1}\right],$$

where $\mathbf{S}_{k|k-1} = \mathbf{H}\mathbf{P}_{k|k-1}\mathbf{H}^T + \mathbf{R}_k$.

3: Output trajectory estimate:

$$\left[\hat{\mathbf{x}}_{k|k}, \mathbf{P}_{k|k}\right] = \mathrm{KF_E}\left[\mathbf{y}_k(i), \hat{\mathbf{x}}_{k|k-1}, \mathbf{P}_{k|k-1}, \mathbf{H}, \mathbf{R}\right].$$

4.3 The probabilistic data association filter

The most successful algorithm in the class of Bayesian all-neighbors filters is the probabilistic data association (PDA) filter. The algorithm updates the object state estimate using all validated measurements and their respective posterior probability weightings. The PDA algorithm is derived by making the assumption that the prediction density in the optimal Bayesain recursion given all past observations is a Gaussian density, although, strictly speaking, it is a Gaussian mixture. This PDA filter approximation is based on the following six strong assumptions:

1. The object being tracked exists and no other object exists.
2. The object motion obeys linear Gaussian statistics.
3. Only one measurement can be from the object of interest.
4. The measurement noise is white Gaussian.
5. The object may or may not be detected all the time and is detected with probability of detection P_D.
6. All non-object originated measurements are assumed to be originated from clutter that is uniformly distributed in space and Poisson distributed in time.
7. Only measurements that fall within a proximity of the expected measurement (i.e., that fall within the validation gate) are considered for processing

The transition and prediction density evaluation

The object dynamics, measurement and noise models are the same as those in the nearest neighbor filter in Section 4.2. Thus the transition and prediction densities are the same as (4.11).

The likelihood function

While the transition and predicted densities of NNF and the PDAF are the same, the key difference is in the way both filters extract information out of the received measurements – i.e., the likelihood function. The likelihood in the PDA is approximated by choosing the subset of measurements from the total measurement set $\mathbf{y}_k$ by gating the measurements based on their statistical distance to the predicted measurements. Since both the measurement noise and model process noise are Gaussian, a chi-square test (Bar-Shalom and Fortmann, 1988) determines the statistical distance and whether a given measurement lies within the minimum volume hyper-ellipsoid that contains a set percentage of the probability distribution of the predicted measurement – the validation gate. The measurements that fall within this ellipsoid are said to be gated. The validation gate is an ellipsoid defined by

$$G = \{\mathbf{y} \in \mathbb{R}^n : [\mathbf{y} - \hat{\mathbf{y}}_{k|k-1}]\mathbf{S}(k)^{-1}[\mathbf{y} - \hat{\mathbf{y}}_{k|k-1}]^T \leq \gamma\},$$

where $\sqrt{\gamma}$ is the gate size, $\hat{\mathbf{y}}_{k|k-1}$ is the predicted measurement and $\mathbf{S}(k)$ is its covariance. A measurement falling inside the validation gate is accepted as a measurement and is indexed as a validated measurement. That is, $\mathbf{y}$ becomes one of the received measurements $\{\mathbf{y}_k(1), \mathbf{y}_k(2), \ldots, \mathbf{y}_k(m_k)\}$. The volume of the n-dimensional gate is

$$V_k = \frac{\pi^{n/2}}{\Gamma(n/2+1)} \sqrt{|\mathbf{S}(k)|} \gamma^{1/2},$$

where $|\mathbf{S}(k)|$ is the determinant of $\mathbf{S}(k)$.

By assuming that the clutter measurements are uniformly distributed in space and have equal probability of being at any point within the gate we can derive the likelihood of the PDA filter. Let P_G denote the probability that the correct measurement falls within the gate, then the likelihood $p(\mathbf{y}_k(1), \mathbf{y}_k(2), \ldots, \mathbf{y}_k(m_k)|\mathbf{x}_k, m_k, \theta_k(i))$ can be evaluated as

$$p(\mathbf{y}_k(1), \mathbf{y}_k(2), \ldots, \mathbf{y}_k(m_k)|\mathbf{x}_k, m_k, \theta_k(i)) = \left(\frac{1}{V_k}\right)^{m_k-1} p(\mathbf{y}_k(i)|\mathbf{x}_k), \quad (4.14)$$

where $p(\mathbf{y}_k(i)|\mathbf{x}_k)$ is the likelihood of the ith measurement being object-originated. It is $N(\mathbf{y}_k(i); \mathbf{Hx}_k, \mathbf{R}_k)$ when the complete measurement space is considered. However, since the effective measurement space is truncated to the validation gate, the likelihood is effectively truncated by the chi-square ellipsoid. This leads to the truncated Gaussian density and does not integrate to 1. By normalizing it with its area in the gate (i.e., the gating probability) it can be re-interpreted as a measurement probability density.

Thus,

$$p(\mathbf{y}_k(i)|\mathbf{x}_k) = \frac{1}{P_G} N(\mathbf{y}_k(i); \mathbf{H}\mathbf{x}_k, \mathbf{R}_k), \tag{4.15}$$

$$p(\mathbf{y}_k(1), \mathbf{y}_k(2), \ldots, \mathbf{y}_k(m_k)|\mathbf{x}_k, m_k, \theta_k(i))$$

$$= \left(\frac{1}{V_k}\right)^{m_k-1} p(\mathbf{y}_k(i)|\mathbf{x}_k)$$

$$= \begin{cases} \left(\frac{1}{V_k}\right)^{m_k-1} P_G^{-1} N(\mathbf{y}_k(i); \mathbf{H}\mathbf{x}_k, \mathbf{R}_k), & \forall i \neq 0, \\ \left(\frac{1}{V_k}\right)^{m_k}, & i = 0. \end{cases} \tag{4.16}$$

Now the first term of (4.2) can be expanded into

$$p(\mathbf{y}_k, m_k|\mathbf{x}_k, \mathbf{y}^{k-1}, m^{k-1})$$

$$= \sum_{i=0}^{m_k} p(\mathbf{y}_k(1), \mathbf{y}_k(2), \ldots, \mathbf{y}_k(m_k)|\mathbf{x}_k, m_k, \theta_k(i)) p(\theta_k(i)|m_k) p(m_k|\mathbf{y}^{k-1}, m^{k-1})$$

$$= \left(\frac{1}{V_k}\right)^{m_k} p(\theta_k(0)|m_k) p(m_k|\mathbf{y}^{k-1}, m^{k-1})$$

$$+ \left(\frac{1}{V_k}\right)^{m_k-1} \sum_{i=1}^{m_k} p(\mathbf{y}_k(i)|\mathbf{x}_k) p(\theta_k(i)|m_k) p(m_k|\mathbf{y}^{k-1}, m^{k-1}). \tag{4.17}$$

The probabilities $p(\theta_k(i)|m_k)$, $i = 0, 1, \ldots, m_k$, can be derived using Bayes' theorem as follows: let P_D stand for the probability that the object is detected and P_G for the probability that the object-originated measurement falls into the gate. Using Bayes' theorem,

$$p(\theta_k(0)|m_k) = \frac{p(m_k|\theta_k(0)) p(\theta_k(0))}{p(m_k)}$$

$$= \frac{p(m_k|\theta_k(0)) p(\theta_k(0))}{p(m_k|\theta_k(0)) p(\theta_k(0)) + p(m_k|\text{NOT } \theta_k(0)) p(\text{NOT } \theta_k(0))}$$

$$= \frac{\mu_F(m_k)(1 - P_D P_G)}{\mu_F(m_k)(1 - P_D P_G) + \mu_F(m_k - 1) P_D P_G},$$

where NOTA is the complement event of event A. Let $\mu_F(m_k)$ be the distribution of m_k clutter measurements that are gated. After some rearranging,

$$p(\theta_k(0)|m_k) = \frac{(1 - P_D P_G)\mu_F(m_k)}{\mu_F(m_k - 1)} \left[P_D P_G + \frac{\mu_F(m_k)}{\mu_F(m_k - 1)}(1 - P_D P_G) \right]^{-1},$$

and for $i = 1, \ldots, m_k$, assuming that all measurements gated are equally likely to be the target-originated measurement,

$$p(\theta_k(i)|m_k) = \frac{1}{m_k} P_D P_G \left[P_D P_G + \frac{\mu_F(m_k)}{\mu_F(m_k - 1)} (1 - P_D P_G) \right]^{-1}.$$

The two most commonly used distributions for $\mu_F(m_k)$ are Poisson and non-parametric (uniform) clutter densities as defined below:

$$\mu_F(m_k) = \frac{\exp(-\lambda V_k)(\lambda V_k)^{m_k}}{m_k!}, \quad \text{for Poisson clutter, where } \lambda V_k \text{ is the mean of}$$
$$\text{the distribution, and}$$

$$\mu_F(m_k) = \frac{1}{N} \; \forall \, m_k = 0, 1 \ldots, N - 1, \text{ for non-parametric clutter.}$$

For example, for the non-parametric model (Bar-Shalom and Fortmann, 1988),

$$P(\theta_k(i)|m_k) = \begin{cases} 1 - P_D P_G, & i = 0, \\ P_D P_G / m_k, & i = 1, 2, \cdots, m_k. \end{cases}$$

Whereas for the Poisson clutter distribution,

$$P(\theta_k(i)|m_k)$$
$$= \begin{cases} (1 - P_D P_G)\lambda V_k [m_k P_D P_G + (1 - P_D P_G)\lambda V_k]^{-1}, & i = 0, \\ P_D P_G [m_k P_D P_G + (1 - P_D P_G)\lambda V_k]^{-1}, & i = 1, 2, \ldots, m_k. \end{cases}$$

The normalization factor

The normalization factor $\delta = p(\mathbf{y}_k, m_k | \mathbf{y}^{k-1}, m^{k-1})$ can be evaluated as follows:

$$\delta = p(\mathbf{y}_k, m_k | \mathbf{y}^{k-1}, m^{k-1}) = \int_{\mathbf{x}_k} p(\mathbf{y}_k, m_k | \mathbf{x}_k, \mathbf{y}^{k-1}, m^{k-1}) p(\mathbf{x}_k | \mathbf{y}^{k-1}, m^{k-1}) d\mathbf{x}_k.$$

Using the result from (4.17) we have

$$\delta = \int_{\mathbf{x}_k} p(\mathbf{y}_k, m_k | \mathbf{x}_k, \mathbf{y}^{k-1}, m^{k-1}) p(\mathbf{x}_k | \mathbf{y}^{k-1}, m^{k-1}) d\mathbf{x}_k$$

$$= \int_{\mathbf{x}_k} p(m_k | \mathbf{y}^{k-1}, m^{k-1}) \left[\frac{p(\theta_k(0)|m_k)}{V_k^{m_k}} + \frac{\sum_{i=1}^{m_k} p(\theta_k(i)|m_k) p(\mathbf{y}_k(i)|\mathbf{x}_k)}{V_k^{m_k-1}} \right]$$

$$\times p(\mathbf{x}_k | \mathbf{y}^{k-1}, m^{k-1}) d\mathbf{x}_k$$

$$= \int_{\mathbf{x}_k} \left[\frac{p(m_k | \mathbf{y}^{k-1}, m^{k-1}) p(\theta_k(0)|m_k)}{V_k^{m_k}} \right.$$

$$\left. + \frac{\sum_{i=1}^{m_k} p(m_k | \mathbf{y}^{k-1}, m^{k-1}) p(\theta_k(i)|m_k) p(\mathbf{y}_k(i)|\mathbf{x}_k)}{V_k^{m_k-1}} \right] N(\mathbf{x}_k; \hat{\mathbf{x}}_{k|k-1}, \mathbf{P}_{k|k-1}) d\mathbf{x}_k$$

$$= \int_{\mathbf{x}_k} \frac{p(m_k|\mathbf{y}^{k-1}, m^{k-1})p(\theta_k(0)|m_k)N(\mathbf{x}_k; \hat{\mathbf{x}}_{k|k-1}, \mathbf{P}_{k|k-1})d\mathbf{x}_k}{V_k^{m_k}}$$

$$+ \int_{\mathbf{x}_k} \frac{\sum_{i=1}^{m_k} p(m_k|\mathbf{y}^{k-1}, m^{k-1})p(\theta_k(i)|m_k)p(\mathbf{y}_k(i)|\mathbf{x}_k)N(\mathbf{x}_k; \hat{\mathbf{x}}_{k|k-1}, \mathbf{P}_{k|k-1})d\mathbf{x}_k}{V_k^{m_k-1}}$$

$$= \frac{p(m_k|\mathbf{y}^{k-1}, m^{k-1})p(\theta_k(0)|m_k)}{V_k^{m_k}} \int_{\mathbf{x}_k} N(\mathbf{x}_k; \hat{\mathbf{x}}_{k|k-1}, \mathbf{P}_{k|k-1})d\mathbf{x}_k$$

$$+ \sum_{i=1}^{m_k} \frac{p(m_k|\mathbf{y}^{k-1}, m^{k-1})p(\theta_k(i)|m_k)}{V_k^{m_k-1}} \int_{\mathbf{x}_k} p(\mathbf{y}_k(i)|\mathbf{x}_k)N(\mathbf{x}_k; \hat{\mathbf{x}}_{k|k-1}, \mathbf{P}_{k|k-1})d\mathbf{x}_k.$$

Since $\int_{\mathbf{x}_k} N(\mathbf{x}_k; \hat{\mathbf{x}}_{k|k-1}, \mathbf{P}_{k|k-1})d\mathbf{x}_k = 1$ and, as per (4.15),

$$p(\mathbf{y}_k(i)|\mathbf{x}_k) = \frac{1}{P_G}N(\mathbf{y}_k(i); H\mathbf{x}_k, \mathbf{R}_k),$$

the normalization factor simplifies to

$$\delta = \frac{p(m_k|\mathbf{y}^{k-1}, m^{k-1})p(\theta_k(0)|m_k)}{V_k^{m_k}} + \sum_{i=1}^{m_k} \frac{p(m_k|\mathbf{y}^{k-1}, m^{k-1})p(\theta_k(i)|m_k)}{P_G V_k^{m_k-1}}$$

$$\times \int_{\mathbf{x}_k} N(\mathbf{y}_k(i); \mathbf{H}\mathbf{x}_k, \mathbf{R}_k)N(\mathbf{x}_k; \hat{\mathbf{x}}_{k|k-1}, \mathbf{P}_{k|k-1})d\mathbf{x}_k.$$

The integral $\int_{\mathbf{x}_k} N(\mathbf{y}_k(i); \mathbf{H}\mathbf{x}_k, \mathbf{R}_k)N(\mathbf{x}_k; \hat{\mathbf{x}}_{k|k-1}, \mathbf{P}_{k|k-1})d\mathbf{x}_k = N(\mathbf{y}_k(i); \hat{\mathbf{y}}_k, \mathbf{S}_k)$, where $\hat{\mathbf{y}}_k = \mathbf{H}\hat{\mathbf{x}}_{k|k-1}$, $\mathbf{S}_k = \mathbf{H}\mathbf{P}_{k|k-1}\mathbf{H}^T + \mathbf{R}_k$. By defining

$$a_0 = \frac{p(\theta_k(0)|m_k)p(m_k|\mathbf{y}^{k-1}, m^{k-1})}{V_k^{m_k}},$$

$$a_1 = \frac{p(m_k|\mathbf{y}^{k-1}, m^{k-1})}{P_G V_k^{m_k-1}\sqrt{2\pi \mathbf{S}_k}},$$

$$e_i = \exp\left\{-\frac{1}{2}[\mathbf{y}_k(i) - \hat{\mathbf{y}}_k]^T \mathbf{S}_k^{-1}[\mathbf{y}_k(i) - \hat{\mathbf{y}}_k]\right\} p(\theta_k(i)|m_k),$$

we can concisely represent the normalization factor in the context of single-object tracking in clutter by the following equation:

$$\delta = a_0 + a_1 \sum_{i=1}^{m_k} e_i. \tag{4.18}$$

The conditional density

Returning to the density of prime interest, the conditional density of the object state, we have

$$p(\mathbf{x}_k|\mathbf{y}^k, m^k) = \frac{1}{\delta} \sum_{i=0}^{m_k} p(\mathbf{y}_k(1), \mathbf{y}_k(2), \ldots, \mathbf{y}_k(m_k), m_k|\mathbf{x}_k, \theta_k(i)) p(\theta_k(i)|m_k)$$

$$\times\ p(\mathbf{x}_k|\mathbf{y}^{k-1}, m^{k-1}),$$

where $\delta = p(\mathbf{y}_k, m_k|\mathbf{y}^{k-1}, m^{k-1})$ is the normalizing factor. Substituting the likelihood of (4.17), and the predicted density of (4.11), the conditional density can be rewritten as

$$p(\mathbf{x}_k|\mathbf{y}^k, m^k) = \frac{1}{\delta}\left(\frac{1}{V_k}\right)^{m_k} N(\mathbf{x}_k; \hat{\mathbf{x}}_{k|k-1}, \mathbf{P}_{k|k-1}) p(\theta_k(0)|m_k) p(m_k|\mathbf{y}^{k-1}, m^{k-1})$$

$$+ \frac{1}{\delta}\left(\frac{1}{V_k}\right)^{m_k-1} \sum_{i=1}^{m_k} p(\mathbf{y}_k(i)|\mathbf{x}_k) N(\mathbf{x}_k; \hat{\mathbf{x}}_{k|k-1}, \mathbf{P}_{k|k-1}) p(\theta_k(i)|m_k)$$

$$\times\ p(m_k|\mathbf{y}^{k-1}, m^{k-1}),$$

where

$$p(\mathbf{y}_k(i)|\mathbf{x}_k) N(\mathbf{x}_k; \hat{\mathbf{x}}_{k|k-1}, \mathbf{P}_{k|k-1}) = P_G^{-1} N(\mathbf{y}_k(i); \mathbf{H}\mathbf{x}_k, \mathbf{R}_k) N(\mathbf{x}_k; \hat{\mathbf{x}}_{k|k-1}, \mathbf{P}_{k|k-1}).$$

$$(4.19)$$

The above expression can be further simplied by multiplying and dividing it by $N(\mathbf{x}_k; \hat{\mathbf{x}}_{k|k-1}, \mathbf{P}_{k|k-1})$. Thus we have,

$$p(\mathbf{y}_k(i)|\mathbf{x}_k) N(\mathbf{x}_k; \hat{\mathbf{x}}_{k|k-1}, \mathbf{P}_{k|k-1})$$

$$= P_G^{-1} N(\mathbf{y}_k(i); \hat{\mathbf{y}}_k, \mathbf{S}_k) \frac{N(\mathbf{y}_k(i); \mathbf{H}\mathbf{x}_k, \mathbf{R}_k) N(\mathbf{x}_k; \hat{\mathbf{x}}_{k|k-1}, \mathbf{P}_{k|k-1})}{N(\mathbf{y}_k(i); \hat{\mathbf{y}}_k, \mathbf{S}_k)}$$

$$= P_G^{-1} N(\mathbf{y}_k(i); \hat{\mathbf{y}}_k, \mathbf{S}_k) N(\mathbf{x}_k; \hat{\mathbf{x}}_{k|k}^i, \mathbf{P}_{k|k}^i),$$

where, using the Gaussian distributions theorem of Appendix A,

$$\left[\hat{\mathbf{x}}_{k|k}^i, \mathbf{P}_{k|k}^i\right] = \mathrm{KF_E}\left[\mathbf{y}_k(i), \hat{\mathbf{x}}_{k|k-1}, \mathbf{P}_{k|k-1}, \mathbf{H}, \mathbf{R}_k\right].$$

These are the update of a Kalman filter using the ith measurement $\mathbf{y}_k(i)$, as in the PDAF solution. Substituting, we have

$$p(\mathbf{x}_k|\mathbf{y}^k, m^k) = \frac{1}{\delta}\left(\frac{1}{V_k}\right)^{m_k} p(\theta_k(0)|m_k)p(m_k|\mathbf{y}^{k-1}, m^{k-1})N(\mathbf{x}_k; \hat{\mathbf{x}}_{k|k-1}, \mathbf{P}_{k|k-1})$$

$$+ \frac{1}{\delta}\left(\frac{1}{V_k}\right)^{m_k-1} P_G^{-1}\sum_{i=1}^{m_k} p(\theta_k(i)|m_k)N(\mathbf{y}_k(i); \hat{\mathbf{y}}_k, \mathbf{S}_k)$$

$$\times p(m_k|\mathbf{y}^{k-1}, m^{k-1})N(\mathbf{x}_k; \hat{\mathbf{x}}_{k|k}^i, \mathbf{P}_{k|k}^i)$$

$$= \beta_k(0)N(\mathbf{x}_k; \hat{\mathbf{x}}_{k|k-1}, \mathbf{P}_{k|k-1}) + \sum_{i=1}^{m_k} \beta_k(i)N(\mathbf{x}_k; \hat{\mathbf{x}}_{k|k}^i, \mathbf{P}_{k|k}^i), \quad (4.20)$$

where, by substituting the normalization factor $\delta = a_0 + a_1 \sum_{i=1}^{m_k} e_i$, we get

$$\beta_k(0) = \frac{1}{\delta}\left(\frac{1}{V_k}\right)^{m_k} p(\theta_k(0)|m_k)p(m_k|\mathbf{y}^{k-1}, m^{k-1})$$

$$= \frac{a_0}{a_0 + a_1 \sum_{i=1}^{m_k} e_i} = \frac{b_k}{b_k + \sum_{i=1}^{m_k} e_i},$$

$$\beta_k(i) = \frac{1}{\delta}\left(\frac{1}{V_k}\right)^{m_k-1} P_G^{-1}p(\theta_k(i)|m_k)N(\mathbf{y}_k(i); \hat{\mathbf{y}}_k, \mathbf{S}_k)$$

$$= \frac{a_1 e_i}{a_0 + a_1 \sum_{i=1}^{m_k} e_i} = \frac{e_i}{b_k + \sum_{i=1}^{m_k} e_i},$$

where

$$b_k = \frac{a_0}{a_1} = p(\theta_k(0)|m_k)\frac{P_G|2\pi\mathbf{S}_k|^{\frac{1}{2}}}{V_k},$$

$$e_i = \exp\left\{-\frac{1}{2}[\mathbf{y}_k(i) - \hat{\mathbf{y}}_k]^T\mathbf{S}_k^{-1}[\mathbf{y}_k(i) - \hat{\mathbf{y}}_k]\right\}p(\theta_k(i)|m_k).$$

Calculating the data association probabilities, $\beta(\cdot)$ is denoted by a pseudo-function

$$\left[\{\beta_k(i)\}_{i=0}^{m_k}\right] = \text{STDA}\left[\{p_k(i)\}_{i=1}^{m_k}\right],$$

where $\beta_k(i)$ are calculated for the ith validated measurement (the measurement within the validation gate V_k).

Thus the posterior density of the target state $p(\mathbf{x}_k|\mathbf{y}^k, m^k)$ is a Gaussian mixture whose mean and covariance are given as follows:

$$\left[\hat{\mathbf{x}}_{k|k}, \mathbf{P}_{k|k}\right] = \text{GMix}\left[\left\{\hat{\mathbf{x}}_{k|k}^i, \mathbf{P}_{k|k}^i, \beta_k(i)\right\}_i\right].$$

The PDA conditional density estimate calculation is denoted by the pseudo-function

$$\left[\hat{\mathbf{x}}_{k|k}, \mathbf{P}_{k|k}\right] = \text{PDA}_{\text{E}}\left[\hat{\mathbf{x}}_{k|k-1}, \mathbf{P}_{k|k-1}, \{\mathbf{y}_k(i)\}_{i=1}^{m_k}, \{\boldsymbol{\beta}_k(i)\}_{i=0}^{m_k}, \mathbf{H}, \mathbf{R}\right].$$

First the a posteriori estimation mean $\hat{\mathbf{x}}_{k|k}(i)$ and covariance $\mathbf{P}_{k|k}(i)$ are calculated given each measurement possibility $i \geq 0$,

$$[\hat{\mathbf{x}}_{k|k}(i), \mathbf{P}_{k|k}^i] = \begin{cases} \left[\hat{\mathbf{x}}_{k|k-1}, \mathbf{P}_{k|k-1}\right], & i = 0, \\ \text{KF}_{\text{E}}\left[\mathbf{y}_k(i), \hat{\mathbf{x}}_{k|k-1}, \mathbf{P}_{k|k-1}, \mathbf{H}, \mathbf{R}\right], & i > 0, \end{cases}$$

after which the mean and covariance of the resulting Gaussian mixture are calculated by

$$\left[\hat{\mathbf{x}}_{k|k}, \mathbf{P}_{k|k}\right] = \text{GMix}\left[\left\{\hat{\mathbf{x}}_{k|k}(i), \mathbf{P}_{k|k}^i, \boldsymbol{\beta}_k(i)\right\}_{i=0}^{m_k}\right].$$

4.3.1 The probability data association filter equations

Algorithm 17 PDA filter recursion equations at time k

1: Prediction:

$$\left[\hat{\mathbf{x}}_{k|k-1}, \mathbf{P}_{k|k-1}\right] = \text{KF}_{\text{P}}\left[\hat{\mathbf{x}}_{k-1|k-1}, \mathbf{P}_{k-1|k-1}, F, Q\right].$$

2: Measurement selection:

$$[\mathbf{y}_k, V_k] = \text{MS}_1\left[\mathbf{Y}_k, \hat{\mathbf{x}}_{k|k-1}, \mathbf{P}_{k|k-1}, \mathbf{H}, \mathbf{R}\right].$$

3: Likelihoods of all selected measurements i:

$$\left[\{p_k(i)\}_i\right] = \text{ML}_1\left[\{\mathbf{y}_k(i)\}_i, \hat{\mathbf{x}}_{k|k-1}, \mathbf{P}_{k|k-1}, \mathbf{H}, \mathbf{R}\right].$$

4: **if** non-parametric tracking **then**
5:    V_k calculated using the equation

$$V_k = \frac{\pi^{n/2}}{\Gamma(n/2+1)}\sqrt{|\mathbf{S}(k)|}\gamma^{1/2},$$

where $|\mathbf{S}(k)|$ is the determinant of $\mathbf{S}(k)$.
6:    Clutter measurement density estimation:

$$\rho = m_k / V_k.$$

7: **end if**
8: Single target data association (sàns target existence):

$$\left[\{\boldsymbol{\beta}_k(i)\}_{i=0}^{m_k}\right] = \text{STDA}\left[\{p_k(i)\}_{i=1}^{m_k}\right].$$

9: Estimation/Merging:

$$\left[\hat{\mathbf{x}}_{k|k}, \mathbf{P}_{k|k}\right] = \text{PDA}_{\text{E}} \left[\hat{\mathbf{x}}_{k|k-1}, \mathbf{P}_{k|k-1}, \{\mathbf{y}_k(i)\}_{i=1}^{m_k}, \{\boldsymbol{\beta}_k(i)\}_{i=0}^{m_k}, \mathbf{H}, \mathbf{R}\right].$$

10: Output trajectory estimate:
   • track mean value $\hat{\mathbf{x}}_{k|k}$ and covariance $\mathbf{P}_{k|k}$.

4.4 Maneuvering object tracking in clutter

4.4.1 Object dynamics and sensor measurements model

As discussed in Chapter 3, the most successful approach in dealing with maneuvering targets is the multiple-model approach where the object motion dynamics was modeled using a jump-Markov process. In this approach the object dynamics are assumed to belong to the set of models defined by (3.6),

$$\mathbf{x}_k = \mathbf{f}_{r_k}(\mathbf{x}_{k-1}) + \mathbf{v}_{r_k} \quad r_k \in \{1, 2, \dots, d\},$$

where the process noise $\mathbf{v}_{r_k}$ is assumed to be additive and r_k is assumed to be a random variable satisfying a homogeneous discrete-time Markov chain with state space $\{1, \dots, d\}$ and transition probability matrix Γ, where

$$\Gamma_{ji} = \Pr(r_k = i | r_{k-1} = j),$$

with initial conditions $\Pr(r_0 = i) = \pi_0(i)$. The measurements, in a fairly general sense, are assumed to be model dependent and related to the true target state through (3.7):

$$\mathbf{y}_k = \mathbf{h}_{r_k}(\mathbf{x}_k) + \mathbf{w}_{r_k}.$$

Due to its success we will adopt the same system dynamics equations for the case of maneuvering target tracking in clutter.

4.4.2 Optimal Bayes' solution for maneuvering object tracking in clutter

Since r_k is a discrete random variable, taking values in a discrete set $\{1, 2, \dots, d\}$, the probability density of interest $p(\mathbf{x}_k | \mathbf{y}^k)$ can be decomposed into d components as follows:

$$p(\mathbf{x}_k | \mathbf{y}^k) = \sum_{i=1}^{d} p(\mathbf{x}_k, r_k = i | \mathbf{y}^k) \quad i = \{1, 2, \dots, d\}.$$

By using the conditional density lemma, it can be further decomposed into

$$p(\mathbf{x}_k|\mathbf{y}^k) = \sum_{i=1}^{d} p(\mathbf{x}_k|r_k = i, \mathbf{y}^k) p(r_k = i|\mathbf{y}^k). \tag{4.21}$$

By applying Bayes' theorem and the Chapman–Kolmogorov lemma to individual components of the above decomposition, we have

$$p(\mathbf{x}_k|r_k = i, \mathbf{y}^k)$$
$$= \frac{p(\mathbf{y}_k|\mathbf{x}_k, r_k = i, \mathbf{y}^{k-1}) \int_{\mathbf{x}_{k-1}} p(\mathbf{x}_k|\mathbf{x}_{k-1}, r_k = i, \mathbf{y}^{k-1}) p(\mathbf{x}_{k-1}|\mathbf{y}^{k-1}) d\mathbf{x}_{k-1}}{p(\mathbf{y}_k|r_k = i, \mathbf{y}^{k-1})}.$$

By defining $p(r_k = i|\mathbf{y}^k) = \mu_{k|k}(i)$ and $p(r_k = i|\mathbf{y}^{k-1}) = \mu_{k|k-1}(i)$, we have

$$\mu_{k|k}(i) = \frac{p(\mathbf{y}_k|r_k = i, \mathbf{y}^{k-1})\mu_{k|k-1}(i)}{\sum_{i=1}^{d} p(\mathbf{y}_k|r_k = i, \mathbf{y}^{k-1}) \sum_{j=1}^{d} \Gamma_{ji} \mu_{k-1|k-1}(j)},$$

and, since we are dealing with object tracking in clutter, the sensor observations at each epoch are a set of measurements $\mathbf{y}_k = \{\mathbf{y}^k(1), \mathbf{y}^k(2), \dots, \mathbf{y}^k(m_k)\}$, where m_k denotes the number of received measurements at time k.

Define the following:

- $\theta_k(0)$ as the association event that none of the measurements in $\mathbf{y}_k$ is object-originated;
- $\theta_k(i)$ as the association event that the ith measurement in $\mathbf{y}_k$ is object-originated, and the rest of the measurements are from clutter, $i = 1, 2, \dots, m(k)$.

We can obtain the likelihood functions as

$$p(\mathbf{y}_k|\mathbf{x}_k, r_k = i, \mathbf{y}^{k-1}) = \sum_{i=0}^{m_k} p(\mathbf{y}_k(1), \mathbf{y}_k(2), \dots, \mathbf{y}_k(m_k)|\mathbf{x}_k, r_k = i, \theta_k(i))$$
$$\times p(\theta_k(i)|\mathbf{x}_k, r_k = i, \mathbf{y}^{k-1}), \tag{4.22}$$

and

$$p(\mathbf{y}_k|r_k = i, \mathbf{y}^{k-1}) = \int_{\mathbf{x}_k} \sum_{i=0}^{m_k} p(\mathbf{y}_k(1), \mathbf{y}_k(2), \dots, \mathbf{y}_k(m_k)|\mathbf{x}_k, r_k = i, \theta_k(i))$$
$$\times p(\theta_k(i)|\mathbf{x}_k, r_k = i, \mathbf{y}^{k-1}) p(\mathbf{x}_k|r_k = i, \mathbf{y}^{k-1}) d\mathbf{x}_k. \tag{4.23}$$

Equation (4.23) is also the normalization factor in the Bayes' recursion associated with the object state's conditional density. The normalization factors associated with the model probability update equation are given by

$$p(\mathbf{y}_k|\mathbf{y}^{k-1}) = \sum_{i=1}^{d} p(\mathbf{y}_k|r_k = i, \mathbf{y}^{k-1}) \, p(r_k = i|\mathbf{y}^{k-1})$$

$$= \sum_{i=1}^{d} p(\mathbf{y}_k|r_k = i, \mathbf{y}^{k-1}) \, \mu_{k|k-1}(i)$$

$$= \sum_{i=1}^{d} p(\mathbf{y}_k|r_k = i, \mathbf{y}^{k-1}) \sum_{j=1}^{d} \Gamma_{ji} \, \mu_{k-1|k-1}(j), \qquad (4.24)$$

where $p(\mathbf{y}_k|r_k = i, \mathbf{y}^{k-1})$ is given by (4.23).

4.4.3 The Bayes' optimal estimates for maneuvering object tracking in clutter

Substituting (3.10) and (3.17) in (3.9), the conditional density of the target state $p(\mathbf{x}_k|\mathbf{y}^k)$ can be determined. From this, the minimum variance target state estimate and associated covariance can be obtained using

$$\hat{\mathbf{x}}_{k|k} = \int_{\mathbf{x}_k} \mathbf{x}_k \, p(\mathbf{x}_k|\mathbf{y}^k) d\mathbf{x}_k,$$

$$P_{k|k} = \int_{\mathbf{x}_k} [\mathbf{x}_k - \hat{x}_{k|k}][\mathbf{x}_k - \hat{x}_{k|k}]^T \, p(\mathbf{x}_k|\mathbf{y}^k) d\mathbf{x}_k.$$

Using (3.9), the mean and covariance equations can be further simplified to

$$\hat{\mathbf{x}}_{k|k} = \int_{\mathbf{x}_k} \mathbf{x}_k \sum_{i=1}^{d} p(\mathbf{x}_k|r_k = i, \mathbf{y}^k) \mu_{k|k}(i) d\mathbf{x}_k,$$

$$P_{k|k} = \int_{\mathbf{x}_k} [\mathbf{x}_k - \hat{x}_k][\mathbf{x}_k - \hat{x}_k]^T \sum_{i=1}^{d} p(\mathbf{x}_k|r_k = i, \mathbf{y}^k) \mu_{k|k}(i) d\mathbf{x}_k.$$

Rearranging the summation and integrals, the conditional mean and covariance are given by

$$\hat{\mathbf{x}}_{k|k} = \sum_{i=1}^{d} \left(\int_{\mathbf{x}_k} \mathbf{x}_k \, p(\mathbf{x}_k|r_k = i, \mathbf{y}^k) d\mathbf{x}_k \right) \mu_{k|k}(i), \qquad (4.25)$$

$$P_{k|k} = \sum_{i=1}^{d} \left(\int_{\mathbf{x}_k} [\mathbf{x}_k - \hat{x}_k][\mathbf{x}_k - \hat{x}_k]^T \, p(\mathbf{x}_k|r_k = i, \mathbf{y}^k) d\mathbf{x}_k \right) \mu_{k|k}(i). \qquad (4.26)$$

Several sub-optimal filters, such as IMMPDA and particle filters for maneuvering object tracking in clutter, can be derived by approximating these Bayes' recursive equations.

4.5 Particle filter for tracking in clutter

In this section the basic particle filtering algorithms described in Section 2.6 will be applied to the problem of single-object tracking in clutter. It will be assumed that the target state evolves according to

$$p(\mathbf{x}_k | \mathbf{x}_{k-1}) = p_{\mathbf{v}_i}(\mathbf{x}_k - \mathbf{f}(\mathbf{x}_{k-1})), \tag{4.27}$$

where $p_{\mathbf{v}_i}$ is the PDF of the process noise. Under the association event $\theta_k(i)$, $i \in \{0, \ldots, m_k\}$, the measurements satisfy

$$p(\mathbf{y}_k | \theta_k(i), \mathbf{x}_k) = \begin{cases} (1/V_k)^{m_k}, & i = 0, \\ (1/V_k)^{m_k-1} p_{\mathbf{w}_k}(\mathbf{y}_k(j) - \mathbf{h}(\mathbf{x}_k)), & i = 1, \ldots, m_k, \end{cases} \tag{4.28}$$

where V_k is the volume of the gate. The number m of measurements in a region of volume V is assumed to be Possion distributed with known mean λV.

4.5.1 The bootstrap filter for object tracking in clutter

Samples of the particle index and target state are drawn from the importance density

$$q(\mathbf{x}_k, t) = w_{k-1}^t p(\mathbf{x}_k | \mathbf{x}_{k-1}^t). \tag{4.29}$$

The marginal importance densities can be found as

$$q(t) = w_{k-1}^t, \tag{4.30}$$

$$q(\mathbf{x}_k | t) = p_{\mathbf{v}_k}(\mathbf{x}_k - \mathbf{f}(\mathbf{x}_{k-1})). \tag{4.31}$$

The sample weights, given by the ratio of the posterior to the importance density, are proportional to the likelihood. The likelihood can be expanded as

$$p(\mathbf{y}_k, m_k | \mathbf{x}_k) = \sum_{j=0}^{m_k} P(\theta_k(j) | m_k) p(\mathbf{y}_k | \theta_k(j), \mathbf{x}_k). \tag{4.32}$$

Substituting (4.28) and

$$p(\theta_k(j) | m_k)$$
$$= \begin{cases} (1 - P_D P_G)\lambda V_k / (m_k P_D P_G + \lambda V_k(1 - P_D P_G)), & j = 0, \\ P_D P_G / (m_k P_D P_G + \lambda V_k(1 - P_D P_G)), & j = 1, \ldots, m_k, \end{cases}$$
$$\tag{4.33}$$

Algorithm 18 Bootstrap filter for object tracking in clutter

1: **for** $i = 1, \ldots, n$ **do**
2:     Draw a mixture index t^i such that $\Pr(t^i = l) = w_{k-1}^l$.
3:     Draw $\mathbf{v}_k^i \sim p_{\mathbf{v}_k}$ and compute the sample target state $\mathbf{x}_k^i = \mathbf{f}(\mathbf{x}_{k-1}^{t^i}) + \mathbf{v}_k^i$.
4:     Compute the un-normalized weight:

$$\tilde{w}_k^i = 1 - P_D P_G + P_D P_G / \lambda \sum_{j=1}^{m_k} p_{\mathbf{w}_k}(\mathbf{y}_k(j) - \mathbf{h}(\mathbf{x}_k^i)). \qquad (4.35)$$

5: **end for**
6: Normalize the weights:

$$w_k^i = \tilde{w}_k^i \bigg/ \sum_{j=1}^{n} \tilde{w}_k^j, \quad i = 1, \ldots, n.$$

7: Compute a state estimate:

$$\hat{\mathbf{x}}_{k|k} = \sum_{i=1}^{n} w_k^i \mathbf{x}_k^i.$$

---

gives

$$p(\mathbf{y}_k, m_k | \mathbf{x}_k) \propto 1 - P_D P_G + P_D P_G / \lambda \sum_{j=1}^{m_k} p_{\mathbf{w}_k}(\mathbf{y}_k(j) - \mathbf{h}(\mathbf{x}_k)). \qquad (4.34)$$

A recursion of the BF for single-object tracking in clutter is given by Algorithm 18. The sampling step is precisely the same as for single-object tracking without clutter. This is to be expected since the bootstrap filter draws samples without consideration of the measurements. The uncertainty in the origin of the measurement is accounted for in the weight calculation. This weight calculation will favor samples which are close to any of the measurements in the validation gate.

4.5.2 *The extended Kalman auxiliary particle filter for object tracking in clutter*

The EK-APF will be derived under the Gaussian assumptions, $\mathbf{v}_k \sim N(0, \mathbf{Q})$ and $\mathbf{w}_k \sim N(0, \mathbf{R})$. The importance density for the EK-APF can be written as

$$q(\mathbf{x}_k, t) = \xi_k^t \, \hat{p}(\mathbf{x}_k | \mathbf{x}_{k-1}^t, \mathbf{y}_{1:k}, m_{1:k}), \qquad (4.36)$$

where

$$\xi_k^t = w_{k-1}^t \hat{p}(\mathbf{y}_k, m_k | \mathbf{x}_{k-1}^t, \mathbf{y}_{1:k-1}) \bigg/ \sum_{i=1}^{n} w_{k-1}^i \hat{p}(\mathbf{y}_k, m_k | \mathbf{x}_{k-1}^i, \mathbf{y}_{1:k-1}). \quad (4.37)$$

Recall that the hat notation is used to denote probability distributions or densities computed with the linearized measurement equation approximation, $\hat{\mathbf{h}}(\mathbf{x}, \hat{\mathbf{x}}) = \mathbf{h}(\hat{\mathbf{x}}) + \mathbf{H}_k(\mathbf{x} - \hat{\mathbf{x}})$, where $\mathbf{H}_k = [\nabla_{\mathbf{x}} \mathbf{h}(\mathbf{x})^T]^T |_{\mathbf{x}=\hat{\mathbf{x}}}$. The update factor $\hat{p}(\mathbf{y}_k | \mathbf{x}_{k-1}^t, \mathbf{y}_{1:k-1})$ for the first-stage weights can be expanded as

$$\hat{p}(\mathbf{y}_k, m_k | \mathbf{x}_{k-1}, \mathbf{y}_{1:k-1}) = \int \hat{p}(\mathbf{y}_k, m_k | \mathbf{x}_k) p(\mathbf{x}_k | \mathbf{x}_{k-1}^t) \, d\mathbf{x}_k. \quad (4.38)$$

Substituting the transition density (4.27) and the likelihood (4.34), with the measurement equation replaced by the linearized approximation, and using the Gaussian assumptions, gives

$$\hat{p}(\mathbf{y}_k, m_k | \mathbf{x}_{k-1}, \mathbf{y}_{1:k-1})$$

$$\propto \int \left[1 - P_D P_G + P_D P_G / \lambda \sum_{j=1}^{m_k} N(\mathbf{y}_k(j); \hat{\mathbf{h}}(\mathbf{x}_k, \mathbf{f}(\mathbf{x}_{k-1})), \mathbf{R}) \right]$$

$$\times N(\mathbf{x}_k; \mathbf{f}(\mathbf{x}_{k-1}), \mathbf{Q}) \, d\mathbf{x}_k \quad (4.39)$$

$$= \sum_{j=0}^{m_k} \tilde{\alpha}_k(j), \quad (4.40)$$

where

$$\tilde{\alpha}_k(j) = \begin{cases} 1 - P_D P_G, & j = 0, \\ P_D P_G N(\mathbf{y}_k(j); \hat{\mathbf{y}}_k, \mathbf{S}_k)/\lambda, & j = 1, \ldots, m_k, \end{cases} \quad (4.41)$$

with $\hat{\mathbf{y}}_k = \mathbf{h}(\mathbf{f}(\mathbf{x}_{k-1}))$ and $\mathbf{S}_k = \mathbf{H}_k \mathbf{Q} \mathbf{H}_k^T + \mathbf{R}$. Equation (4.40) was obtained using Theorem 2.1. The sampling density for the state vector can be written, using Bayes' rules, as

$$\hat{p}(\mathbf{x}_k | \mathbf{x}_{k-1}, \mathbf{y}_{1:k}, m_{1:k}) \propto \hat{p}(\mathbf{y}_k, m_k | \mathbf{x}_k) p(\mathbf{x}_k | \mathbf{x}_{k-1}). \quad (4.42)$$

The same procedure used to derive the first-stage weights can be used to find the sampling density as

$$\hat{p}(\mathbf{x}_k | \mathbf{x}_{k-1}, \mathbf{y}_{1:k}, m_{1:k}) = \sum_{j=0}^{m_k} \alpha_k(j) N(\mathbf{x}_k; \boldsymbol{\mu}_k(j), \boldsymbol{\Sigma}_k(j)), \quad (4.43)$$

where

$$\alpha_k(j) = \tilde{\alpha}_k(j) \Big/ \sum_{i=0}^{m_k} \tilde{\alpha}_k(i), \tag{4.44}$$

$$\mu_k(j) = \begin{cases} \mathbf{f}(\mathbf{x}_{k-1}), & j = 0, \\ \mathbf{f}(\mathbf{x}_{k-1}) + \mathbf{K}_k(\mathbf{y}_k(j) - \hat{\mathbf{y}}_k), & j = 1, \dots, m_k, \end{cases} \tag{4.45}$$

$$\Sigma_k(j) = \begin{cases} \mathbf{Q}, & j = 0, \\ \mathbf{Q} - \mathbf{K}_k \mathbf{H}_k \mathbf{Q}, & j = 1, \dots, m_k, \end{cases} \tag{4.46}$$

with $\mathbf{K}_k = \mathbf{Q}\mathbf{H}_k \mathbf{S}_k^{-1}$. It can be seen from (4.40) that sample indexes drawn according to the first-stage weights will tend to correspond to samples with a predicted measurement close to at least one of the validated measurements. The sampling density for the state vector is a mixture density with each mixture component corresponding to an association hypothesis. The procedure for drawing a sample of the target state consists of selecting an association hypothesis, according to the probabilities $\alpha_k(0), \dots, \alpha_k(m_k)$, and then drawing from the corresponding component distribution. The sampling density corresponding to $\theta_k(j)$, $j > 0$, is the sampling density for the EK-APF without clutter assuming that the jth measurement is target-originated. The sampling density for $\theta_k(0)$ is just the transition density, since this hypothesis states that all measurements are clutter. A recursion of the EK-APF is given by Algorithm 19.

Algorithm 19 Extended Kalman auxiliary particle filter for target tracking in clutter

---

1: **for** $i = 1, \dots, n$ **do**
2:     Compute the Jacobian $\mathbf{H}_k^i = [\nabla_\mathbf{x}\mathbf{h}(\mathbf{x})^T]^T|_{\mathbf{x}=\mathbf{f}(\mathbf{x}_{k-1}^i)}$.
3:     Compute:

$$\mathbf{x}_{k|k-1}^i = \mathbf{f}(\mathbf{x}_{k-1}^i), \qquad\qquad \hat{\mathbf{y}}_k^i = \mathbf{h}(\mathbf{x}_{k|k-1}^i),$$

$$\mathbf{S}_k^i = \mathbf{H}_k^i \mathbf{Q}(\mathbf{H}_k^i)^T + \mathbf{R}, \qquad \mathbf{K}_k^i = \mathbf{Q}(\mathbf{H}_k^i)^T(\mathbf{S}_k^i)^{-1}.$$

4:     For $j = 0, \dots, m_k$, compute $\tilde{\alpha}_k^i(j)$ using (4.41).
5:     Compute the first-stage weight update:

$$a_k^i = \sum_{j=0}^{m_k} \tilde{\alpha}_k^i(j).$$

6:     For $j = 0, \dots, m_k$, compute the association probabilities $\alpha_k^i(j) = \tilde{\alpha}_k^i(j)/a_k^i$.
7: **end for**

8: Compute the first-stage weights:

$$\xi_k^t = w_{k-1}^t a_k^t \left/ \sum_{i=1}^{n} w_{k-1}^i a_k^i \right., \quad t = 1, \ldots, n.$$

9: **for** $i = 1, \ldots, n$ **do**

10:    Draw a mixture index t^i such that $\Pr(t^i = l) = \xi_k^l$.

11:    Draw an association hypothesis j^i such that $\Pr(j^i = l) = \alpha_k^{t^i}(l)$.

12:    Compute the conditional mean $\boldsymbol{\mu}_k^{t^i}(j^i)$ and covariance matrix $\boldsymbol{\Sigma}_k^{t^i}(j^i)$ using (4.45) and (4.46).

13:    Draw the sample target state $\mathbf{x}_k^i \sim N(\boldsymbol{\mu}_k^{t^i}(j^i), \boldsymbol{\Sigma}_k^{t^i}(j^i))$.

14:    Compute the un-normalized weight:

$$\tilde{w}_k^i = \frac{1 - P_D P_G + P_D P_G / \lambda \sum_{j=1}^{m_k} N(\mathbf{y}_k(j); \mathbf{h}(\mathbf{x}_k^i), \mathbf{R})}{1 - P_D P_G + P_D P_G / \lambda \sum_{j=1}^{m_k} N(\mathbf{y}_k(j); \hat{\mathbf{h}}(\mathbf{x}_k^i, \mathbf{f}(\mathbf{x}_{k-1}^{t^i}), \mathbf{R})}.$$

15: **end for**

16: Normalize the weights:

$$w_k^i = \tilde{w}_k^i \left/ \sum_{j=1}^{n} \tilde{w}_k^j \right., \quad i = 1, \ldots, n.$$

17: Compute a state estimate:

$$\hat{\mathbf{x}}_{k|k} = \sum_{i=1}^{n} w_k^i \mathbf{x}_k^i.$$

4.6 Performance bounds

The performances of the sub-optimal tracking algorithms described in this chapter can be assessed by comparing their mean square errors (MSEs) with the posterior Cramér–Rao bound (PCRB). Recall that the PCRB is a lower bound on the MSE of random parameter estimators (Van Trees, 1968). Thus, an estimator $\hat{\mathbf{x}}_{k|k}$ of the state $\mathbf{x}_k$ based on the measurement sequence $\mathbf{y}^k$ has a MSE which satisfies

$$\text{mse}(\hat{\mathbf{x}}_{k|k}) \geq \mathbf{J}_k(\mathbf{x}_k)^{-1}, \tag{4.47}$$

where the information matrix $\mathbf{J}(\mathbf{x}_k)$ can be computed using the recursion

$$\mathbf{J}_k(\mathbf{x}_k) = \mathbf{W}_k - \mathbf{V}_k^T [\mathbf{J}_{k-1}(\mathbf{x}_{k-1}) + \mathbf{U}_k]^{-1} \mathbf{V}_k, \tag{4.48}$$

with

$$\mathbf{U}_k = -\mathsf{E}\left[\nabla_{\mathbf{x}_{k-1}} \nabla_{\mathbf{x}_{k-1}}^T \log p(\mathbf{x}_k|\mathbf{x}_{k-1})\right], \tag{4.49}$$

$$\mathbf{V}_k = -\mathsf{E}\left[\nabla_{\mathbf{x}_{k-1}} \nabla_{\mathbf{x}_k}^T \log p(\mathbf{x}_k|\mathbf{x}_{k-1})\right], \tag{4.50}$$

$$\mathbf{W}_k = -\mathsf{E}\left[\nabla_{\mathbf{x}_k} \nabla_{\mathbf{x}_k}^T \log p(\mathbf{x}_k|\mathbf{x}_{k-1})\right] - \mathsf{E}\left[\nabla_{\mathbf{x}_k} \nabla_{\mathbf{x}_k}^T \log p(\mathbf{y}_k|\mathbf{x}_k)\right]. \tag{4.51}$$

Inspection of (4.49)–(4.51) shows that only the matrix $\mathbf{W}_k$, which includes the contribution of the current measurement $\mathbf{y}_k$, is affected by measurement origin uncertainty. We will therefore concentrate on this matrix, and in particular the second term of (4.51), which we denote

$$\mathbf{Z}_k = \mathsf{E}\left[\nabla_{\mathbf{x}_k} \nabla_{\mathbf{x}_k}^T \log p(\mathbf{y}_k|\mathbf{x}_k)\right]. \tag{4.52}$$

An important characteristic of object tracking in clutter is that we cannot know beforehand how many measurements will be received. The expectation in (4.52) must therefore be over the measurement number m_k in addition to the current measurement $\mathbf{y}_k$ and the current state $\mathbf{x}_k$. We can then write (Hernandez *et al.*, 2005; Mušicki *et al.*, 2005b)

$$\mathbf{Z}_k = \sum_{m_k=0}^{\infty} \mathsf{P}(m_k)\mathbf{Z}_{k,m_k}, \tag{4.53}$$

where

$$\mathbf{Z}_{k,m_k} = \mathsf{E}\left[\nabla_{\mathbf{x}_k} \nabla_{\mathbf{x}_k}^T \log p(\mathbf{y}_k|\mathbf{x}_k, m_k)\right]. \tag{4.54}$$

Computation of the PCRB requires the measurement number probability $\mathsf{P}(m_k)$, $m_k = 0, 1, \ldots$, and the information increment $\mathbf{Z}_{k,m_k}$ conditional on receiving m_k measurements. It is assumed that the target moves in a surveillance region S with volume V. All measurements are independently generated random variables. A measurement $\mathbf{y}$ from a target with state $\mathbf{x}$ satisfies

$$\mathbf{y} = \mathbf{h}(\mathbf{x}) + \mathbf{e},$$

where $\mathbf{e} \sim N(\mathbf{0}, \mathbf{R})$. The usual assumptions are made regarding the clutter, i.e., clutter measurements are uniformly distributed in the surveillance region S and the number of clutter measurements in S is Poisson distributed with mean λV.

Consider first the measurement number probability. There are two possible events which can produce m_k measurements: m_k clutter measurements or $m_k - 1$ clutter measurements and one object measurement. It follows that

$$\mathsf{P}(m_k) = \frac{(\lambda V)^{m_k} \exp(-\lambda V)}{m_k!} \left[1 - P_D + m_k P_D/(\lambda V)\right], \tag{4.55}$$

The conditional measurement PDF must also account for the two different ways in which m_k measurements can be received. Let $\tau(m_k)$ denote the probability of receiving an object measurement given that m_k measurements are received. It can be shown from (4.55) that

$$\tau(m_k) = m_k P_D / [\lambda V (1 - P_D) + m_k P_D]. \tag{4.56}$$

The measurement PDF conditional on receiving m_k measurements is then

$$p(\mathbf{y}_k | \mathbf{x}_k, m_k) = [1 - \tau(m_k)] / V^{m_k} + \frac{\tau(m_k)}{m_k V^{m_k-1}} \sum_{j=1}^{m_k} N(\mathbf{y}_{k,j}; \mathbf{h}(\mathbf{x}_k), \mathbf{R}). \tag{4.57}$$

The gradient of the measurement PDF with respect to the object state is

$$\nabla_{\mathbf{x}_k} \log p(\mathbf{y}_k | \mathbf{x}_k, m_k)$$

$$= \frac{\tau(m_k) V / m_k \mathbf{H}(\mathbf{x}_k)^T \mathbf{R}^{-1} \sum_{j=1}^{m_k} [\mathbf{y}_{k,j} - \mathbf{h}(\mathbf{x}_k)] N(\mathbf{y}_{k,j}; \mathbf{h}(\mathbf{x}_k), \mathbf{R})}{1 - \tau(m_k) + \tau(m_k) V / m_k \sum_{i=1}^{m_k} N(\mathbf{y}_{k,i}; \mathbf{h}(\mathbf{x}_k), \mathbf{R})}$$

$$= \mathbf{B}(\mathbf{y}_k, \mathbf{x}_k, m_k) \sum_{j=1}^{m_k} \boldsymbol{\gamma}(\mathbf{y}_{k,j}, \mathbf{x}_k), \tag{4.58}$$

where

$$\mathbf{B}(\mathbf{y}, \mathbf{x}, m) = \frac{\tau(m) V / m \mathbf{H}(\mathbf{x})^T \mathbf{R}^{-1}}{1 - \tau(m) + \tau(m) V / m \sum_{j=1}^{m} N(\mathbf{y}_j; \mathbf{h}(\mathbf{x}), \mathbf{R})}, \tag{4.59}$$

$$\boldsymbol{\gamma}(\mathbf{y}, \mathbf{x}) = [\mathbf{y} - \mathbf{h}(\mathbf{x})] N(\mathbf{y}; \mathbf{h}(\mathbf{x}), \mathbf{R}), \tag{4.60}$$

where $\mathbf{H}(\mathbf{x}_k) = [\nabla_{\mathbf{x}} \mathbf{h}(\mathbf{x})^T |_{\mathbf{x}=\mathbf{x}_k}]^T$. Then

$$\mathbf{Z}_{k,m_k} = -\mathsf{E}[\nabla_{\mathbf{x}_k} \log p(\mathbf{y}_k | \mathbf{x}_k, m_k) \nabla_{\mathbf{x}_k}^T \log p(\mathbf{y}_k | \mathbf{x}_k, m_k)]$$

$$= \mathsf{E}[\mathsf{E}[\mathbf{C}(\mathbf{y}_k, \mathbf{x}_k, m_k) | \mathbf{x}_k, m_k]], \tag{4.61}$$

where

$$\mathbf{C}(\mathbf{y}, \mathbf{x}, m) = \mathbf{B}(\mathbf{y}, \mathbf{x}, m) \sum_{j_1=1}^{m} \sum_{j_2=1}^{m} \boldsymbol{\gamma}(\mathbf{y}_{j_1}, \mathbf{x}) \boldsymbol{\gamma}(\mathbf{y}_{j_2}, \mathbf{x})^T \mathbf{B}(\mathbf{y}, \mathbf{x}, m)^T. \tag{4.62}$$

The outer expectation in (4.61) is over the object state while the inner expectation is over the measurement conditional on the object state. We concentrate on the inner expectation:

$$
\mathsf{E}[\mathbf{C}(\mathbf{y}, \mathbf{x}, m)|\mathbf{x}, m] = [\tau(m)/m]^2 / V^{m-2} \mathbf{H}(\mathbf{x})^T \mathbf{R}^{-1} \sum_{j_1=1}^{m} \sum_{j_2=1}^{m} \mathbf{A}_{j_1, j_2}(m) \mathbf{R}^{-1} \mathbf{H}(\mathbf{x}),
$$

(4.63)

where

$$
\mathbf{A}_{j_1, j_2}(m, \mathbf{x})
$$

$$
= \int_{S^m} \frac{[\mathbf{y}_{j_1} - \mathbf{h}(\mathbf{x})][\mathbf{y}_{j_2} - \mathbf{h}(\mathbf{x})]^T N(\mathbf{y}_{j_1}; \mathbf{h}(\mathbf{x}), \mathbf{R}) N(\mathbf{y}_{j_2}; \mathbf{h}(\mathbf{x}), \mathbf{R})}{a(m) + b(m) \sum_{i=1}^{m_k} N(\mathbf{y}_i; \mathbf{h}(\mathbf{x}), \mathbf{R})} d\mathbf{y}_1 \cdots d\mathbf{y}_m,
$$

(4.64)

with $a(m) = 1 - \tau(m)$, $b(m) = \tau(m)V/m$. The integral (4.68) cannot be evaluated exactly but can be approximated using, for instance, Monte Carlo methods. Before proceeding with an approximation it is useful to simplify the integral. We apply the change of variable $\mathbf{z} = \mathbf{G}^{-1}[\mathbf{y} - \mathbf{h}(\mathbf{x})]$, where $\mathbf{G}$ is the matrix square root of the measurement noise covariance matrix, i.e. $\mathbf{R} = \mathbf{G}\mathbf{G}^T$. Gating is also applied so that the region of integration is reduced to

$$
\{\mathbf{z} = [z_1, \ldots, z_n]^T : |z_i| < g, i = 1, \ldots, n\}, \quad (4.65)
$$

where g determines the size of the gate. Note that after this gating the surveillance volume V should be replaced by the gate volume $|\mathbf{G}|(2g)^n$. Then, for $j_1, j_2 = 1, \ldots, m$,

$$
\mathbf{A}_{j_1, j_2}(m) = |\mathbf{G}|^{m-2} \int_{[-g,g]^{nm}} \frac{\mathbf{G} \mathbf{z}_{j_1} \mathbf{z}_{j_2}^T \mathbf{G}^T N(\mathbf{z}_{j_1}; \mathbf{0}, \mathbf{I}) N(\mathbf{z}_{j_2}; \mathbf{0}, \mathbf{I})}{a(m) + b(m)/|\mathbf{G}| \sum_{i=1}^{m} N(\mathbf{z}_i; \mathbf{0}, \mathbf{I})} d\mathbf{z}_1 \cdots d\mathbf{z}_m.
$$

(4.66)

It can be seen that $\mathbf{A}_{j_1, j_2} = \mathbf{0}$ for $j_1 \neq j_2$ since (4.68) involves the integration of odd-symmetric integrands over a symmetric region. Thus, in (4.63) it is necessary to consider only the terms for which $j_1 = j_2 = j$. These terms can be written as

$$
\mathbf{A}_{j, j}(m) = |\mathbf{G}|^{m-2} \mathbf{G} \mathbf{D}(m) \mathbf{G}^T, \quad (4.67)
$$

where

$$
\mathbf{D}(m) = \int_{[-g,g]^{nm}} \frac{\mathbf{z}_1 \mathbf{z}_1^T N(\mathbf{z}_1; \mathbf{0}, \mathbf{I})^2}{a(m) + b(m)/|\mathbf{G}| \sum_{i=1}^{m} N(\mathbf{z}_i; \mathbf{0}, \mathbf{I})} d\mathbf{z}_1 \cdots d\mathbf{z}_m. \quad (4.68)
$$

The symmetry of the integration region means that only the diagonal elements of $\mathbf{D}$ are non-zero. We obtain

$$\mathbf{D}(m) = \kappa(m)\mathbf{I}, \tag{4.69}$$

where

$$\kappa(m) = \int_{[-g,g]^{nm}} \frac{z_k(1)^2 N(\mathbf{z}_1; \mathbf{0}, \mathbf{I})^2}{a(m) + b(m)/|\mathbf{G}| \sum_{i=1}^{m} N(\mathbf{z}_i; \mathbf{0}, \mathbf{I})} \, d\mathbf{z}_1 \cdots d\mathbf{z}_m. \tag{4.70}$$

Substituting (4.69) into (4.67) and the result into (4.63) gives

$$\mathsf{E}[\mathbf{C}(\mathbf{y}, \mathbf{x}, m)|\mathbf{x}, m] = [\tau(m)V/m]^2 m|\mathbf{G}|^{m-2}\kappa(m)\mathbf{H}(\mathbf{x})^T \mathbf{R}^{-1}\mathbf{H}(\mathbf{x}). \tag{4.71}$$

It follows that

$$\mathbf{Z}_k = v_k \mathsf{E}[\mathbf{H}(\mathbf{x})^T \mathbf{R}^{-1}\mathbf{H}(\mathbf{x})], \tag{4.72}$$

where

$$v_k = \sum_{m=1}^{\infty} \frac{\mathsf{P}(m)\tau(m)^2|\mathbf{G}|^{m-2}\kappa(m)}{mV^{m-2}}. \tag{4.73}$$

Recall that, for the case of no clutter and perfect detection, $\mathsf{E}[\mathbf{H}(\mathbf{x})^T \mathbf{R}^{-1}\mathbf{H}(\mathbf{x})]$ is the contribution of the current measurement to the information matrix. Equation (4.72) indicates that the effect of clutter and missed detections on the PCRB is a scaling of this measurement contribution by the quantity v_k of (4.73). This quantity is referred to as the information reduction factor (IRF) (Mušicki *et al.*, 2005b).

The bound described here is not the tightest bound available for tracking in clutter. The measurement sequence conditioning approach of Hernandez *et al.* (2006) and the measurement existence conditioning (MSC) approach of McGinnity and Irwin (2001) are both provably tighter. The IRF bound has been chosen here because of the interesting structure it provides. Empirical evidence suggests that the difference between the IRF bound and the tightest of the three bounds, the MSC bound, decreases as measurements are acquired (Hernandez *et al.*, 2006).

It should be noted that estimation accuracy is not the only performance criteria for tracking in clutter. For sufficiently dense clutter, accurate tracking becomes impossible for any algorithm. In such cases the MSE is not particularly useful as a measure of performance. Instead it is useful to consider the probability of an algorithm remaining 'in track'. This essentially means that the algorithm continues to provide state estimates which are, in some sense, close to the true state.

The problem of bounding performance in this sense does not seem to have been considered.

4.7 Illustrative example

In this section the angle tracking example described in Chapter 2 and depicted in Figure 2.2 is used to assess several clutter tracking algorithms. In this example an object of interest moving along the ground is observed by an airborne sensor which measures elevation and azimuth. The scenario is complicated here by occasional missed target detections and the addition of non-object related clutter measurements.

The particular object and sensor trajectories used here are shown in Figure 2.3. The object is detected with probability $P_D = 0.8$. Object measurements are affected by additive noise with covariance matrix $\mathbf{R} = (\pi/360)^2 \mathbf{I}_2$. Simulations are performed for several values of the clutter density. In particular, we use $\lambda = 5, 25$ and 100 points/rad^2. These values of clutter density roughly correspond to light, moderate and heavy clutter, respectively.

The algorithms used in the performance analysis are the nearest neighbour filter (NNF), the probabilistic data association filter (PDAF) and the extended Kalman auxiliary particle filter (EK-APF). State estimation for the NNF and PDAF is implemented using the UKF. The EK-APF is implemented with a sample size of 5000. As discussed above, MSE is not the only useful performance measure for tracking in clutter. In addition it is necessary to consider the ability of the algorithm to maintain reliable in adverse environments. We thus consider the track loss probability where track loss is defined as the event that the error of algorithm's state estimate exceeds the PCRB by an order of magnitude for ten consecutive scans. The MSE is computed by averaging only over those realizations for which track loss does not occur. The performance measures are computed over 1000 realizations for each clutter density.

The time-averaged RMS position errors and the track loss percentages are shown in Tables 4.1 and 4.2. The EK-APF clearly outperforms the PDAF, which in turn is far superior to the NNF. The main differences in performance are observed in the track loss percentages rather than the RMSE. These results highlight the ability of PFs to accurately approximate a multi-modal posterior density, a task which is rather difficult for Gaussian approximations such as the PDAF and NNF. The results suggest that, in this scenario, the PCRB does not give a good indication of the achievable MSE, even for low clutter densities. The PCRB changes very little as the clutter density is increased from 5 to 100 points/rad^2 despite the dramatically increased difficulty of the scenario. These observations should be kept in mind when considering the use of the PCRB in sensor scheduling applications.

Table 4.1 *Time-averaged RMS position errors for single object tracking in clutter.*

Clutter density (points/rad^2)	NNF	PDAF	EK-APF	PCRB
5	16.2	17.1	13.3	7.3
25	20.3	22.6	17.0	7.4
100	25.7	29.8	25.5	7.7

Table 4.2 *Track loss probability for single object tracking in clutter.*

Clutter density (points/rad^2)	NNF	PDAF	EK-APF
5	0.232	0.075	0.027
25	0.518	0.188	0.070
100	0.846	0.661	0.165

4.8 Summary

In this chapter the problem of tracking a single object in the presence of clutter measurements was considered. The principal difficulty introduced by clutter measurements is that it is no longer possible to know which of the available measurements originated from the object of interest. In the optimal Bayesian solution measurement origin uncertainty is overcome by enumerating and evaluating all possible measurement origin hypotheses. The practical unfeasibility of doing this exactly has led to a large number of approximations. Several of the more popular approximations, such as the nearest neighbour filter, the probabilistic data association filter and particle filters, have been described in this chapter. These filters were chosen for the impact they have had on the research community and are far from an exhaustive list of algorithms. An important class of filters which have not been considered in detail here are those filters which approximate the posterior by a Gaussian mixture with a fixed and finite number of components (Ross, 2003; Wang *et al.*, 2008). The basis of these filters are mixture-reduction algorithms which attempt to approximate a Gaussian mixture with a large number of components by another mixture with a much smaller number of components. Also interesting is a generalization of the PDAF which involves merging mixture components over multiple scans (Sidenbladh, 2003).

5

Single- and multiple-object tracking in clutter: object-existence-based approach

In many practical situations, the number and existence of objects that are supposed to be tracked are a priori unknown. This information is an important part of the tracking output. In this chapter we include the object existence in the track state. As in previous chapters, the track state pdf propagates between scans as a Markov process, and is updated using the Bayes formula.

Object existence is particularly important in the cluttered environment, when the origin of each measurement is a priori unknown. This chapter reveals the close relationship (generalization/specialization) of a number of object-existence-based target tracking filters, which have a common derivation and common update cycle.

Some of the algorithms mentioned here also appear in other chapters of this book. These include probabilistic data association (PDA) (Section 4.3), integrated PDA (IPDA) (Sections 5.4.4 and 6.4.4) and joint IPDA (JIPDA) (Section 6.4.5). The derivations of this chapter follow a different track, and the results are more general as they also cater for non-homogeneous clutter.

5.1 Introduction

Object tracking aims to estimate the states of a (usually moving) unknown number of objects, using measurements received from sensors, and based on assumptions and models of the objects and measurements.

The object tracking algorithms presented in this chapter are based on the following assumptions, unless stated otherwise:

- Object:
  - There are zero or more objects in the surveillance area. The number and the position of the objects are a priori unknown.
  - Possible object trajectory models are assumed known, and the trajectory models are assumed to propagate as a Markov chain.

– At each scan one measurement per object may be present, described by the probability of detection, P_D, which may be different for each object. In other words, we assume point objects and each object produces zero or one measurement per scan randomly.

There are a large number of estimation algorithms for processes with Markovian switching coefficients, which is the model for object trajectory described here. We limit ourselves here to using interacting multiple models (IMM) (Blom and Bar-Shalom, 1988), detailed in Section 3.4. When the number of trajectory models equals one, IMM propagation and estimation collapses into the Kalman filter propagation and estimation respectively.

• Clutter:
– The number and location of clutter measurements are random and follow a Poisson distribution with non-homogeneous measurement density. This is a more general model than the usual one (Chapter 4), which assumes homogeneous density of clutter measurements within the surveillance area. It is also a more realistic model, as in practice the clutter measurement density is NOT uniform in the surveillance area in a vast majority of applications.
– Clutter measurement density in the surveillance area is a priori known, otherwise we estimate it from the received measurements.

• Measurements
– Measurements are produced by the sensor with infinite resolution. In other words, each measurement can have only one source. When a measurement is used to update a track, it can be one of
   * measurement (detection) of the object being tracked;
   * clutter measurement; or
   * measurement (detection) of an object being followed by some other track. This possibility defines the multi-object tracking, and is ignored by the single-object tracking.
– An additional possibility for each measurement is that it may be a detection of an object not being followed by an existing track. Such an object is termed here a "new object." Due to the assumed lack of prior knowledge of new objects' distribution, this possibility is handled separately by the track initialization process, described in Section 9.4.
– We consider here only position measurements. Additional measurement attributes, such as amplitude, Doppler speed, etc., can be included in a straightforward manner using a priori probability density functions of the measurement attributes (Lerro and Bar-Shalom, 1990, 1993; Mušicki and Evans, 2008; Wang *et al.*, 2008).

These assumptions are usually not completely satisfied in practical situations. For example, sensors usually do not have infinite resolution and close measurements will be merged. In the opposite direction, with high-resolution sensors, one object may have more than one measurement per scan. However, assumptions listed here simplify the object tracking problem considerably and are instrumental to derivations of a majority of object tracking algorithms. As a consequence the practical implementation of object tracking may fall short of expectations in situations when these assumptions are violated to a large extent (Mušicki and Evans, 1995).

In an automatic object tracking system, tracks are initialized and updated using measurements. If a track is initialized using one or more clutter measurements, or measurements from more than one object, the initialized track (usually) does not follow any object. Also, a track may "lose" its object due to an unfavorable detection sequence, object maneuvers, clutter measurements, measurement noise, multi-object effects or some combination thereof. Finally, the object that was followed by a true track may physically disappear from the surveillance area due to terminating its trajectory ("aircraft landing"), entering an observation shadow, actually disappearing ("successful" defense action) or simply departing the surveillance space. Therefore, in most tracking situations, each existing track may be either a true track which follows an object, or a false track which does not. The existence of an object being followed by a track is uncertain, i.e., object existence is a random event and is best described in probabilistic terms. Thus, each track has an associated probability of existence of the underlying object. The object existence event propagates as a Markov event, and its probability is recursively updated using measurements in a Bayesian fashion by algorithms presented in this chapter. This object existence paradigm is introduced in Mušicki *et al.* (1994).

The probability of object existence is usually used as a track quality measure for the false track discrimination procedure. False track discrimination tries to recognize and confirm true tracks, and recognize and terminate false tracks. The exact use of the probability of object existence is not part of the algorithms themselves, but rather the algorithms provide the tool for the job. A simple false track discrimination scheme may be described by the following track status propagation (Figure 5.1):

- each new track has the tentative status, which may be changed by confirmation or termination using subsequent measurements;
- if the probability of object existence rises above a predetermined track confirmation threshold t_c, the track becomes confirmed, and stays confirmed until termination; and
- when the probability of object existence falls below a predetermined track termination threshold t_t, the track is terminated and removed from memory.

In this case all existing tracks are either tentative or confirmed. Confirmed tracks are used by the operators, fusion centers or higher processing stages, as

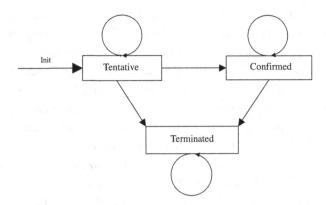

Figure 5.1 Track status propagation.

object existence is assumed with a high confidence. Tentative tracks are usually just propagated and updated by the object tracking algorithm (also named here "tracker," as a term of endearment), whilst being ignored by operators and higher levels of information processing.

False track discrimination is often the most important task of the surveillance system; it is usually more important to know precisely the existence and the number of objects in the surveillance area than to increase the precision of trajectory estimation by a small factor.

The algorithms presented in this chapter may be classified according to multiple criteria:

- Single-object tracking/multi-object tracking. Single-object tracking filters update each track separately, ignoring the possible presence of objects (and their detections) being followed by other tracks.

  Multi-object tracking filters look at the "global" allocations of measurements to tracks.
- Single trajectory model (non-maneuvering objects)/multiple trajectory models (maneuvering objects) trackers.
- Single-scan/multi-scan filters. Single-scan tracking filters approximate the a posteriori object trajectory pdf by one Gaussian pdf per trajectory model. The effects of all measurements in the current scan are merged together.

  Multi-scan filters ideally create one trajectory pdf for each possible measurement sequence in time, and each such trajectory pdf is termed a component. The complete trajectory pdf is a mixture of components. As the number of components grows exponentially in time, practical implementations are almost always sub-optimal, as the number of components must be kept to a level consistent with available computational resources.

Table 5.1 *Object-existence-based tracker classification.*

		Multi-scan	Single-scan	Deterministic existence
Multi-object	Multi-model	IMM-JITS	IMM-JIPDA	IMM-JPDA
	Single model	JITS	JIPDA	JPDA
Single object	Multi-model	IMM-ITS	IMM-IPDA	IMM-PDA
	Single model	ITS	IPDA	PDA

- Random object existence/deterministic object existence. Filters with the deterministic object existence assume object existence for each track under consideration (with probability one).

  Random object existence algorithms for each track calculate the probability that there is an object being followed. The probability of object existence is meant for use in the false track discrimination procedure.

The object tracking algorithms presented in this chapter are listed in Table 5.1.

In Table 5.1, "IMM" stands for the "interacting multiple models" algorithm, which is incorporated in the tracker. The letter "J" stands for "joint", or multi-object data association. "ITS" stands for "integrated track splitting", "IPDA" stands for "integrated probabilistic data association" and "PDA" stands for "probabilistic data association."

The most general algorithms in Table 5.1 are IMM-JITS and IMM-ITS for multi-object and single-object tracking respectively. Moving to the right from column to column or down from row to row increases the specialization of the algorithms. For example, JITS is in fact IMM-JITS with the number of object trajectory models equal to one; IMM-JIPDA is IMM-JITS with the number of measurement sequences (or track components as elucidated below) equal to one. JPDA is identical to JIPDA with the probability of object existence fixed to one (deterministic object existence).

Generally speaking, the more specialized the algorithms are, the less accurate the object trajectory state pdf becomes, with a corresponding detriment to the object tracking performance. On the other hand, more specialized algorithms have significantly smaller computational requirements.

Strictly speaking, algorithms in the rightmost column of Table 5.1, namely PDA, JPDA, IMM-PDA and IMM-JPDA, do not share the object existence paradigm as they assume all objects to be present (deterministic object existence). They are included in this chapter due to seamless derivation by specialization, with a fringe benefit of obtaining somewhat more general formulae than their original publications (Bar-Shalom and Tse, 1975; Bar-Shalom and Fortmann, 1988; Bar-Shalom and Li, 1993; Blom and Bloem, 2002), and derivations in Chapter 4. In particular, these algorithms now conform to the non-uniform clutter measurement density assumption.

The algorithms listed in Table 5.1 share a common track state, common operations and common (Bayesian) derivations. While the single-object tracking algorithms are specializations of the multi-object tracking algorithms, we separate their data association expressions and derivations for reasons of clarity and reader sanity. Other operations are common to both classes of algorithms. This is utilized to make this chapter more consistent and more readable with a single thread to follow. Common (Bayesian) derivations for the most general algorithm are provided, from which specialized algorithms follow by applying constraints.

A common track state also creates a possibility to use different object tracking algorithms in the same application. An example used in Mušicki and Evans (2004b, 2008) and Mušicki and La Scala (2008) is to apply IPDA (or ITS) on the set of tentative tracks, and JIPDA (or JITS) on the set of confirmed tracks in difficult multi-object scenarios. A common false track discrimination procedure (based on the updated probability of object existence) may be used for all tracks. This combination significantly reduces the computational requirements, whilst still providing the multi-object tracking benefits.

In this chapter a superscript (e.g., τ) is sometimes used to denote track. The symbol τ is also used to denote the (possible) object being followed by track τ. For example, $p(\chi_k^\tau)$ denotes the probability of existence of the object τ. Whenever the meaning is obvious from the context, the track superscript will be omitted for reasons of clarity. For example, the track prediction operation is performed on all tracks independently, and the track superscript is omitted.

The problem statement and models used by algorithms presented in this chapter are presented in Section 5.2. The hybrid track state, which consists of the object existence event and object trajectory state is defined and detailed in Section 5.3. The optimal Bayes derivation is presented in Section 5.4. The track update cycle, which is common to all algorithms of this chapter, is presented in Section 5.5. Various techniques to the number of components are presented in Section 5.6. Finally, Sections 5.7 and 5.8 detail concrete algorithms for single-object and multi-object tracking respectively.

For each tracker presented in this chapter, only one update cycle is included. Issues of track initialization, track merging and clutter measurement density estimation are relegated to Chapter 9. Illustrative examples are presented in Section 9.6.

5.2 Problem statement/models

Braving the risk of repetition, this section defines and details the elements of the problem statements or models that are assumed.

5.2.1 Sensors

A batch of measurements is received at each sampling time k, without a priori information on the origin of each measurement. Zero or one measurement of each object is received at time k randomly, with the probability of detection denoted by P_D. A random number of clutter measurements is also received at time k.

Denote by $\mathbf{Y}_k$ the set of measurements delivered by the sensor at time k. As detailed in Section 5.5.2, a subset of $\mathbf{Y}_k$, denoted here by $\mathbf{y}_k$, is selected for track update at time k. The selection process selects the object detection with a probability P_G, if the detection is placed within a selection gate part of the surveillance area of volume V_k.

Symbol m_k denotes the (random) number of measurements in (cardinality of) set $\mathbf{y}_k$, and $\mathbf{y}_k(i)$ denotes the ith element of $\mathbf{y}_k$. Finally, $\mathbf{Y}^k$ denotes the sequence of all measurement sets used for track update up to and including time k:

$$\mathbf{Y}^k = \mathbf{y}_k \bigcup \mathbf{Y}^{k-1}; \qquad \mathbf{Y}^0 = \{\}. \tag{5.1}$$

In principle, and especially with intelligent sensors including the Electronically Steered Antenna radars, the tracker has no a priori knowledge of the scan time of sampling the next batch of measurements. The time interval between previous, $k - 1$, and current, k, scan time denoted here by Δt_k may be different from scan to scan and is only known at the moment of arrival of measurement set $\mathbf{Y}_k$.

5.2.2 Objects

Single-object tracking algorithms with random object existence (IMM-ITS, ITS, IMM-IPDA and IPDA) assume the existence of either zero or one object, and single-object tracking algorithms with deterministic object existence (PDA and IMM-PDA) assume the existence of one object.

Multi-object tracking algorithms with random object existence (IMM-JITS, JITS, IMM-JIPDA and JIPDA) allow for the existence of any number of objects between zero and the number of tracks, whilst multi-object tracking algorithms with deterministic object existence (JPDA and IMM-JPDA) assume the existence of one object per existing track.

Maneuvering objects may change their trajectory model at any time. The model is simplified by assuming that the trajectory model change occurs only at the sampling time. Further simplification limits the number of models to M a priori known models (as in Section 3.2.1), and the object trajectory propagation is modeled as

$$\mathbf{x}_k = \mathbf{f}(k, r_k, \mathbf{x}_{k-1}, \nu(k, r_k)), \tag{5.2}$$

where $\mathbf{x}_k$ denotes the object trajectory at time k, $r_k = 1, \ldots, M$ denotes the trajectory model in the time interval between $k - 1$ and k, ν denotes the plant noise (random element) of trajectory propagation and $\mathbf{f}$ is a possibly non-linear function.

In this chapter we consider the linear object trajectory model, defined by

$$\mathbf{x}_k = \mathbf{F}_{r_k}\mathbf{x}_{k-1} + \mathbf{v}(k, r_k), \tag{5.3}$$

where $\mathbf{F}_{r_k}$ is the trajectory state propagation matrix of the maneuvering model r_k, and $\mathbf{v}(k, r_k)$ is assumed to be zero mean, white Gaussian noise with covariance $\mathbf{Q}_{r_k}$, which is also independent of any measurement noise sample. The transition probability density function follows directly as

$$p(\mathbf{x}_k|\mathbf{x}_{k-1}, r_k) = \mathcal{N}(\mathbf{x}_k; \mathbf{F}_{r_k}\mathbf{x}_{k-1}, \mathbf{Q}_{r_k}). \tag{5.4}$$

This model also contains the non-maneuvering object trajectory, which is obtained by setting the number of models equal to one, $M = 1$.

When propagating and updating track estimates, the exact value of object trajectory model index r_k is unknown, and state is estimated conditioned on event $r_k = \sigma$, where $\sigma = 1, \ldots, M$. For reasons of clarity, shorthand σ denotes the event $r_k = \sigma$ in this chapter. Events $r_k = \sigma$ are mutually exclusive, as an object may follow only one trajectory model at one time. The event set is also assumed complete, i.e., an object must follow one of the trajectory models from the set.

Propagation of the object trajectory model r_k is modeled by a Markov process with transition probability matrix by $\boldsymbol{\Gamma}$, where $\boldsymbol{\Gamma}_{\sigma,\omega} = p(r_k = \omega|r_{k-1} = \sigma)$ denotes transitional probability from trajectory model σ to model ω, and r_k denotes the object trajectory model at time k. To a large extent, this repeats the maneuvering model of Chapter 3.

5.2.3 Object measurements

Object measurements in practice do not depend on the object trajectory model, and may be expressed by

$$\mathbf{y}_k(i) = \mathbf{h}(k, \mathbf{x}_k, \boldsymbol{v}(k)), \tag{5.5}$$

where $\mathbf{y}_k(i)$ is the object measurement, $\mathbf{h}$ is a possibly non-linear function and $\boldsymbol{v}(k)$ denotes the measurement noise. The object measurement exists only if the object exists, and if the object is detected, and if the object measurement is selected. Both the existence of the object measurement, and the possible index i of the object measurement are not known a priori.

In this chapter we consider the linear measurement model, where object measurement equals

$$\mathbf{y}_k(i) = \mathbf{H}\mathbf{x}_k + \boldsymbol{v}_k, \tag{5.6}$$

where $\mathbf{H}$ denotes the measurement matrix, and v_k denotes a sample of zero mean white Gaussian noise with covariance $\mathbf{R}$, which is uncorrelated with the plant noise sequence.

5.2.4 Clutter measurements

In this chapter we generalize the clutter model of Chapter 4. The number of clutter measurements follows a Poisson distribution with somehow known intensity, which we call the clutter measurement density.

The number of clutter measurements in surveillance space is modeled by a non-homogeneous Poisson process, with clutter measurement density at point $\mathbf{y}$ denoted by $\rho(\mathbf{y})$ dependent on $\mathbf{y}$. Denote by $N(\mathbf{y}, V)$ the number of clutter measurements in surveillance space of volume V centered at $\mathbf{y}$. The necessary and sufficient conditions that process N is a Poisson process with intensity $\rho(\mathbf{y})$ are (Kingman, 1992; Ross, 2003):

1. the numbers of clutter measurements in non-overlapping regions of surveillance space are statistically independent;
2. $p(N(\mathbf{y}, V) \geq 2) = o(V)$; and
3. $p(N(\mathbf{y}, V) = 1) = \rho(\mathbf{y})V + o(V)$,

where $o(V)$ denotes the higher order of V. Property 1 is a direct consequence of the infinite-resolution sensor assumption. In this chapter we use the shorthand notation $\rho_k(i) \overset{\triangle}{=} \rho(\mathbf{y}_k(i))$.

Assume that we observe measurement space V_k at time k; this measurement space is either the track selection gate (for single-object trackers), or the cluster area (for multi-object trackers). The statistically mean number of clutter measurements in V_k is

$$\bar{m}_k = \int_{V_k} \rho(\mathbf{y}) \mathrm{d}V.$$

The likelihood (a priori pdf) of a measurement $\mathbf{y}_k(i)$, given that it is a clutter measurement, is

$$p_{\rho,k}(i) = \rho_k(i)/\bar{m}_k. \tag{5.7}$$

The probability that the number of clutter measurements equals m in a measurement space V_k at time k follows the Poisson distribution

$$\mu_F(m) = \exp(-\bar{m}_k)\frac{\bar{m}_k^m}{m!}, \tag{5.8}$$

where $\exp(\cdot)$ denotes $e^{(\cdot)}$, with e being the base of the natural logarithm.

The non-homogeneous clutter measurement model is both more general (it also includes the homogeneous clutter measurement model as a special case), and more realistic. The clutter measurement density is generally non-homogeneous, i.e., will be different at different points in the observation space. If the clutter measurement density is a priori known, we call this "parametric" object tracking. In some applications, e.g., when the main source of clutter is the thermal noise, the clutter measurement density can be calculated. In other applications the clutter measurement density can be estimated by using measurements from prior scans, e.g., using clutter mapping.

If the clutter measurement density is a priori unknown, we call this "non-parametric" object tracking. In this case the clutter measurement density is estimated in each scan using the received measurements. One approach to clutter measurement estimation is presented in Section 9.3. All the trackers in Chapter 4 are "non-parametric" object trackers. The non-parametric object tracking assumption is that the clutter measurement density is uniform within the surveillance area covered by the track selection gate.

Some of the object tracking algorithms presented in this chapter, namely PDA, JPDA, IMM-PDA and IMM-JPDA, have been derived in Chapters 4 and 6 using the homogeneous clutter model; they are extended here for more general and more realistic non-homogeneous clutter measurement models.

5.3 Track state

Here we consider a hybrid track state. Track state has a discrete component, which is the object existence at time k, and a continuous component, which is the object trajectory state pdf at time k.

5.3.1 Object existence

Two models for object existence propagation have been identified in Mušicki *et al.* (1994). In one, termed Markov Chain One, object existence has two possible states:

- the object either exists; or
- the object does not exist.

If it exists, the object is detectable (generates a measurement) with the probability of detection P_D. This is the default model which is used in this chapter. The other object existence model is termed Markov Chain Two in Mušicki *et al.* (1994), and has three possible object existence states:

- the object exists and is detectable; or
- the object exists and is temporarily not detectable; or
- the object does not exist.

The object may exist and be temporarily not detectable if, for example, the object gets temporarily concealed by an obstacle between the sensor and the object. If the object exists and is detectable, its measurements are present in each scan with the probability P_D. The Markov Chain Two model is adaptable to unknown or fluctuating probability of detection (Wang and Mušicki, 2007; Mušicki and Wang, 2004), at the expense of a slightly more complex model. For reasons of clarity and simplicity, the rest of this chapter follows the Markov Chain One model.

The Markov Chain One object existence model defines two mutually exclusive and exhaustive events modeled by a random variable E_k:

$$\chi_k \overset{\triangle}{=} E_k = 1 \quad \text{the event that an object exists,}$$
$$\bar{\chi}_k \overset{\triangle}{=} E_k = 0 \quad \text{the event that the object does not exist.}$$

The probability of propagated object existence is obtained by applying the Markov chain propagation formula,

$$\begin{bmatrix} p(\chi_k|\mathbf{Y}^{k-1}) \\ p(\bar{\chi}_k|\mathbf{Y}^{k-1}) \end{bmatrix} = \begin{bmatrix} p(\chi_k|\mathbf{Y}^{k-1}) \\ 1 - p(\chi_k|\mathbf{Y}^{k-1}) \end{bmatrix}$$
$$= \gamma^T \begin{bmatrix} p(\chi_{k-1}|\mathbf{Y}^{k-1}) \\ p(\bar{\chi}_{k-1}|\mathbf{Y}^{k-1}) \end{bmatrix} = \gamma^T \begin{bmatrix} p(\chi_{k-1}|\mathbf{Y}^{k-1}) \\ 1 - p(\chi_{k-1}|\mathbf{Y}^{k-1}) \end{bmatrix}, \tag{5.9}$$

with the elements of matrix γ being the transitional probabilities between object existence states:

$$\gamma_{ij} \overset{\triangle}{=} p(E_k = 2 - j | E_{k-1} = 2 - i), i, j \in \{1, 2\},$$

and

$$\gamma_{11} + \gamma_{12} = \gamma_{21} + \gamma_{22} = 1.$$

This operation denotes the Markov Chain One model for object existence prediction (propagation), and its pseudo-function is

$$p(\chi_k|\mathbf{Y}^{k-1}) = \text{TEX}_\text{P}[p(\chi_{k-1}|\mathbf{Y}^{k-1}), \gamma].$$

As discussed in Mušicki *et al.* (1994, 2007), the value of γ_{21} should be zero, from which $\gamma_{22} = 1$. The value of γ_{21} is the transitional probability of a false track becoming a true track. While such an event may happen in practice when the false track starts to follow an object, it also renders the object trajectory state pdf $p(\mathbf{x}_k|\chi_k)$ meaningless. The authors recommend using $\gamma_{21} = 0$, and treating the emergence of new tracks as part of the track "birth" process or track initialization, detailed in Section 9.4. Equation (5.9) therefore reduces to

$$p(\chi_k|\mathbf{Y}^{k-1}) = \gamma_{11} p(\chi_{k-1}|\mathbf{Y}^{k-1}), \tag{5.10}$$

where γ_{11} denotes the probability that an object will continue to exist at time k, given that it exists at time $k-1$. Value of γ_{11} is calculated as (Bar-Shalom and Li, 1993)

$$\gamma_{11} = 1 - \frac{\Delta t_k}{T_\chi},$$

where T_χ denotes the average object lifetime in the surveillance region, and Δt_k denotes the time between measurement scans $k-1$ and k, with the assumption that $\Delta t_k \ll T_\chi$.

5.3.2 Object trajectory state

The object trajectory state at time k is denoted by $\mathbf{x}_k$. The a priori track state probability density function is given by

$$p(\mathbf{x}_k, \chi_k|\mathbf{Y}^{k-1}) = p(\mathbf{x}_k|\chi_k, \mathbf{Y}^{k-1}) p(\chi_k|\mathbf{Y}^{k-1}), \tag{5.11}$$

and the a posteriori track state probability density function is given by

$$p(\mathbf{x}_k, \chi_k|\mathbf{Y}^k) = p(\mathbf{x}_k|\chi_k, \mathbf{Y}^k) p(\chi_k|\mathbf{Y}^k). \tag{5.12}$$

The object trajectory state pdf, $p(\mathbf{x}_k)$, is always calculated conditioned on the object existence event χ_k. The object trajectory state pdf conditioned on the object non-existence event is undefined and, indeed, does not make sense.

Denote by $\theta_k(i_k)$, $i_k \geq 0$, the event that measurement $\mathbf{y}_k (i_k)$ is the detection of the object being tracked by the track. Event $\theta_k(0)$ denotes the event that no selected measurement is the detection of the object, which can happen because either the object does not exist, or the object was not detected at time k, or its detection was not selected. Assuming that the track was initialized at time $k = 1$, each measurement sequence

$$\xi_k(c_k) = \{i_1, \ldots, i_k\}; \quad i_\ell = 0, \ldots, m_\ell; \quad \ell = 1, \ldots, k, \tag{5.13}$$

denotes one possible object detection sequence; the index c_k is the past measurement sequence index at time k, $c_k = 1, \ldots, C_k$ and

$$C_k = \prod_{\ell=1}^{k} (1 + m_\ell). \tag{5.14}$$

Table 5.2 *Measurement sequences/track components.*

c_1	$\xi_1(c_1)$	c_2	$\xi_2(c_2)$	c_3	$\xi_3(c_3)$
		1	$\{1, 0\}$	1	$\{1, 0, 0\}$
				2	$\{1, 0, 1\}$
1	$\{1\}$	2	$\{1, 1\}$	3	$\{1, 1, 0\}$
				4	$\{1, 1, 1\}$
		3	$\{1, 2\}$	5	$\{1, 2, 0\}$
				6	$\{1, 2, 1\}$

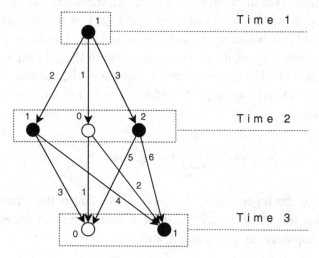

Figure 5.2 Measurement sequences/track components.

This is illustrated in Figure 5.2 and Table 5.2. At time $k = 1$ there is only one measurement which starts the track. There is only one measurement sequence $\xi_1(c_1)$ at time $k = 1$. At time $k = 2$, there are two measurements, indexed as 1 and 2. The "null" measurement indexed as 0 corresponds to object non-detection or object non-existence events. Each measurement at time $k = 2$ associates with the measurement sequence at time $k = 1$ to create a measurement sequence $\xi_2(c_2)$ at time $k = 2$. The process repeats at time $k = 3$ where we have one measurement indexed as 1 and one "null" measurement indexed as 0. Each measurement sequence $\xi_2(c_2)$ at time $k = 2$ associates with each measurement at time $k = 3$ to create one measurement sequence $\xi_3(c_3)$ at time $k = 3$.

The object trajectory pdf conditioned on $\xi_k(c_k)$, or in other words assuming that $\xi_k(c_k)$ is the sequence of object detections of measurement set $\mathbf{Y}^k$, is denoted by shorthand $p(\mathbf{x}_k|c_k)$:

$$p(\mathbf{x}_k|c_k) \overset{\triangle}{=} p(\mathbf{x}_k|\xi_k(c_k), \chi_k, \mathbf{Y}^k),$$

and

$$p(c_k) \overset{\triangle}{=} p(\xi_k(c_k)|\chi_k, \mathbf{Y}^k)$$

is the a posteriori probability that measurement sequence $\xi_k(c_k)$ consists entirely of object detections. Events $\{\xi_k(c_k)|\chi_k, \mathbf{Y}^k\}$ are mutually exclusive, as there is one and only one measurement sequence, which consists of object detections:

$$\sum_{c_k} p(c_k) = \sum_{c_k} p(\xi_k(c_k)|\chi_k, \mathbf{Y}^k) = 1.$$

Given a linear system with a known object trajectory model, $p(\mathbf{x}_k|c_k)$ is obtained by applying measurement sequence $\xi_k(c_k)$ to the Kalman filter with corresponding object trajectory model parameters. In that case $p(\mathbf{x}_k|c_k)$ is a Gaussian pdf defined by its mean and covariance. In this chapter we consider a more general case of M possible object trajectory models, and use the interacting multiple models (IMM) estimator. Therefore, $p(\mathbf{x}_k|c_k)$ is a Gaussian mixture of M Gaussian pdfs

$$p(\mathbf{x}_k|c_k) = \sum_{\sigma=1}^{M} p(\mathbf{x}_k|c_k, \sigma)\mu_{k|k}(c_k, \sigma), \tag{5.15}$$

where σ indexes the trajectory models, and $\mu_{k|k}(c_k, \sigma)$ is the a posteriori probability of object trajectory model σ at time k conditioned on the event that the measurement sequence $\xi_k(c_k)$ is correct, and

$$p(\mathbf{x}_k|c_k, \sigma) = \mathcal{N}\left(\mathbf{x}_k; \hat{\mathbf{x}}_{k|k}(c_k, \sigma), \mathbf{P}_{k|k}(c_k, \sigma)\right). \tag{5.16}$$

Applying the total probability theorem,

$$p(\mathbf{x}_k|\chi_k, \mathbf{Y}^k) = \sum_{c_k=1}^{C_k} p(\mathbf{x}_k|c_k)p(c_k) \tag{5.17}$$

$$= \sum_{i_1=0}^{m_1} \cdots \sum_{i_k=0}^{m_k} p(\mathbf{x}_k|\chi_k, \theta_1(i_1), \ldots, \theta_k(i_k), \mathbf{Y}^k)$$

$$\times p(\theta_1(i_1), \ldots, \theta_k(i_k)|\chi_k, \mathbf{Y}^k). \tag{5.18}$$

Each trajectory state pdf at time k, given measurement history $\xi_k(c_k)$, is called a track component, indexed by c_k. The number of track components, as defined by (5.14), grows exponentially with k. This soon exhausts the available reasonable computational resources. Any practical implementation has to limit the number of components using the various methods discussed in Section 5.5.5. Due to this component management, the strict definition of components as defined by (5.13) and (5.18) often is not valid; however, the track trajectory state pdf remains a set of mutually exclusive components and (5.15), (5.16) and (5.17) remain valid.

As both trajectory models and track components are mutually exclusive and exhaustive events, the following relations hold:

$$\sum_{c_k=1}^{C_k} p(c_k) = \sum_{\sigma=1}^{M} \mu_{k|k}(c_k, \sigma) = 1,$$

$$\sum_{c_k=1}^{C_k} \sum_{\sigma=1}^{M} p(c_k)\mu_{k|k}(c_k, \sigma) = 1,$$

$$p(c_k), \mu_{k|k}(c_k, \sigma) \in [0\ 1], \quad \forall c_k, \sigma.$$

This object trajectory state model (a set of track components, each having an IMM block) is a general and generic model for all algorithms presented in this chapter. This model is directly used for the IMM-ITS and IMM-JITS algorithms, and various degrees of specialization, one motion model and/or one track component will be applied for other algorithms.

5.4 Optimal Bayes' recursion

In this section we present optimal prediction and Bayes' update. These operations are part of every algorithm subsequently presented in this chapter. The optimal update of the single-object trackers differ substantially from the optimal update of the multi-object trackers, thus they are presented separately.

5.4.1 Track prediction

Track state prediction involves propagating the track state at time $k-1$, defined by

$$p(\mathbf{x}_{k-1}, \chi_{k-1}|\mathbf{Y}^{k-1}) = p(\chi_{k-1}|\mathbf{Y}^{k-1})p(\mathbf{x}_{k-1}|\chi_{k-1}, \mathbf{Y}^{k-1}),$$

to the track state at time k defined by

$$p(\mathbf{x}_k, \chi_k|\mathbf{Y}^{k-1}) = p(\chi_k|\mathbf{Y}^{k-1})p(\mathbf{x}_k|\chi_k, \mathbf{Y}^{k-1}).$$

The probability of object existence propagates as a Markov chain (Section 5.3.1):

$$p(\chi_k|\mathbf{Y}^{k-1}) = \text{TEX}_\text{P}[p(\chi_{k-1}|\mathbf{Y}^{k-1}), \boldsymbol{\gamma}]. \tag{5.19}$$

Barring some exotic situations (Mušicki *et al.*, 2007), relative component probabilities do not change when propagating from time $k-1$ to k, and

$$p(\mathbf{x}_k|\chi_k, \mathbf{Y}^{k-1}) = \sum_{c_{k-1}=1}^{C_{k-1}} p(\mathbf{x}_k|c_{k-1}, \chi_{k-1}, \mathbf{Y}^{k-1})p(c_{k-1}).$$

In other words, trajectory state pdf propagation consists of propagating each trajectory state component pdf individually. Each state component c_{k-1} is an IMM block. Therefore, object trajectory state pdf propagation consists of IMM mixing and prediction operations for each component c_{k-1} of each track (Section 3.4.1),

$$[\{\mu_{k|k-1}(c_{k-1}, \sigma), \hat{\mathbf{x}}_{k|k-1}(c_{k-1}, \sigma), \mathbf{P}_{k|k-1}(c_{k-1}, \sigma)\}_\sigma]$$

$$= \text{IMM}_{\text{MP}}[\{\mu_{k-1|k-1}(c_{k-1}, \sigma), \hat{\mathbf{x}}_{k-1|k-1}(c_{k-1}, \sigma),$$

$$\mathbf{P}_{k-1|k-1}(c_{k-1}, \sigma), \mathbf{F}_\sigma, \mathbf{Q}_\sigma\}_\sigma, \boldsymbol{\Gamma}].$$

IMM mixing and prediction collapses into Kalman filter prediction if the number of object trajectory models (IMM models) equals one, $M = 1$, as is the case for JITS, JIPDA, JPDA, ITS, IPDA and PDA. Then, for each component of each track (Section 2.2.1),

$$\left[\hat{\mathbf{x}}_{k|k-1}(c_{k-1}), \mathbf{P}_{k|k-1}(c_{k-1})\right] = \text{KF}_{\text{P}}\left[\hat{\mathbf{x}}_{k-1|k-1}(c_{k-1}), \mathbf{P}_{k-1|k-1}(c_{k-1}), \mathbf{F}, \mathbf{Q}\right].$$

The predicted pdf of the object trajectory state at time k is given by a mixture of the predicted pdfs of the object trajectory state component pdfs,

$$p(\mathbf{x}_k|\chi_k, \mathbf{Y}^{k-1}) = \sum_{c_{k-1}=1}^{C_{k-1}} p(\mathbf{x}_k|c_{k-1}, \chi_k, \mathbf{Y}^{k-1})p(c_{k-1}), \qquad (5.20)$$

and each pdf of the object trajectory component state is a mixture of the object trajectory state component pdfs given individual object trajectory models σ,

$$p(\mathbf{x}_k|c_{k-1}, \chi_k, \mathbf{Y}^{k-1}) = \sum_{\sigma=1}^{M} \mu_{k|k-1}(c_{k-1}, \sigma)p(\mathbf{x}_k|c_{k-1}, \sigma, \chi_k, \mathbf{Y}^{k-1}), \qquad (5.21)$$

$$p(\mathbf{x}_k|c_{k-1}, \sigma, \chi_k, \mathbf{Y}^{k-1}) = \mathcal{N}(\mathbf{x}_k; \hat{\mathbf{x}}_{k|k-1}(c_{k-1}, \sigma), \mathbf{P}_{k|k-1}(c_{k-1}, \sigma)).$$

The track state propagation parameters, γ, $\boldsymbol{\Gamma}$, $\mathbf{F}$ and $\mathbf{Q}$ depend on the time interval Δt_k between sampling times $k - 1$ and k. An example of values for $\mathbf{F}$ and $\mathbf{Q}$ matrices is presented in Section 2.8; γ and $\boldsymbol{\Gamma}$ are discussed in Sections 5.3.1 and 3.4.1 respectively.

5.4.2 Object measurement likelihood

We use the total probability theorem to obtain the prior pdf $p(\mathbf{y}|\chi_k, \mathbf{Y}^{k-1})$ of the object detection position for track,

$$p(\mathbf{y}|\chi_k, \mathbf{Y}^{k-1}) = \int_{\mathbf{x}_k} p(\mathbf{y}|\mathbf{x}_k) \, p(\mathbf{x}_k|\chi_k, \mathbf{Y}^{k-1}) \, d\mathbf{x}_k,$$

where we use the fact that the measurement noise in (5.6) is white.

Applying (5.20) and interchanging summation and integration, we first obtain the prior pdf $p(\mathbf{y}|c_{k-1}, \chi_k, \mathbf{Y}^{k-1})$ of the object detection position given track component c_{k-1},

$$p(\mathbf{y}|\chi_k, \mathbf{Y}^{k-1}) = \sum_{c_{k-1}=1}^{C_{k-1}} p(c_{k-1})p(\mathbf{y}|c_{k-1}, \chi_k, \mathbf{Y}^{k-1}), \quad (5.22)$$

$$p(\mathbf{y}|c_{k-1}, \chi_k, \mathbf{Y}^{k-1}) = \int_{\mathbf{x}_k} p(\mathbf{y}|\mathbf{x}_k) \, p(\mathbf{x}_k|c_{k-1}, \chi_k, \mathbf{Y}^{k-1}) \, d\mathbf{x}_k.$$

Applying (5.21) and interchanging summation and integration again, we obtain the prior pdf $p(\mathbf{y}|c_{k-1}, \sigma, \chi_k, \mathbf{Y}^{k-1})$ of the object detection position given track component c_{k-1} and object trajectory model σ,

$$p(\mathbf{y}|c_{k-1}, \chi_k, \mathbf{Y}^{k-1}) = \sum_{\sigma=1}^{M} \mu_{k|k-1}(c_{k-1}, \sigma)p(\mathbf{y}|c_{k-1}, \sigma, \chi_k, \mathbf{Y}^{k-1}), \quad (5.23)$$

$$p(\mathbf{y}|c_{k-1}, \sigma, \chi_k, \mathbf{Y}^{k-1}) = \int_{\mathbf{x}_k} p(\mathbf{y}|\mathbf{x}_k) \, p(\mathbf{x}_k|c_{k-1}, \sigma, \chi_k, \mathbf{Y}^{k-1}) \, d\mathbf{x}_k,$$

with

$$p(\mathbf{y}|c_{k-1}, \sigma, \chi_k, \mathbf{Y}^{k-1})$$
$$= \int_{\mathbf{x}_k} \mathcal{N}\left(\mathbf{y}; \mathbf{Hx}_k, \mathbf{R}_k\right) \mathcal{N}\left(\mathbf{x}_k; \hat{\mathbf{x}}_{k|k-1}(c_{k-1}, \sigma), \mathbf{P}_{k|k-1}(c_{k-1}, \sigma)\right) d\mathbf{x}_k$$
$$= \mathcal{N}\left(\mathbf{y}; \hat{\mathbf{y}}_{k|k-1}(c_{k-1}, \sigma), \mathbf{S}_k(c_{k-1}, \sigma)\right). \quad (5.24)$$

In (5.24) the predicted measurement position mean $\hat{\mathbf{y}}_{k|k-1}(c_{k-1}, \sigma)$ and covariance $\mathbf{S}_k(c_{k-1}, \sigma)$, given component c_{k-1} and trajectory model σ, are calculated by the pseudo-function

$$[\hat{\mathbf{y}}_{k|k-1}(c_{k-1}, \sigma), \mathbf{S}_k(c_{k-1}, \sigma)] = \text{MP}[\hat{\mathbf{x}}_{k|k-1}(c_{k-1}, \sigma), \mathbf{P}_{k|k-1}(c_{k-1}, \sigma), \mathbf{H}, \mathbf{R}],$$
$$(5.25)$$

which is defined in Section 2.2.1 by

$$\hat{\mathbf{y}}_{k|k-1}(c_{k-1}, \sigma) = \mathbf{H}\hat{\mathbf{x}}_{k|k-1}(c_{k-1}, \sigma),$$
$$\mathbf{S}_k(c_{k-1}, \sigma) = \mathbf{HP}_{k|k-1}(c_{k-1}, \sigma)\mathbf{H}^T + \mathbf{R}.$$

Equation (5.24) calculates the prior measurement pdf for all surveillance space. The prior measurement pdfconditioned on measurement being selected is given by

$$p(\mathbf{y}|\mathbf{y} \in V_k(c_{k-1}, \sigma), c_{k-1}, \sigma, \chi_k, \mathbf{Y}^{k-1})$$

$$= \frac{1}{P_G} \mathcal{N}(\mathbf{y}; \hat{\mathbf{y}}_{k|k-1}(c_{k-1}, \sigma), \mathbf{S}_k(c_{k-1}, \sigma)).$$

Equations (5.22) and (5.23) remain valid.

The likelihoods of selected measurements are equal to their prior pdfs. The following shortcut is observed in this chapter:

$$p_k(i, c_{k-1}, \sigma) \overset{\triangle}{=} p(\mathbf{y}_k(i)|c_{k-1}, \sigma, \chi_k, \mathbf{Y}^{k-1}) \tag{5.26}$$

$$= \begin{cases} \frac{1}{P_G} \mathcal{N}\left(\mathbf{y}_k(i); \hat{\mathbf{y}}_{k|k-1}(c_{k-1}, \sigma), \mathbf{S}_k(c_{k-1}, \sigma)\right), & \mathbf{y}_k(i) \in V_k(c_{k-1}, \sigma), \\ 0, & \mathbf{y}_k(i) \notin V_k(c_{k-1}, \sigma). \end{cases}$$

A pseudo-function comprising equations (5.25)–(5.26), which calculate the object measurement likelihoods for one Gaussian state pdf, is denoted by:

$$\left[\{p_k(i, c_{k-1}, \sigma)\}_i\right] = \mathrm{ML}_1\left[\{\mathbf{y}_k(i)\}_i, \hat{\mathbf{x}}_{k|k-1}(c_{k-1}, \sigma), \mathbf{P}_{k|k-1}(c_{k-1}, \sigma), \mathbf{H}, \mathbf{R}\right]. \tag{5.27}$$

We can proceed to the level of component and the level of track by

$$p_k(i, c_{k-1}) \overset{\triangle}{=} p(\mathbf{y}_k(i)|c_{k-1}, \chi_k, \mathbf{Y}^{k-1}) = \sum_{\sigma=1}^{M} \mu_{k|k-1}(c_{k-1}, \sigma) p_k(i, c_{k-1}, \sigma),$$

$$p_k(i) \overset{\triangle}{=} p(\mathbf{y}_k(i)|\chi_k, \mathbf{Y}^{k-1}) = \sum_{c_{k-1}=1}^{C_{k-1}} p(c_{k-1}) p_k(i, c_{k-1}).$$

The likelihoods of all measurements not selected by a track are assumed to be zero with respect to the track.

5.4.3 Optimal track update

The derivation is presented according to the template followed by the book. An alternative derivation is presented in Mušicki *et al.* (2007) and Mušicki and Evans (2008).

The track state is hybrid and consists of the object existence event and the object trajectory state as detailed in Section 5.3. The a priori track state probability density function is given by (5.11),

$$p(\mathbf{x}_k, \chi_k|\mathbf{Y}^{k-1}) = p(\mathbf{x}_k|\chi_k, \mathbf{Y}^{k-1}) p(\chi_k|\mathbf{Y}^{k-1}), \tag{5.28}$$

and is updated by the measurement information $\mathbf{y}_k$. The information provided by set $\mathbf{y}_k$ includes the number of selected measurements m_k, as well as the position of the selected measurements. The Bayes' equation becomes

$$p(\mathbf{x}_k, \chi_k | \mathbf{Y}^k) = \frac{p(\mathbf{y}_k, m_k | \mathbf{x}_k, \chi_k, \mathbf{Y}^{k-1})}{p(\mathbf{y}_k, m_k | \mathbf{Y}^{k-1})} p(\mathbf{x}_k, \chi_k | \mathbf{Y}^{k-1}),$$

where we choose to explicitly include m_k. Using (5.28) and multiplying and dividing the right-hand side by $p(\mathbf{y}_k, m_k | \chi_k, \mathbf{Y}^{k-1})$,

$$p(\mathbf{x}_k, \chi_k | \mathbf{Y}^k) = \frac{p(\mathbf{y}_k, m_k | \chi_k, \mathbf{Y}^{k-1})}{p(\mathbf{y}_k, m_k | \mathbf{Y}^{k-1})} p(\chi_k | \mathbf{Y}^{k-1})$$

$$\times \frac{p(\mathbf{y}_k, m_k | \mathbf{x}_k, \chi_k, \mathbf{Y}^{k-1})}{p(\mathbf{y}_k, m_k | \chi_k, \mathbf{Y}^{k-1})} p(\mathbf{x}_k | \chi_k, \mathbf{Y}^{k-1}).$$

Applying the Bayes' equation separately,

$$p(\chi_k | \mathbf{Y}^k) = \frac{p(\mathbf{y}_k, m_k | \chi_k, \mathbf{Y}^{k-1})}{p(\mathbf{y}_k, m_k | \mathbf{Y}^{k-1})} p(\chi_k | \mathbf{Y}^{k-1}), \tag{5.29}$$

$$p(\mathbf{x}_k | \chi_k, \mathbf{Y}^k) = \frac{p(\mathbf{y}_k, m_k | \mathbf{x}_k, \chi_k, \mathbf{Y}^{k-1})}{p(\mathbf{y}_k, m_k | \chi_k, \mathbf{Y}^{k-1})} p(\mathbf{x}_k | \chi_k, \mathbf{Y}^{k-1}), \tag{5.30}$$

we obtain (5.12):

$$p(\mathbf{x}_k, \chi_k | \mathbf{Y}^k) = p(\mathbf{x}_k | \chi_k, \mathbf{Y}^k) p(\chi_k | \mathbf{Y}^k).$$

The likelihoods needed to calculate the updated track state using (5.29) and (5.30) differ substantially for single-object and multi-object tracking, and are considered separately.

5.4.4 Single-object update

Measurement likelihoods

Each measurement may be a clutter measurement, or a measurement originated by the object. To recap, the likelihood of measurement $\mathbf{y}_k(i)$, given that it is a clutter measurement, is (5.7):

$$p_{\rho,k}(i) = \rho_k(i)/\bar{m}_k. \tag{5.31}$$

Let $\theta_k(0)$ denote the event that none of the selected measurements is the object detection, and let $\theta_k(i)$, $i > 0$, denote the event that measurement $\mathbf{y}_k(i)$ is the object detection. The event that measurement $\mathbf{y}_k(i)$ is the object detection can exist either conditional on the object existence $\{\theta_k(i)|\chi_k\}$, or as a subset of the object existence event $\{\theta_k(i), \chi_k\}$. The prior probabilities of event $\theta_k(0)$ are

$$p(\theta_k(0)|\mathbf{Y}^{k-1}) = 1 - P_D P_G p(\chi_k|\mathbf{Y}^{k-1}),$$

$$p(\theta_k(0)|\chi_k, \mathbf{Y}^{k-1}) = 1 - P_D P_G,$$

as the presence of the object measurement requires the object existence (in the first equation), object detection and object measurement selection simultaneously. Denote by $\bar{\theta}_k(0)$ the complement of event $\theta_k(0)$, i.e., the event that the measurement detection is present in $\mathbf{y}_k$. Then

$$p(\bar{\theta}_k(0)|\cdot) = 1 - p(\theta_k(0)|\cdot),$$

and, for $i > 0$,

$$p(\theta_k(i)|m_k, \bar{\theta}_k(0), \chi_k) = 1/m_k,$$

as the object measurement is equally likely to be any of the selected measurements.

The likelihood of measurement $\mathbf{y}_k(i)$, given that it is the object detection, is

$$p(\mathbf{y}_k(i)|\theta_k(i), \chi_k, \mathbf{Y}^{k-1}) = p_k(i), \tag{5.32}$$

defined by (5.26). Based on (5.6), the likelihood of measurement $\mathbf{y}_k(i)$, given that it is an object detection, and given the object state $\mathbf{x}_k$, is

$$f_k(i) \overset{\Delta}{=} p(\mathbf{y}_k(i)|\mathbf{x}_k, \theta_k(i), \chi_k, \mathbf{Y}^{k-1}) = \tfrac{1}{P_G}\mathcal{N}(\mathbf{y}_k(i); \mathbf{H}_k\mathbf{x}_k, \mathbf{R}_k). \tag{5.33}$$

The positions of the Poisson clutter measurements are mutually independent, and they are also independent from the position of the object measurement. Thus the likelihood of the measurement set $\mathbf{y}_k$ is the product of the likelihoods of the individual measurements $\mathbf{y}_k(i)$, $i = 1, \ldots, m_k$.

All conditions which result in all measurements in $\mathbf{y}_k$ to be the clutter measurements have the same likelihood of $\mathbf{y}_k$:

$$p(\mathbf{y}_k|\bar{\chi}_k) = p(\mathbf{y}_k|\theta_k(0), \chi_k) = p_{\rho,k} \overset{\Delta}{=} \prod_{i=1}^{m_k} p_{\rho,k}(i) = \frac{1}{\bar{m}_k^{m_k}} \prod_{i=1}^{m_k} \rho_k(i),$$

where $p_{\rho,k}$ is used as the appropriate shortcut. Given the same conditions, the number of selected clutter measurements equals m_k, with the prior probability

$$\mu_F(m_k) = \exp(-\bar{m}_k)\frac{\bar{m}_k^{m_k}}{m_k!}.$$

The likelihood of the measurement set $\mathbf{y}_k$, given that $\mathbf{y}_k(i)$ is the object detection, is

$$p(\mathbf{y}_k|\theta_k(i), \chi_k, \mathbf{Y}^{k-1}) = p_k(i) \prod_{\substack{j=1 \\ j \neq i}}^{m_k} P_{\rho,k}(j) = p_{\rho,k}\bar{m}_k \frac{p_k(i)}{\rho_k(i)}.$$

Also,

$$p(\mathbf{y}_k|\mathbf{x}_k, \theta_k(i), \chi_k, \mathbf{Y}^{k-1}) = f_k(i) \prod_{\substack{j=1 \\ j \neq i}}^{m_k} P_{\rho,k}(j) = p_{\rho,k}\bar{m}_k \frac{f_k(i)}{\rho_k(i)}.$$

In this case the number of selected clutter measurements equals $m_k - 1$, with prior probability

$$\mu_F(m_k - 1) = \exp(-\bar{m}_k)\frac{\bar{m}_k^{m_k-1}}{(m_k - 1)!} = \mu_F(m_k)\frac{m_k}{\bar{m}_k}.$$

Using the total probability theorem,

$$\begin{aligned} p(\mathbf{y}_k, m_k|\chi_k, \mathbf{Y}^{k-1}) &= p(\mathbf{y}_k, m_k|\theta_k(0), \chi_k, \mathbf{Y}^{k-1})p(\theta_k(0)|\chi_k, \mathbf{Y}^{k-1}) \\ &\quad + p(\mathbf{y}_k, m_k|\bar{\theta}_k(0), \chi_k, \mathbf{Y}^{k-1})p(\bar{\theta}_k(0)|\chi_k, \mathbf{Y}^{k-1}), \end{aligned}$$

with

$$\begin{aligned} p(\mathbf{y}_k, m_k|\theta_k(0), \chi_k, \mathbf{Y}^{k-1}) &= p(\mathbf{y}_k|m_k, \theta_k(0), \chi_k, \mathbf{Y}^{k-1})p(m_k|\theta_k(0), \chi_k, \mathbf{Y}^{k-1}) \\ &= p(\mathbf{y}_k|\theta_k(0), \chi_k, \mathbf{Y}^{k-1})\mu_F(m_k) \\ &= p_{\rho,k}\mu_F(m_k). \end{aligned}$$

Event $\bar{\theta}_k(0)$ is a union of mutually exclusive events $\theta_k(i)$, $i = 1, \ldots, m_k$, and

$$p(\mathbf{y}_k, m_k|\bar{\theta}_k(0), \chi_k, \mathbf{Y}^{k-1}) = \sum_{i=1}^{m_k} p(\mathbf{y}_k, m_k, \theta_k(i)|\bar{\theta}_k(0), \chi_k, \mathbf{Y}^{k-1}),$$

and, as $i > 0$,

$$\begin{aligned} p(\mathbf{y}_k, &m_k, \theta_k(i)|\bar{\theta}_k(0), \chi_k, \mathbf{Y}^{k-1}) \\ &= p(\mathbf{y}_k|m_k, \theta_k(i), \bar{\theta}_k(0), \chi_k, \mathbf{Y}^{k-1}) \\ &\quad \times p(\theta_k(i)|m_k, \bar{\theta}_k(0), \chi_k, \mathbf{Y}^{k-1})p(m_k|\bar{\theta}_k(0), \chi_k, \mathbf{Y}^{k-1}) \\ &= p(\mathbf{y}_k|\theta_k(i), \chi_k, \mathbf{Y}^{k-1})p(\theta_k(i)|m_k, \bar{\theta}_k(0), \chi_k)\mu_F(m_k - 1) \\ &= p_{\rho,k}\mu_F(m_k)p_k(i)/\rho_k(i). \end{aligned}$$

Bringing it all together,

$$p(\mathbf{y}_k, m_k | \chi_k, \mathbf{Y}^{k-1}) = p_{\rho,k} \mu_F(m_k) \left(1 - P_D P_G + P_D P_G \sum_{i=1}^{m_k} \frac{p_k(i)}{\rho_k(i)}\right).$$

Define the measurement likelihood ratio Λ_k by

$$\Lambda_k = \frac{p(\mathbf{y}_k, m_k | \chi_k, \mathbf{Y}^{k-1})}{p(\mathbf{y}_k, m_k | \bar{\chi}_k, \mathbf{Y}^{k-1})}$$

$$= 1 - P_D P_G + P_D P_G \sum_{i=1}^{m_k} \frac{p_k(i)}{\rho_k(i)}, \tag{5.34}$$

then

$$p(\mathbf{y}_k, m_k | \chi_k, \mathbf{Y}^{k-1}) = p_{\rho,k} \mu_F(m_k) \Lambda_k. \tag{5.35}$$

Following a similar path,

$$p(\mathbf{y}_k, m_k | \mathbf{x}_k, \chi_k, \mathbf{Y}^{k-1}) = p(\mathbf{y}_k, m_k | \theta_k(0), \chi_k, \mathbf{x}_k, \mathbf{Y}^{k-1}) p(\theta_k(0) | \chi_k, \mathbf{Y}^{k-1})$$
$$+ p(\mathbf{y}_k, m_k | \bar{\theta}_k(0), \chi_k, \mathbf{x}_k, \mathbf{Y}^{k-1}) p(\bar{\theta}_k(0) | \chi_k, \mathbf{Y}^{k-1}),$$

and noting that

$$p(\mathbf{y}_k, m_k | \theta_k(0), \chi_k, \mathbf{x}_k, \mathbf{Y}^{k-1}) = p(\mathbf{y}_k, m_k | \theta_k(0), \chi_k, \mathbf{Y}^{k-1}) = \rho_k \mu_F(m_k),$$

and

$$p(\mathbf{y}_k, m_k | \bar{\theta}_k(0), \chi_k, \mathbf{x}_k, \mathbf{Y}^{k-1}) = \sum_{i=1}^{m_k} p(\mathbf{y}_k, m_k, \theta_k(i) | \bar{\theta}_k(0), \chi_k, \mathbf{x}_k, \mathbf{Y}^{k-1}),$$

with

$$p(\mathbf{y}_k, m_k, \theta_k(i) | \bar{\theta}_k(0), \chi_k, \mathbf{x}_k, \mathbf{Y}^{k-1})$$
$$= p(\mathbf{y}_k | \mathbf{x}_k, m_k, \theta_k(i), \bar{\theta}_k(0), \chi_k, \mathbf{Y}^{k-1})$$
$$\times p(\theta_k(i) | m_k, \bar{\theta}_k(0), \chi_k, \mathbf{x}_k, \mathbf{Y}^{k-1}) p(m_k | \bar{\theta}_k(0), \chi_k, \mathbf{x}_k, \mathbf{Y}^{k-1})$$
$$= p(\mathbf{y}_k | \mathbf{x}_k, \theta_k(i), \chi_k, \mathbf{Y}^{k-1}) p(\theta_k(i) | m_k, \bar{\theta}_k(0), \chi_k) \mu_F(m_k - 1)$$
$$= p_{\rho,k} \mu_F(m_k) f_k(i) / \rho_k(i),$$

thus

$$p(\mathbf{y}_k, m_k | \mathbf{x}_k, \chi_k, \mathbf{Y}^{k-1}) = p_{\rho,k} \mu_F(m_k) \left(1 - P_D P_G + P_D P_G \sum_{i=1}^{m_k} \frac{f_k(i)}{\rho_k(i)}\right). \tag{5.36}$$

Finally, the unconditional measurement likelihood $p(\mathbf{y}_k, m_k|\mathbf{Y}^{k-1})$ equals

$$p(\mathbf{y}_k, m_k|\mathbf{Y}^{k-1}) = p(\mathbf{y}_k, m_k|\bar{\chi}_k, \mathbf{Y}^{k-1})p(\bar{\chi}_k|\mathbf{Y}^{k-1})$$
$$+ p(\mathbf{y}_k, m_k|\chi_k, \mathbf{Y}^{k-1})p(\chi_k|\mathbf{Y}^{k-1}),$$

where

$$p(\mathbf{y}_k, m_k|\bar{\chi}_k, \mathbf{Y}^{k-1}) = p(\mathbf{y}_k, m_k|\theta_k(0), \chi_k, \mathbf{Y}^{k-1}),$$
$$p(\bar{\chi}_k|\mathbf{Y}^{k-1}) = 1 - p(\chi_k|\mathbf{Y}^{k-1}),$$

and rearranging we obtain

$$p(\mathbf{y}_k, m_k|\mathbf{Y}^{k-1}) = p_{\rho,k}\mu_F(m_k)(1 - p(\chi_k|\mathbf{Y}^{k-1})(1 - \Lambda_k)). \quad (5.37)$$

Optimal update for $m_k = 0$

When $m_k = 0$, the only information that $\mathbf{y}_k$ provides is the number of selected measurements. Thus (5.29) becomes

$$p(\chi_k|m_k = 0, \mathbf{Y}^{k-1}) = \frac{p(m_k = 0|\chi_k, \mathbf{Y}^{k-1})}{p(m_k = 0|\mathbf{Y}^{k-1})}p(\chi_k|\mathbf{Y}^{k-1})$$

$$= \frac{p(\theta_k(0)|\chi_k, \mathbf{Y}^{k-1})}{p(\theta_k(0)|\mathbf{Y}^{k-1})}p(\chi_k|\mathbf{Y}^{k-1}),$$

which equals

$$p(\chi_k|m_k = 0, \mathbf{Y}^{k-1}) = \frac{(1 - P_D P_G)p(\chi_k|\mathbf{Y}^{k-1})}{1 - P_D P_G p(\chi_k|\mathbf{Y}^{k-1})}. \quad (5.38)$$

Equation (5.30) becomes

$$p(\mathbf{x}_k|\chi_k, m_k = 0, \mathbf{Y}^{k-1}) = \frac{p(m_k = 0|\mathbf{x}_k, \chi_k, \mathbf{Y}^{k-1})}{p(m_k = 0|\chi_k, \mathbf{Y}^{k-1})}p(\mathbf{x}_k|\chi_k, \mathbf{Y}^{k-1})$$
$$\approx p(\mathbf{x}_k|\chi_k, \mathbf{Y}^{k-1}), \quad (5.39)$$

where the approximation is more valid with higher P_G (Li and Bar-Shalom, 1997).

Optimal update for $m_k > 0$

Equation (5.29) is used with the measurement set likelihoods of (5.35) and (5.37) and becomes

$$p(\chi_k|\mathbf{Y}^k) = \frac{\Lambda_k p(\chi_k|\mathbf{Y}^{k-1})}{1 - (1 - \Lambda_k)p(\chi_k|\mathbf{Y}^{k-1})}. \quad (5.40)$$

Equation (5.30) is used with the measurement set likelihoods of (5.35) and (5.36):

$$p(\mathbf{x}_k|\chi_k, \mathbf{Y}^k) = \frac{1 - P_D P_G + P_D P_G \sum_{i=1}^{m_k} \frac{f_k(i)}{\rho_k(i)}}{\Lambda_k} p(\mathbf{x}_k|\chi_k, \mathbf{Y}^{k-1}).$$

Please note that

$$f_k(i)p(\mathbf{x}_k|\chi_k, \mathbf{Y}^{k-1}) = f_k(i)p(\mathbf{x}_k|\chi_k, \mathbf{Y}^{k-1})\frac{p_k(i)}{p_k(i)}$$
$$= p_k(i)p(\mathbf{x}_k|\chi_k, \mathbf{y}_k(i), \mathbf{Y}^{k-1}),$$

(5.41)

where the last equation is obtained by using the Bayes' equation, and $p(\mathbf{x}_k|\chi_k, \mathbf{y}_k(i), \mathbf{Y}^{k-1})$ is the object state pdf after measurement $\mathbf{y}_k(i)$ was used to update the propagated object trajectory state pdf.

Define

$$\beta_k(i) = \frac{1}{\Lambda_k} \cdot \begin{cases} 1 - P_D P_G, & i = 0, \\ P_D P_G p_k(i)/\rho_k(i), & i > 0. \end{cases}$$

(5.42)

Note that

$$p(\mathbf{x}_k|\theta_k(0), \chi_k, \mathbf{Y}^k) = p(\mathbf{x}_k|\chi_k, \mathbf{Y}^{k-1}),$$
$$p(\mathbf{x}_k|\theta_k(i), \chi_k, \mathbf{Y}^k) = p(\mathbf{x}_k|\chi_k, \mathbf{y}_k(i), \mathbf{Y}^{k-1}),$$

(5.43)

then

$$p(\mathbf{x}_k|\chi_k, \mathbf{Y}^k) = \sum_{i=0}^{m_k} \beta_k(i)p(\mathbf{x}_k|\theta_k(i), \chi_k, \mathbf{Y}^k).$$

(5.44)

It is easy to confirm that

$$\beta_k(i) = p(\theta_k(i)|\chi_k, \mathbf{Y}^k),$$

(5.45)

and (5.44) is just another form of the total probability theorem.

Equations (5.34), (5.40) and (5.42) are derived without making any assumptions on the shape of the prior object measurement pdf $p(\mathbf{y}|\chi_k, \mathbf{Y}^{k-1})$ (or, for that matter, on the shape of the clutter measurement density). These equations are therefore valid for $p(\mathbf{y}|\chi_k, \mathbf{Y}^{k-1})$ being a single Gaussian pdf, Gaussian mixture, or some non-linear pdf represented by a particle filter.

Noting that the sum operation on an empty set returns zero, it is easy to show that (5.34), (5.40) and (5.42) are valid in both the $m_k = 0$ and $m_k > 0$ cases.

5.4.5 Multi-object update

A multi-object situation arises when tracks share measurements, i.e., some measurements are selected (gated) by more than one track. Multi-object tracking differs from single-object tracking by the measurement origin assumption. When updating a track, each measurement has only one origin, which is one of:

- clutter;
- the object whose track is being updated;
- objects followed by other tracks.

As the event that a measurement is a track detection is mutually exclusive across tracks, and not mutually independent, the optimal Bayes' update operations have to consider all tracks in a cluster (defined below) simultaneously. The optimal Bayes' track update equations in Section 5.4.3 are still valid; however, a significant amount of work is required to adapt them to the multi-object situation. Otherwise, track prediction and likelihoods presented in Section 5.4.1 are still valid, as these operations are performed on a track-by-track basis, and are thus identical for both single- and multi-object tracking.

In this section, track superscripts are used (usually τ or η) to distinguish between tracks.

Clusters

In the (possible) presence of multiple objects, the measurement origins are no longer independent. If a measurement is a detection of one object, it changes the possibilities of the measurement outcomes for other (possible) objects, as they have "lost" one candidate for their detections. Thus, the allocation of measurements to possible objects (tracks) must be considered jointly or globally. A joint measurement allocation is an allocation of all measurements to all possible objects and to clutter.

The optimal Bayes approach to multi-object tracking consists of enumerating all (feasible, defined below) joint measurement allocations, and then evaluating them. The number of feasible joint measurement allocations grows combinatorially with the number of tracks and the number of selected measurements. For this computational reason tracks are first grouped into mutually exclusive clusters, as detailed below:

A cluster at time k is a set of tracks and the measurements these tracks select, defined by its complement: tracks not belonging to the cluster must not select any of the cluster measurements.

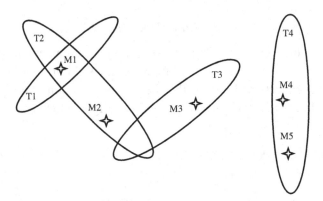

Figure 5.3 Clusters.

In other words, any track which selects cluster measurements also belongs to the cluster, and adds its set of selected measurements to the set of cluster measurements. Please note that this definition allows a track to belong to a cluster even if it does not share the selected measurements with other cluster tracks. A trivial cluster is a set of all existing tracks. For reasons of computational complexity minimization, clusters are always formed to contain the minimum number of tracks conforming to the definition. A track which does not share measurements with any other track can form a cluster.

Also note that clusters are formed anew at each scan. Two tracks which share a cluster in one scan may be in different clusters in subsequent scans, and vice versa. This may cause confusion, as the term cluster is used in a different manner in the MHT (Reid, 1979) literature.

Consider the situation depicted in Figure 5.3. There are four tracks, labeled T1 to T4, and five measurements, labeled M1 to M5. Each track is depicted by its selection gate border in the two-dimensional surveillance space. In this situation there are three clusters with a minimum number of tracks:

- Tracks T1 and T2 both select (share) measurement M1.
- Track T3, which does not share its selected measurement M3 with any other track, forms a cluster on its own. The fact that the selection gate of track T3 intersects the selection gate of track T2 does not count.
- Track T4 forms a cluster on its own.

Any union of two or more clusters is also a cluster.

The optimal Bayes multi-object tracking approach suffers from the curse of complexity. The number of possible allocations of all measurements to all tracks ("feasible joint measurement allocations" or "feasible joint events") grows combinatorially with the number of measurements and the number of tracks involved.

The number of unique assignments of m measurements to T tracks, assuming that all tracks select all measurements (including the possibilities of object non-detections), is (Mušicki and La Scala, 2008)

$$T! \sum_{t=0}^{T} \frac{1}{t!} \binom{m}{T-t} \geq (m+1)T!; \quad m \geq T \geq 1,$$

$$(5.46)$$

$$m! \sum_{i=0}^{m} \frac{1}{i!} \binom{T}{m-i} \geq (T+1)m!; \quad T \geq m \geq 1.$$

The number of feasible joint events depends only on the number of measurements, the number of tracks and the measurement selection outcomes. The number of feasible joint events does not depend on the number of track components.

The purpose of clustering is to minimize the number of operations by separating the tracks into groups or clusters. Assume $T = 6$ and $m = 6$ in (5.46). Then the total number of feasible joint assignments is 13 327. If we divide the set of $T = 6$ tracks into two smaller clusters of three tracks each, each cluster selecting three measurements, the number of feasible joint events for each cluster becomes 34, for a total of 68.

Joint multi-object tracking is applied to all tracks in a cluster simultaneously, while ignoring tracks which do not belong to the cluster. Thus, for the rest of Section 5.4.5 we assume that all tracks and measurements mentioned belong to the same cluster.

The cluster area is a union of the selection gate areas of all tracks which form the cluster, and it effectively becomes a new selection gate for all tracks belonging to the cluster. Denote by $\mathbf{y}_k^\tau$ the set of measurements selected by track τ at time k. Then the measurement set selected by the cluster becomes

$$\mathbf{y}_k = \bigcup_\tau \mathbf{y}_k^\tau.$$

Feasible joint data association events

A joint event is one allocation of all measurements to all tracks in the cluster, i.e., a joint event is a hypothesis of the origin of all measurements within the cluster. Mathematically, a joint event ε is defined by $\varepsilon \overset{\Delta}{=} \bigcap_{\tau=1}^{T} \theta_k^\tau(i(\tau, \varepsilon))$, with T denoting the number of tracks in the cluster, and $i(\tau, \varepsilon)$ denoting the index of the measurement allocated to track τ by the joint event ε.

Based on the infinite-resolution sensor and the point object assumptions, a feasible joint event is a joint event which satisfies the following:

- each track is assigned zero or one measurement; and
- each measurement is allocated to zero or one track.

Table 5.3 *Multi-object joint events.*

Joint event ε	Track T1	Track T2	Track T3
(1)	0	0	0
(2)	M1	0	0
(3)	M2	0	0
$\vdots$	$\vdots$	$\vdots$	$\vdots$
(21)	M2	M3	M4

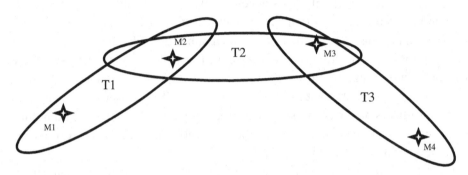

Figure 5.4 Multi-object situation.

Two feasible joint events are different if the assignment of at least one measurement is different. Different feasible joint events are mutually exclusive, and we enumerate all feasible joint events. Consider the one-cluster multi-object situation shown in Figure 5.4, with three tracks, T1–T3, selecting four measurements, M1–M4. The selection gate area limits of the tracks are shown as ellipsoids. In this one cluster, we can identify 21 feasible joint events, some of which are listed in Table 5.3. For example, feasible joint event (3) allocates measurement M2 to track T1, whereas no measurements are allocated to tracks T2 and T3. Measurements M1, M3 and M4 are declared clutter. Feasible joint event (21) allocates measurement M2, M3 and M4 to track T1, T2 and T3 respectively, whereas measurement M1 is declared clutter.

The feasible joint event space is "finely" divided into a set of mutually exclusive feasible joint events. It is also more "coarsely" divided into a set of mutually exclusive joint track detection events. A joint track detection event $\mathcal{D}$ divides the set of existing tracks into two sets:

- $T_0(\mathcal{D})$: set of tracks allocated no measurement;
- $T_1(\mathcal{D})$: set of tracks allocated one measurement.

The detection event of each track is a priori independent. The prior probabilities of no selected detection of each track $\eta \in T_0(\mathcal{D})$ equal

$$p(\theta_k^\eta(0)|\mathbf{Y}^{k-1}) = 1 - P_D^\eta P_G^\eta p(\chi_k^\eta|\mathbf{Y}^{k-1}),$$

$$p(\theta_k^\eta(0)|\chi_k^\eta, \mathbf{Y}^{k-1}) = p(\theta_k^\eta(0)|\chi_k^\eta, \mathbf{x}_k^\eta, \mathbf{Y}^{k-1}) = 1 - P_D^\eta P_G^\eta,$$

the prior probabilities of the selected detection of each track $\eta \in T_1(\mathcal{D})$ equal

$$p(\bar{\theta}_k^\eta(0)|\mathbf{Y}^{k-1}) = P_D^\eta P_G^\eta p(\chi_k^\eta|\mathbf{Y}^{k-1}),$$

$$p(\bar{\theta}_k^\eta(0)|\chi_k^\eta, \mathbf{Y}^{k-1}) = p(\bar{\theta}_k^\eta(0)|\chi_k^\eta, \mathbf{x}_k^\eta, \mathbf{Y}^{k-1}) = P_D^\eta P_G^\eta,$$

thus the prior probability of a joint detection event $\mathcal{D}$ is

$$p(\mathcal{D}|\mathbf{Y}^{k-1}) = \prod_{\eta \in T_0(\mathcal{D})} \left(1 - P_D^\eta P_G^\eta p(\chi_k^\eta|\mathbf{Y}^{k-1})\right) \prod_{\eta \in T_1(\mathcal{D})} P_D^\eta P_G^\eta p(\chi_k^\eta|\mathbf{Y}^{k-1}).$$

Following the same argument, only conditioned on the object τ existence,

$$p(\mathcal{D}|\chi_k^\tau, \mathbf{Y}^{k-1}) = p(\mathcal{D}|\chi_k^\tau, \mathbf{x}_k^\tau, \mathbf{Y}^{k-1})$$

$$= p(\mathcal{D}|\mathbf{Y}^{k-1}) \begin{cases} \dfrac{1 - P_D^\tau P_G^\tau}{1 - P_D^\tau P_G^\tau p(\chi_k^\tau|\mathbf{Y}^{k-1})}, & \tau \in T_0(\varepsilon), \\[4mm] \dfrac{1}{p(\chi_k^\tau|\mathbf{Y}^{k-1})}, & \tau \in T_1(\varepsilon). \end{cases} \qquad (5.47)$$

Each feasible joint event ε corresponds to one joint detection event $\mathcal{D}(\varepsilon)$. Each joint detection event $\mathcal{D}$ potentially corresponds to a number of feasible joint events. Thus, it is a "one-to-many" correspondence. In Table 5.3, joint events (2) and (3) correspond to the same detection event: they both imply detection of track T1, and non-detection of tracks T2 and T3. Number of joint detection events equals 2^T, with T denoting here the number of tracks. In this example, $2^3 = 8$ joint detection events partition the feasible joint event space of 21 feasible joint events.

As $\varepsilon \in \mathcal{D}(\varepsilon)$, then

$$\varepsilon = \varepsilon \bigcap \mathcal{D}(\varepsilon),$$

and

$$p(\cdot, \varepsilon|\mathbf{Y}^{k-1}) = p(\cdot, \varepsilon, \mathcal{D}(\varepsilon)|\mathbf{Y}^{k-1}) = p(\cdot, \varepsilon|\mathcal{D}(\varepsilon), \mathbf{Y}^{k-1}) p(\mathcal{D}(\varepsilon)|\mathbf{Y}^{k-1}).$$

Denote by $\varepsilon(\mathcal{D})$ the set of all feasible joint events corresponding to the joint detection event $\mathcal{D}$. The set of all joint events $\mathcal{D}$ encompasses the feasible joint event space, and then

$$\sum_{\mathcal{D}} \sum_{\varepsilon \in \varepsilon(\mathcal{D})} \cdot = \sum_{\varepsilon} \cdot.$$

Measurement likelihoods

Each track τ is updated using the Bayes' equations, i.e., Section 5.4.3 still applies. For convenience, we repeat (5.29) and (5.30), where superscript τ denotes track τ:

$$p\left(\chi_k^\tau | \mathbf{Y}^k\right) = \frac{p(\mathbf{y}_k, m_k | \chi_k^\tau, \mathbf{Y}^{k-1})}{p(\mathbf{y}_k, m_k | \mathbf{Y}^{k-1})} p\left(\chi_k^\tau | \mathbf{Y}^{k-1}\right), \qquad (5.48)$$

$$p\left(\mathbf{x}_k^\tau | \chi_k^\tau, \mathbf{Y}^k\right) = \frac{p(\mathbf{y}_k, m_k | \mathbf{x}_k^\tau, \chi_k^\tau, \mathbf{Y}^{k-1})}{p(\mathbf{y}_k, m_k | \chi_k^\tau, \mathbf{Y}^{k-1})} p\left(\mathbf{x}_k^\tau | \chi_k^\tau, \mathbf{Y}^{k-1}\right). \qquad (5.49)$$

Equation (5.12) is also still valid:

$$p\left(\mathbf{x}_k^\tau, \chi_k^\tau | \mathbf{Y}^k\right) = p\left(\mathbf{x}_k^\tau | \chi_k^\tau, \mathbf{Y}^k\right) p\left(\chi_k^\tau | \mathbf{Y}^k\right).$$

In multi-object situations, we need to consider the exhaustive set of mutually exclusive feasible joint events ε and summing out ε or using the total probability theorem

$$p(\mathbf{y}_k, m_k | \cdot, \mathbf{Y}^{k-1}) = \sum_\varepsilon p(\mathbf{y}_k, m_k, \varepsilon, \mathcal{D}(\varepsilon) | \cdot, \mathbf{Y}^{k-1})$$

$$= \sum_\varepsilon p(\mathbf{y}_k | m_k, \varepsilon, \mathcal{D}(\varepsilon), \cdot, \mathbf{Y}^{k-1}) \, p(\varepsilon | m_k, \mathcal{D}(\varepsilon), \cdot, \mathbf{Y}^{k-1}) \ (5.50)$$

$$\times p(m_k | \mathcal{D}(\varepsilon), \cdot, \mathbf{Y}^{k-1}) \, p(\mathcal{D}(\varepsilon) | \cdot, \mathbf{Y}^{k-1}).$$

For the joint event ε define:

- $M_0(\varepsilon)$: set of measurements not allocated to any track;
- $T_0(\varepsilon) = T_0(\mathcal{D}(\varepsilon))$: set with cardinality $t_0(\varepsilon)$ of tracks allocated no measurement;
- $T_1(\varepsilon) = T_1(\mathcal{D}(\varepsilon))$: set with cardinality $t_1(\varepsilon)$ of tracks allocated one measurement;
- $i(\tau, \varepsilon)$: index of measurement allocated to track τ.

For joint event (2) listed in Table 5.3, $T_0(2) = \{T2, T3\}$, $T_1(2) = \{T1\}$, and $i(T1, 2) = 1$, $i(T2, 2) = 0$, $i(T3, 2) = 0$. For joint event (21) listed in Table 5.3, $T_0(21) = \{ \ \}$ (empty set), $T_1(21) = \{T1, T2, T3\}$, and $i(T1, 21) = 2$, $i(T2, 21) = 3$, $i(T3, 21) = 4$.

Denote by $t_c(\varepsilon)$ the number of measurements assumed to be clutter by joint event $\mathcal{D}(\varepsilon)$,

$$t_c(\varepsilon) + t_1(\varepsilon) = m_k.$$

The number of possible allocations of m_k measurements among $t_1(\varepsilon)$ tracks equals

$$V_{t_1(\varepsilon)}^{m_k} = t_1(\varepsilon)! C_{t_1(\varepsilon)}^{m_k} = \frac{m_k!}{t_c(\varepsilon)!},$$

where each allocation is equally (a priori) probable and therefore

$$p(\varepsilon|m_k, \mathcal{D}(\varepsilon), \mathbf{Y}^{k-1}) = \frac{t_c(\varepsilon)!}{m_k!}. \tag{5.51}$$

The prior probability of $t_c(\varepsilon)$ selected clutter measurements equals

$$p(m_k|\mathcal{D}(\varepsilon), \mathbf{Y}^{k-1}) = \mu_F(t_c(\varepsilon)) = \exp(-\bar{m}_k)\frac{\bar{m}_k^{t_c(\varepsilon)}}{t_c(\varepsilon)!}$$

$$= \mu_F(m_k)\bar{m}_k^{-t_1(\varepsilon)}\frac{m_k!}{t_c(\varepsilon)!}, \tag{5.52}$$

which does not depend on the object existence event, given the joint detection event, thus

$$p(m_k|\mathcal{D}(\varepsilon), \chi_k^\tau, \mathbf{Y}^{k-1}) = p(m_k|\mathcal{D}(\varepsilon), \chi_k^\tau, \mathbf{x}_k^\tau, \mathbf{Y}^{k-1}) = p(m_k|\mathcal{D}(\varepsilon), \mathbf{Y}^{k-1}). \tag{5.53}$$

Following the derivation of Section 5.4.4, given the measurement allocation, the likelihoods of individual measurements $\mathbf{y}_k(i)$ are mutually independent and the likelihood of measurement set $\mathbf{y}_k$ is the product of the individual likelihoods:

$$p(\mathbf{y}_k|\varepsilon, m_k, \mathcal{D}(\varepsilon), \mathbf{Y}^{k-1}) = \prod_{\mathbf{y}_k(i)\in M_0} \frac{\rho_k(i)}{\bar{m}_k} \prod_{\eta\in T_1} p_k^\eta(i(\eta, \varepsilon))$$

$$= p_{\rho,k}\bar{m}_k^{-t_1(\varepsilon)} \prod_{\eta\in T_1} \frac{p_k^\eta(i(\eta, \varepsilon))}{\rho_k(i(\eta, \varepsilon))}. \tag{5.54}$$

The object existence event, given ε, does not change the value of the likelihoods, although knowledge of object τ state may, and

$$p(\mathbf{y}_k|\varepsilon, m_k, \mathcal{D}(\varepsilon), \chi_k^\tau, \mathbf{Y}^{k-1}) = p(\mathbf{y}_k|\varepsilon, m_k, \mathcal{D}(\varepsilon), \mathbf{Y}^{k-1}), \tag{5.55}$$

$$p(\mathbf{y}_k|\varepsilon, m_k, \mathcal{D}(\varepsilon), \chi_k^\tau, \mathbf{x}_k^\tau, \mathbf{Y}^{k-1}) = p(\mathbf{y}_k|\varepsilon, m_k, \mathcal{D}(\varepsilon), \mathbf{Y}^{k-1}) \begin{cases} 1, & \tau \in T_0(\varepsilon), \\ \frac{f_k^\tau(i)}{p_k^\tau(i)}, & \tau \in T_1(\varepsilon). \end{cases} \tag{5.56}$$

Putting together (5.50), (5.51), (5.52) and (5.54) we obtain

$$p(\mathbf{y}_k, m_k|\mathbf{Y}^{k-1}) = p_{\rho,k}\mu_F(m_k) \sum_\varepsilon \prod_{\eta\in T_0(\varepsilon)} \left(1 - P_D^\eta P_G^\eta P\{c_k, \sigma\eta|\mathbf{Y}^{k-1}\}\right)$$

$$\times \prod_{\eta\in T_1(\varepsilon)} P_D^\eta P_G^\eta P\{\chi_k^\eta|\mathbf{Y}^{k-1}\}\frac{p_k^\eta(i(\eta, \varepsilon))}{\rho_k(i(\eta, \varepsilon))}.$$

For convenience, we define shortcut $A(\varepsilon)$ as

$$A(\varepsilon) = \prod_{\eta \in T_0(\varepsilon)} (1 - P_D^\eta P_G^\eta P\{c_k, \sigma \eta | \mathbf{Y}^{k-1}\}) \prod_{\eta \in T_1(\varepsilon)} P_D^\eta P_G^\eta P\{\chi_k^\eta | \mathbf{Y}^{k-1}\} \frac{p_k^\eta(i(\eta, \varepsilon))}{\rho_k(i(\eta, \varepsilon))},$$

thus

$$p(\mathbf{y}_k, m_k | \mathbf{Y}^{k-1}) = p_{\rho,k} \mu_F(m_k) \sum_\varepsilon A(\varepsilon). \tag{5.57}$$

Comparing (5.50) and (5.57),

$$p(\mathbf{y}_k, m_k, \varepsilon | \mathbf{Y}^{k-1}) = p_{\rho,k} \mu_F(m_k) A(\varepsilon),$$

the a posteriori probability of the feasible joint event ε equals

$$\begin{aligned} p(\varepsilon | \mathbf{Y}^k) &= \frac{p(\mathbf{y}_k, m_k, \varepsilon | \mathbf{Y}^{k-1})}{p(\mathbf{y}_k, m_k | \mathbf{Y}^{k-1})} \\ &= \frac{A(\varepsilon)}{\sum_\epsilon A(\epsilon)}. \end{aligned} \tag{5.58}$$

Denote by $\Xi(\tau, i)$ the set of feasible joint events which allocate measurement i to track τ. Set $\Xi(\tau, 0)$ contains all feasible joint events ε for which $\tau \in T_0(\varepsilon)$. For $i \neq j$ sets $\Xi(\tau, i)$ and $\Xi(\tau, j)$ are mutually disjoint, and union of all sets $\Xi(\tau, i)$, $i = 0, \ldots, m_k$, is the set of all feasible joint events.

Putting together (5.50), (5.47), (5.53) and (5.55) we obtain

$$\begin{aligned} p(\mathbf{y}_k, m_k | \chi_k^\tau, \mathbf{Y}^{k-1}) &= p_{\rho,k} \mu_F(m_k) \\ &\times \left(\frac{1 - P_D^\tau P_G^\tau}{1 - P_D^\tau P_G^\tau p(\chi_k^\tau | \mathbf{Y}^{k-1})} \sum_{\varepsilon \in \Xi(\tau,0)} A(\varepsilon) + \frac{1}{p(\chi_k^\tau | \mathbf{Y}^{k-1})} \sum_{i=1}^{m_k} \sum_{\varepsilon \in \Xi(\tau,i)} A(\varepsilon) \right), \end{aligned} \tag{5.59}$$

and using (5.50), (5.47), (5.53) and (5.56) we obtain

$$\begin{aligned} p(\mathbf{y}_k, m_k | \chi_k^\tau, \mathbf{x}_k^\tau, \mathbf{Y}^{k-1}) &= p_{\rho,k} \mu_F(m_k) \left(\frac{1 - P_D^\tau P_G^\tau}{1 - P_D^\tau P_G^\tau p(\chi_k^\tau | \mathbf{Y}^{k-1})} \sum_{\varepsilon \in \Xi(\tau,0)} A(\varepsilon) \right. \\ &\left. + \frac{1}{p(\chi_k^\tau | \mathbf{Y}^{k-1})} \sum_{i=1}^{m_k} \frac{f_k^\tau(i)}{p_k^\tau(i)} \sum_{\varepsilon \in \Xi(\tau,i)} A(\varepsilon) \right). \end{aligned} \tag{5.60}$$

Placing the likelihood values defined by (5.57) and (5.59) into (5.48), the a posteriori probability of object existence for track τ is

$$p(\chi_k^\tau|\mathbf{Y}^k) = \frac{\frac{(1-P_D^\tau P_G^\tau)p(\chi_k^\tau|\mathbf{Y}^{k-1})}{1-P_D^\tau P_G^\tau p(\chi_k^\tau|\mathbf{Y}^{k-1})}\sum_{\varepsilon\in\Xi(\tau,0)} A(\varepsilon) + \sum_{i=1}^{m_k}\sum_{\varepsilon\in\Xi(\tau,i)} A(\varepsilon)}{\sum_\varepsilon A(\varepsilon)},$$

which, after inclusion of (5.58), becomes

$$p(\chi_k^\tau|\mathbf{Y}^k) = \frac{(1-P_D^\tau P_G^\tau)p(\chi_k^\tau|\mathbf{Y}^{k-1})}{1-P_D^\tau P_G^\tau p(\chi_k^\tau|\mathbf{Y}^{k-1})}\sum_{\varepsilon\in\Xi(\tau,0)} p(\varepsilon|\mathbf{Y}^k) + \sum_{i=1}^{m_k}\sum_{\varepsilon\in\Xi(\tau,i)} p(\varepsilon|\mathbf{Y}^k).$$

$$(5.61)$$

Remark 5.1 *Event $\theta_k^\tau(0)$ equals the union of joint events which are elements of $\Xi(\tau,0)$:*

$$\theta_k^\tau(0) = \bigcup_{\varepsilon\in\Xi(\tau,0)} \varepsilon \qquad (5.62)$$

Proof Each feasible joint event ε which is an element of set $\Xi(\tau,0)$ implies that τ is allocated no measurements, i.e., implies event $\theta_k^\tau(0)$. Thus

$$\varepsilon \in \Xi(\tau,0) \Rightarrow \theta_k^\tau(0),$$

or

$$\theta_k^\tau(0) \subseteq \bigcup_{\varepsilon\in\Xi(\tau,0)} \varepsilon. \qquad (5.63)$$

On the other hand, each joint event that is not an element of $\Xi(\tau,0)$ implies the event that track τ is allocated some measurement, meaning that $\theta_k^\tau(0)$ is not true, or

$$\varepsilon \notin \Xi(\tau,0) \Rightarrow \bar\theta_k^\tau(0),$$

which implies that

$$\theta_k^\tau(0) \Rightarrow \varepsilon \in \Xi(\tau,0),$$

or

$$\bigcup_{\varepsilon\in\Xi(\tau,0)} \varepsilon \subseteq \theta_k^\tau(0). \qquad (5.64)$$

From (5.63) and (5.64),

$$\bigcup_{\varepsilon\in\Xi(\tau,0)} \varepsilon = \theta_k^\tau(0).$$

$$\square$$

Remark 5.2 *For $1 \le i \le m_k$ event $< \theta_k^\tau(i), \chi_k^\tau >$ equals the union of joint events which are elements of $\Xi(\tau, i)$:*

$$< \theta_k^\tau(i), \chi_k^\tau > = \bigcup_{\varepsilon \in \Xi(\tau,i)} \varepsilon. \tag{5.65}$$

Proof The proof closely follows the proof of Remark 5.1. $\square$

As a consequence of Remark 5.1 and the mutual exclusiveness of feasible joint events,

$$p(\theta_k^\tau(0)|\mathbf{Y}^k) = \sum_{\varepsilon \in \Xi(\tau,0)} p(\varepsilon|\mathbf{Y}^k). \tag{5.66}$$

As a consequence of Remark 5.2 and the mutual exclusiveness of feasible joint events, for $i > 0$,

$$p(\theta_k^\tau(i), \chi_k^\tau|\mathbf{Y}^k) = \sum_{\varepsilon \in \Xi(\tau,i)} p(\varepsilon|\mathbf{Y}^k). \tag{5.67}$$

Including these results in (5.61),

$$p(\chi_k^\tau|\mathbf{Y}^k) = \frac{(1 - P_D^\tau P_G^\tau)p(\chi_k^\tau|\mathbf{Y}^{k-1})}{1 - P_D^\tau P_G^\tau p(\chi_k^\tau|\mathbf{Y}^{k-1})} p(\theta_k^\tau(0)|\mathbf{Y}^k) + \sum_{i=1}^{m_k} p(\theta_k^\tau(i), \chi_k^\tau|\mathbf{Y}^k).$$

Following a derivation similar to that of (5.38) (replacing the event $m_k = 0$ by the event $\theta_k^\tau(0)$, and $\mathbf{Y}^{k-1}$ by $\mathbf{Y}^k$), it is easy to show that

$$p(\theta_k^\tau(0), \chi_k^\tau|\mathbf{Y}^k) = \frac{(1 - P_D^\tau P_G^\tau)p(\chi_k^\tau|\mathbf{Y}^{k-1})}{1 - P_D^\tau P_G^\tau p(\chi_k^\tau|\mathbf{Y}^{k-1})} p(\theta_k^\tau(0)|\mathbf{Y}^k), \tag{5.68}$$

and the updated probability of target existence becomes

$$p(\chi_k^\tau|\mathbf{Y}^k) = \sum_{i=0}^{m_k} p(\theta_k^\tau(i), \chi_k^\tau|\mathbf{Y}^k). \tag{5.69}$$

To obtain the posterior object trajectory state pdf, we place the likelihood values defined by (5.59) and (5.60) into (5.49), and use (5.66) and (5.67):

$$p(\mathbf{x}_k^\tau|\chi_k^\tau, \mathbf{Y}^k)$$
$$= p(\mathbf{x}_k^\tau|\chi_k^\tau, \mathbf{Y}^{k-1})$$
$$\times \frac{\frac{1 - P_D^\tau P_G^\tau}{1 - P_D^\tau P_G^\tau p(\chi_k^\tau|\mathbf{Y}^{k-1})} p(\theta_k^\tau(0)|\mathbf{Y}^k) + \frac{1}{p(\chi_k^\tau|\mathbf{Y}^{k-1})} \sum_{i=1}^{m_k} \frac{f_k^\tau(i)}{p_k^\tau(i)} p(\theta_k^\tau(i), \chi_k^\tau|\mathbf{Y}^k)}{\frac{1 - P_D^\tau P_G^\tau}{1 - P_D^\tau P_G^\tau p(\chi_k^\tau|\mathbf{Y}^{k-1})} p(\theta_k^\tau(0)|\mathbf{Y}^k) + \frac{1}{p(\chi_k^\tau|\mathbf{Y}^{k-1})} \sum_{i=1}^{m_k} p(\theta_k^\tau(i), \chi_k^\tau|\mathbf{Y}^k)}$$

which, using (5.68) and (5.69), becomes

$$p(\mathbf{x}_k^\tau | \chi_k^\tau, \mathbf{Y}^k) = \frac{p(\theta_k^\tau(0), \chi_k^\tau | \mathbf{Y}^k) + \sum_{i=1}^{m_k} \frac{f_k^\tau(i)}{p_k^\tau(i)} p(\theta_k^\tau(i), \chi_k^\tau | \mathbf{Y}^k)}{p(\chi_k^\tau | \mathbf{Y}^k)} p(\mathbf{x}_k^\tau | \chi_k^\tau, \mathbf{Y}^{k-1}).$$

As in the case of the optimal single-object trajectory update, we define

$$\beta_k^\tau(i) = p(\theta_k^\tau(i) | \chi_k^\tau, \mathbf{Y}^k) = \frac{p(\theta_k^\tau(i), \chi_k^\tau | \mathbf{Y}^k)}{p(\chi_k^\tau | \mathbf{Y}^k)}, \qquad (5.70)$$

and note that, as in (5.41),

$$f_k^\tau(i) p(\mathbf{x}_k^\tau | \chi_k^\tau, \mathbf{Y}^{k-1}) = p_k^\tau(i) p(\mathbf{x}_k^\tau | \chi_k^\tau, \mathbf{y}_k(i), \mathbf{Y}^{k-1}),$$

where $p(\mathbf{x}_k^\tau | \chi_k^\tau, \mathbf{y}_k(i), \mathbf{Y}^{k-1})$ is the object trajectory state pdf after measurement $\mathbf{y}_k(i)$ was used to update the propagated object trajectory state pdf.

We can now immediately write

$$p(\mathbf{x}_k^\tau | \chi_k^\tau, \mathbf{Y}^k) = \sum_{i=0}^{m_k} \beta_k^\tau(i) p(\mathbf{x}_k^\tau | \theta_k^\tau(i), \chi_k^\tau, \mathbf{Y}^k), \qquad (5.71)$$

which is identical to the single-object trajectory state pdf equation (5.44). Once the appropriate conditional data association probabilities $\beta_k^\tau(i)$ are calculated, both single- and multi-object trajectory updates proceed in an identical manner.

5.4.6 Track trajectory update

This procedure is common to both single- and multi-target tracking and is performed on each track separately. Here we omit the track superscript. Both single-object update (Section 5.4.4) and multi-object update (Section 5.4.5) define the conditional a posteriori data association probabilities

$$\beta_k(i) = p(\theta_k(i) | \chi_k, \mathbf{Y}^k) = \frac{p(\theta_k(i), \chi_k | \mathbf{Y}^k)}{p(\chi_k | \mathbf{Y}^k)},$$

and calculate and use them to update the trajectory state pdf as (5.44), (5.71),

$$p(\mathbf{x}_k | \chi_k, \mathbf{Y}^k) = \sum_{i=0}^{m_k} \beta_k(i) p(\mathbf{x}_k | \theta_k(i), \chi_k, \mathbf{Y}^k). \qquad (5.72)$$

Given that measurement $\mathbf{y}_k(i)$ is the object detection,

$$p(\mathbf{x}_k | \theta_k(i), \chi_k, \mathbf{Y}^k) = p(\mathbf{x}_k | \chi_k, \mathbf{y}_k(i), \mathbf{Y}^{k-1}),$$

and given that no measurement in $\mathbf{y}_k$ is the object detection, the updated pdf is still the propagated pdf, $i = 0$,

$$p(\mathbf{x}_k | \theta_k(0), \chi_k, \mathbf{Y}^k) = p(\mathbf{x}_k | \chi_k, \mathbf{Y}^{k-1}).$$

The propagated object trajectory pdf at time k is

$$p(\mathbf{x}_k | \chi_k, \mathbf{Y}^{k-1}) = \sum_{c_{k-1}=1}^{C_{k-1}} p(c_{k-1}) \cdot \sum_{\sigma=1}^{M} \mu_{k|k-1}(c_{k-1}, \sigma) p(\mathbf{x}_k | c_{k-1}, \sigma, \chi_k, \mathbf{Y}^{k-1}),$$

with

$$p(\mathbf{x}_k | c_{k-1}, \sigma, \chi_k, \mathbf{Y}^{k-1}) = \mathcal{N}(\mathbf{x}_k; \hat{\mathbf{x}}_{k|k-1}(c_{k-1}, \sigma), \mathbf{P}_{k|k-1}(c_{k-1}, \sigma)).$$

Element $i = 0$ of the summation (5.72) is

$$\beta_k(0) p(\mathbf{x}_k | \chi_k, \mathbf{Y}^{k-1}) = \beta_k(0) \sum_{c_{k-1}=1}^{C_{k-1}} p(c_{k-1})$$

$$\times \sum_{\sigma=1}^{M} \mu_{k|k-1}(c_{k-1}, \sigma) p(\mathbf{x}_k | c_{k-1}, \sigma, \chi_k, \mathbf{Y}^{k-1}).$$

Element i, $i > 0$, of the summation (5.72) is

$$\beta_k(i) p(\mathbf{x}_k | \chi_k, \theta_k(i), \mathbf{Y}^k) = \beta_k(i) p(\mathbf{x}_k | \chi_k, \mathbf{y}_k(i), \mathbf{Y}^{k-1}).$$

The object trajectory state pdf $p(\mathbf{x}_k | \chi_k, \mathbf{y}_k(i), \mathbf{Y}^{k-1})$ is obtained by applying measurement $\mathbf{y}_k(i)$ to the propagated object trajectory state pdf, and using the Bayes' formula,

$$p(\mathbf{x}_k | \chi_k, \mathbf{y}_k(i), \mathbf{Y}^{k-1})$$

$$= \frac{p(\mathbf{y}_k(i) | \chi_k, \mathbf{x}_k, \mathbf{Y}^{k-1}) p(\mathbf{x}_k | \chi_k, \mathbf{Y}^{k-1})}{p(\mathbf{y}_k(i) | \chi_k, \mathbf{Y}^{k-1})}$$

$$= \frac{p(\mathbf{y}_k(i) | \mathbf{x}_k)}{p_k(i)} \sum_{c_{k-1}=1}^{C_{k-1}} \sum_{\sigma=1}^{M} p(c_{k-1}) \mu_{k|k-1}(c_{k-1}, \sigma) p(\mathbf{x}_k | c_{k-1}, \sigma, \chi_k, \mathbf{Y}^{k-1}).$$

Further applying the Bayes' formula we obtain

$$p(\mathbf{y}_k(i) | \mathbf{x}_k) p(\mathbf{x}_k | c_{k-1}, \sigma, \chi_k, \mathbf{Y}^{k-1})$$

$$= p(\mathbf{y}_k(i) | \mathbf{x}_k) p(\mathbf{x}_k | c_{k-1}, \sigma, \chi_k, \mathbf{Y}^{k-1}) \frac{p(\mathbf{y}_k(i) | c_{k-1}, \sigma, \chi_k, \mathbf{Y}^{k-1})}{p(\mathbf{y}_k(i) | c_{k-1}, \sigma, \chi_k, \mathbf{Y}^{k-1})}$$

$$= p_k(i, c_{k-1}, \sigma) p(\mathbf{x}_k | c_{k-1}, \sigma, \chi_k, \mathbf{y}_k(i), \mathbf{Y}^{k-1}).$$

Thus, for $i > 0$,

$$\beta_k(i) p(\mathbf{x}_k | \chi_k, \theta_k(i), \mathbf{Y}^k)$$

$$= \beta_k(i) \sum_{c_{k-1}=1}^{C_{k-1}} p(c_{k-1})$$

$$\times \sum_{\sigma=1}^{M} \mu_{k|k-1}(c_{k-1}, \sigma) \frac{p_k(i, c_{k-1}, \sigma)}{p_k(i)} p(\mathbf{x}_k | \chi_k, c_{k-1}, \sigma, \theta_k(i), \mathbf{Y}^k),$$

where $p(\mathbf{x}_k | \chi_k, c_{k-1}, \sigma, \theta_k(i), \mathbf{Y}^k)$ is obtained by applying measurement $\mathbf{y}_k(i)$ to the propagated trajectory state pdf $p(\mathbf{x}_k | \chi_k, c_{k-1}, \sigma, \mathbf{Y}^{k-1})$ (Kalman filter update).

Each pair of a priori component c_{k-1} and selected measurement i, $i \geq 0$, at time k creates a new a posteriori tentative component denoted by $\tilde{c}_k \overset{\Delta}{=} \{c_{k-1}, \theta_k(i)\}$. Equation (5.72) may be rewritten as

$$p(\mathbf{x}_k | \chi_k, \mathbf{Y}^k) = \sum_{\tilde{c}_k=1}^{\tilde{C}_k} p(\tilde{c}_k) \cdot \sum_{\sigma=1}^{M} \mu_{k|k}(\tilde{c}_k, \sigma) p(\mathbf{x}_k | \tilde{c}_k, \sigma, \chi_k, \mathbf{Y}^k),$$

with

$$p(\mathbf{x}_k | \tilde{c}_k, \sigma, \chi_k, \mathbf{Y}^k) = \mathcal{N}\left(\hat{\mathbf{x}}_{k|k}(\tilde{c}_k, \sigma), \mathbf{P}_{k|k}(\tilde{c}_k, \sigma)\right).$$

Thus, for $\tilde{c}_k = \{c_{k-1}, \theta_k(0)\}$,

$$p(\tilde{c}_k) \mu_{k|k}(\tilde{c}_k, \sigma) = \beta_k(0) p(c_{k-1}) \mu_{k|k-1}(c_{k-1}, \sigma),$$

and

$$\hat{\mathbf{x}}_{k|k}(\tilde{c}_k, \sigma) = \hat{\mathbf{x}}_{k|k-1}(c_{k-1}, \sigma),$$

$$\mathbf{P}_{k|k}(\tilde{c}_k, \sigma) = \mathbf{P}_{k|k-1}(c_{k-1}, \sigma).$$

Given that $\tilde{c}_k = \{c_{k-1}, i\}$, $i > 0$,

$$p(\tilde{c}_k | \chi_k, \mathbf{Y}^k) \mu_{k|k}(\tilde{c}_k, \sigma) = \beta_k(i) p(c_{k-1}) \mu_{k|k-1}(c_{k-1}, \sigma) \frac{p_k(i, c_{k-1}, \sigma)}{p_k(i)},$$

and

$$p(\mathbf{x}_k | \tilde{c}_k, \sigma, \chi_k, \mathbf{Y}^k) \overset{\Delta}{=} p(\mathbf{x}_k | c_{k-1}, \sigma, \chi_k, \mathbf{y}_k(i), \mathbf{Y}^{k-1})$$

$$= \mathcal{N}(\mathbf{x}_k; \hat{\mathbf{x}}_{k|k}(\tilde{c}_k, \sigma), \mathbf{P}_{k|k}(\tilde{c}_k, \sigma)).$$

Values of $\hat{\mathbf{x}}_{k|k}(\tilde{c}_k, \sigma)$ and $\mathbf{P}_{k|k}(\tilde{c}_k, \sigma)$ are obtained by applying measurement $\mathbf{y}_k(i)$ to the Kalman filter update with predicted mean $\hat{\mathbf{x}}_{k|k-1}(c_{k-1}, \sigma)$ and covariance

$\mathbf{P}_{k|k-1}(c_{k-1}, \sigma)$:

$$[\hat{\mathbf{x}}_{k|k}(\tilde{c}_k, \sigma), \mathbf{P}_{k|k}(\tilde{c}_k, \sigma)] = \text{KF}_{\text{E}}[\mathbf{y}_k(i), \hat{\mathbf{x}}_{k|k-1}(c_{k-1}, \sigma), \mathbf{P}_{k|k-1}(c_{k-1}, \sigma), \mathbf{H}, \mathbf{R}],$$

where KF$_{\text{E}}$ is the Kalman filter estimate, as defined in Section 2.2.1.

Therefore, for each component $\tilde{c}_k = \{c_{k-1}, i \geq 0\}$,

$$p(\tilde{c}_k | \chi_k, \mathbf{Y}^k) \mu_{k|k}(\tilde{c}_k, \sigma)$$

$$= \beta_k(i) p(c_{k-1}) \mu_{k|k-1}(c_{k-1}, \sigma) \begin{cases} 1, & i = 0, \\ \dfrac{p_k(i, c_{k-1}, \sigma)}{p_k(i)}, & i > 0. \end{cases}$$

By summing out the probability of the trajectory model, we obtain the a posteriori probability of a new component:

$$p(\tilde{c}_k | \chi_k, \mathbf{Y}^k) = \sum_{\sigma=1}^{M} p(\tilde{c}_k | \chi_k, \mathbf{Y}^k) \mu_{k|k}(\tilde{c}_k, \sigma)$$

$$= \sum_{\sigma=1}^{M} \beta_k(i) p(c_{k-1}) \mu_{k|k-1}(c_{k-1}, \sigma) \begin{cases} 1, & i = 0 \\ \dfrac{p_k(i, c_{k-1}, \sigma)}{p_k(i)}, & i > 0 \end{cases},$$

$$= \beta_k(i) p(c_{k-1}) \begin{cases} 1, & i = 0 \\ \dfrac{p_k(i, c_{k-1})}{p_k(i)}, & i > 0 \end{cases},$$

and then directly obtain the a posteriori probability of the trajectory model, given the a posteriori component

$$\mu_{k|k}(\tilde{c}_k, \sigma) \stackrel{\triangle}{=} p(M_k = \sigma | \tilde{c}_k, \chi_k, \mathbf{Y}^k)$$

$$= \frac{p(\tilde{c}_k | \chi_k, \mathbf{Y}^k) \mu_{k|k}(\tilde{c}_k, \sigma)}{p(\tilde{c}_k | \chi_k, \mathbf{Y}^k)}$$

$$= \mu_{k|k-1}(c_{k-1}, \sigma) \begin{cases} 1, & i = 0, \\ \dfrac{p_k(i, c_{k-1}, \sigma)}{p_k(i, c_{k-1})}, & i > 0. \end{cases}$$

The a posteriori object trajectory state pdf for each tentative component $\tilde{c}_k$ is

$$p(\mathbf{x}_k | \tilde{c}_k, \chi_k, \mathbf{Y}^k) = \sum_{\sigma=1}^{M} \mu_{k|k}(\tilde{c}_k, \sigma) p(\mathbf{x}_k | \tilde{c}_k, \sigma, \chi_k, \mathbf{Y}^k),$$

where each object trajectory state pdf conditioned on the tentative a posteriori component $\tilde{c}_k$ and object trajectory model σ is a Gaussian pdf,

$$p(\mathbf{x}_k | \tilde{c}_k, \sigma, \chi_k, \mathbf{Y}^k) = \mathcal{N}(\mathbf{x}_k; \hat{\mathbf{x}}_{k|k}(\tilde{c}_k, \sigma), \mathbf{P}_{k|k}(\tilde{c}_k, \sigma)),$$

with mean and covariance defined by

$$\left[\hat{\mathbf{x}}_{k|k}\left(\tilde{c}_k, \sigma\right), \mathbf{P}_{k|k}\left(\tilde{c}_k, \sigma\right)\right]$$

$$= \begin{cases} \left[\hat{\mathbf{x}}_{k|k-1}\left(c_{k-1}, \sigma\right), \mathbf{P}_{k|k-1}\left(c_{k-1}, \sigma\right)\right], & i = 0, \\ \mathrm{KF_E}\left[\mathbf{y}_k\left(i\right), \hat{\mathbf{x}}_{k|k-1}\left(c_{k-1}, \sigma\right), \mathbf{P}_{k|k-1}\left(c_{k-1}, \sigma\right), \mathbf{H}, \mathbf{R}\right], & i > 0. \end{cases}$$

Given that

$$\sum_{i=0}^{m_k} \beta_k(i) = \sum_{c_{k-1}=1}^{C_{k-1}} p(c_{k-1}) = \sum_{\sigma=1}^{M} \mu_{k|k-1}(c_{k-1}, \sigma) = 1,$$

it is straightforward to verify that

$$\sum_{\tilde{c}_k=1}^{\tilde{C}_k} p(\tilde{c}_k) = \sum_{\sigma=1}^{M} \mu_{k|k}(\tilde{c}_k, \sigma) = 1.$$

5.5 Optimal track update cycle

The object tracking algorithms presented in this chapter are recursive. A single update cycle is presented in this section. This cycle is repeated for each existing track. Track initialization issues are presented in Section 9.4.

All algorithms presented in this chapter follow the same update cycle. This section presents the optimal track update cycle for two cases, single-object tracking and multi-object tracking. The most general trajectory pdf shape is used, which is multi-component, where each component is using an IMM estimator for multiple trajectory models. Both single- and multi-object tracking share all track cycle operations, with the exception of the data association operation.

The single-object data association may be derived from the multi-object data association formulae by limiting the number of tracks to one. However, for reasons of clarity and the reader's sanity, the data association operation is presented separately for single-object tracking and multi-object tracking.

The individual single-object tracking algorithms detailed in Section 5.7 and multi-object tracking algorithms detailed in Section 5.8 are obtained by introducing various degrees of sub-optimality, or specialization, to the optimal track update cycle formulae presented in this section.

The track update cycle at time k begins with the arrival of the $\mathbf{Y}_k$ batch of measurements. The a posteriori state of each track τ at time $k - 1$, $p(\mathbf{x}_{k-1}^\tau, \chi_{k-1}^\tau | \mathbf{Y}^{k-1})$ is known, and is propagated (predicted) from time $k - 1$ to time k to obtain $p(\mathbf{x}_k^\tau, \chi_k^\tau | \mathbf{Y}^{k-1})$.

Table 5.4 *Cycle k object – existence-based track update.*

Operation	Input	Output
Scan $k-1$		$p(\mathbf{x}_{k-1}, \chi_{k-1}\|\mathbf{Y}^{k-1})$
Prediction	$p(\chi_{k-1}\|\mathbf{Y}^{k-1})$, $p(\mathbf{x}_{k-1}\|\chi_{k-1}, \mathbf{Y}^{k-1})$	$p(\chi_k\|\mathbf{Y}^{k-1})$, $p(\mathbf{x}_k\|\chi_k, \mathbf{Y}^{k-1})$
Selection	$\mathbf{Y}_k$, $p(\mathbf{x}_k\|\chi_k, \mathbf{Y}^{k-1})$	$\mathbf{y}_k$, $p(\mathbf{y}_k(i)\|\chi_k, \mathbf{Y}^{k-1})$
Data association	$p(\chi_k\|\mathbf{Y}^{k-1})$, $p(\mathbf{y}_k(i)\|\chi_k, \mathbf{Y}^{k-1})$	$p(\chi_k\|\mathbf{Y}^k)$, $\boldsymbol{\beta}_k(i)$
Estimation	$\mathbf{y}_k$, $\boldsymbol{\beta}_k(i)$, $p(\mathbf{x}_k\|\chi_k, \mathbf{Y}^{k-1})$	$p(\mathbf{x}_k\|\chi_k, \mathbf{Y}^k)$
Track output	$p(\chi_k\|\mathbf{Y}^k)$, $p(\mathbf{x}_k\|\chi_k, \mathbf{Y}^k)$	status, $\hat{\mathbf{x}}_{k\|k}$, $\mathbf{P}_{k\|k}$

Each track at time k selects a subset of measurements $\mathbf{y}_k^\tau$ which is used to update the track state. At this stage, the measurement likelihoods $p(\mathbf{y}_k(i)\|\chi_k^\tau, \mathbf{Y}^{k-1})$, $i = 1, \ldots, m_k$, are also calculated.

The data association operation uses the measurement likelihood information to update the probability of object existence (calculate $p(\chi_k^\tau\|\mathbf{Y}^k)$), as well as to calculate the a posteriori data association probabilities conditioned on the object existence, $\boldsymbol{\beta}_k^\tau(i) \overset{\Delta}{=} p(\theta_k^\tau(i)\|\chi_k^\tau, \mathbf{Y}^k)$, $i = 0, \ldots, m_k^\tau$, for each track τ.

The track trajectory state pdf $p(\mathbf{x}_k^\tau\|\chi_k^\tau, \mathbf{Y}^k)$ is obtained by using the predicted object trajectory pdf $p(\mathbf{x}_k^\tau\|\chi_k^\tau, \mathbf{Y}^{k-1})$, selected measurements $\mathbf{y}_k^\tau$ and a posteriori data association probabilities $\boldsymbol{\beta}_k^\tau$.

The track output at time k is calculated using a posteriori track state $p(\mathbf{x}_k^\tau, \chi_k^\tau\|\mathbf{Y}^k)$. It usually includes track status (such as tentative, confirmed or terminated), and object trajectory state mean $\hat{\mathbf{x}}_{k\|k}^\tau$ and covariance $\mathbf{P}_{k\|k}^\tau$. Track outputs are used by external clients of the tracker (operators, fusion center, etc.), and are not used in subsequent track update cycles.

One track cycle update is presented in Table 5.4. Please note that the prediction, selection, estimation and track output operations are performed on all tracks separately (or in parallel if the hardware allows it). Data association for single-object tracking is also performed on all tracks separately, whereas data association for multi-object tracking is performed on all tracks (within a cluster, as discussed in Section 5.5.4) simultaneously. Thus we omit the track superscript when describing all operations, except for the multi-object data association.

5.5.1 Track state prediction

Track state prediction involves propagating the probability of object existence, as well as propagating the object trajectory state pdf from time $k-1$ to time k.

The inputs to the track state prediction operation at time k are:

- the a posteriori object existence probability at time $k-1$, $p(\chi_{k-1}\|\mathbf{Y}^{k-1})$; and

- the a posteriori pdf of object trajectory state at time $k-1$, $p(\mathbf{x}_{k-1}|\chi_{k-1}, \mathbf{Y}^{k-1})$, which is parameterized by:
  - the set of track components indexed by c_{k-1}; and for each component c_{k-1},
  - the relative component probability, $p(c_{k-1}) \overset{\triangle}{=} p(\xi_{k-1}(c_{k-1})|\chi_{k-1}, \mathbf{Y}^{k-1})$; and
  - the a posteriori IMM pdf parameterized by the trajectory model probabilities $\mu_{k-1|k-1}(c_{k-1}, \sigma)$, mean values $\hat{\mathbf{x}}_{k-1|k-1}(c_{k-1}, \sigma)$ and covariances $\mathbf{P}_{k-1|k-1}(c_{k-1}, \sigma)$ for all object trajectory models σ.

The outputs of the track state prediction operation at time k are:

- the a priori object existence probability at time k, $p(\chi_k|\mathbf{Y}^{k-1})$; and
- the a priori pdf of object trajectory state at time k, $p(\mathbf{x}_k|\chi_k, \mathbf{Y}^{k-1})$ parameterized by:
  - the set of track components indexed by c_{k-1} (not changed by this operation); and for each component c_{k-1},
  - the relative component probabilities $p(c_{k-1})$ (not changed by this operation), $p(\xi_{k-1}(c_{k-1})|\chi_k, \mathbf{Y}^{k-1}) = p(c_{k-1})$; and
  - the a priori IMM pdf parameterized by the trajectory model probabilities $\mu_{k|k-1}(c_{k-1}, \sigma)$, mean values $\hat{\mathbf{x}}_{k|k-1}(c_{k-1}, \sigma)$ and covariances $\mathbf{P}_{k|k-1}(c_{k-1}, \sigma)$ for all object trajectory models σ.

Here we use the results of Section 5.4.1. The probability of object existence propagates as a Markov chain (Section 5.3.1):

$$p(\chi_k|\mathbf{Y}^{k-1}) = \text{TEX}_\text{P}[p(\chi_{k-1}|\mathbf{Y}^{k-1}), \gamma].$$

The propagated object trajectory state pdf at time k is a mixture of the propagated component pdfs,

$$p(\mathbf{x}_k|\chi_k, \mathbf{Y}^{k-1}) = \sum_{c_{k-1}=1}^{C_{k-1}} p(c_{k-1})p(\mathbf{x}_k|c_{k-1}, \chi_k, \mathbf{Y}^{k-1}), \tag{5.73}$$

where $p(c_{k-1})$ is the relative probability of the component and $p(\mathbf{x}_k|c_{k-1}, \chi_k, \mathbf{Y}^{k-1})$ is the propagated component pdf. Each propagated component pdf is a mixture of the propagated object trajectory state pdfs, given the component and individual object trajectory models σ:

$$p(\mathbf{x}_k|c_{k-1}, \chi_k, \mathbf{Y}^{k-1}) = \sum_{\sigma=1}^{M} \mu_{k|k-1}(c_{k-1}, \sigma)p(\mathbf{x}_k|c_{k-1}, \sigma, \chi_k, \mathbf{Y}^{k-1}), \tag{5.74}$$

$$p(\mathbf{x}_k|c_{k-1}, \sigma, \chi_k, \mathbf{Y}^{k-1}) = \mathcal{N}(\mathbf{x}_k; \hat{\mathbf{x}}_{k|k-1}(c_{k-1}, \sigma), \mathbf{P}_{k|k-1}(c_{k-1}, \sigma)).$$

In other words, trajectory state propagation consists of the propagation of each trajectory state component individually. Each trajectory component state c_{k-1} is an IMM block and propagates by the IMM mixing and prediction operations (Section 3.4.1):

$$[\{\boldsymbol{\mu}_{k|k-1}(c_{k-1}, \sigma), \hat{\mathbf{x}}_{k|k-1}(c_{k-1}, \sigma), \mathbf{P}_{k|k-1}(c_{k-1}, \sigma)\}_\sigma]$$
$$= \text{IMM}_{\text{MP}}[\{\boldsymbol{\mu}_{k-1|k-1}(c_{k-1}, \sigma), \hat{\mathbf{x}}_{k-1|k-1}(c_{k-1}, \sigma),$$
$$\mathbf{P}_{k-1|k-1}(c_{k-1}, \sigma), \mathbf{F}_\sigma, \mathbf{Q}_\sigma\}_\sigma, \boldsymbol{\Gamma}].$$

IMM mixing and prediction collapses into Kalman filter prediction if the number of object trajectory models (IMM models) equals one, $M = 1$, as is the case for JITS, JIPDA, JPDA, ITS, IPDA and PDA. Then, for each component of each track (Section 2.2.1),

$$\left[\hat{\mathbf{x}}_{k|k-1}(c_{k-1}), \mathbf{P}_{k|k-1}(c_{k-1})\right] = \text{KF}_{\text{P}}\left[\hat{\mathbf{x}}_{k-1|k-1}(c_{k-1}), \mathbf{P}_{k-1|k-1}(c_{k-1}), \mathbf{F}, \mathbf{Q}\right].$$

The track state propagation parameters, $\boldsymbol{\gamma}$, $\boldsymbol{\Gamma}$, $\mathbf{F}$ and $\mathbf{Q}$, depend on the time interval Δt_k between sampling times $k - 1$ and k. An example of values for $\mathbf{F}$ and and $\mathbf{Q}$ matrices is presented in Section 2.8; $\boldsymbol{\gamma}$ and $\boldsymbol{\Gamma}$ are discussed in Sections 5.3.1 and 3.4.1 respectively.

5.5.2 Measurement selection (gating)

Measurement selection (gating) selects a subset of measurements which, given that the object exists and is detected, contains the object detection with high probability. This probability is termed the "gating probability" and is denoted here by P_G. For computational and software organization purposes, measurement likelihoods are in practice usually calculated as part of this step.

This operation is performed on each track separately. The inputs to the measurement selection operation at time k are:

- the set $\mathbf{Y}_k$ of measurements delivered by the sensor; and
- the a priori pdf of object trajectory state at time k, $p(\mathbf{x}_k | \chi_k, \mathbf{Y}^{k-1})$. In the case of a linear system, it is parameterized by:
  - the set of track components indexed by c_{k-1}; and for each component c_{k-1},
  - the relative component probabilities $p(c_{k-1})$; and
  - the a priori IMM pdf parameterized by the model probabilities $\boldsymbol{\mu}_{k|k-1}(c_{k-1}, \sigma)$, mean values $\hat{\mathbf{x}}_{k|k-1}(c_{k-1}, \sigma)$ and covariances $\mathbf{P}_{k|k-1}(c_{k-1}, \sigma)$ for all trajectory models σ.

The outputs of the measurement selection operation at time k are:

- the set of measurements $\mathbf{y}_k$ used to update track at time k, and for each selected measurement $\mathbf{y}_k(i)$;
- the measurement likelihood $p(\mathbf{y}_k(i)|\chi_k, \mathbf{Y}^{k-1})$. In the case of a linear system, this is parameterized by:
  - for each track component c_{k-1}, likelihood $p(\mathbf{y}_k(i)|c_{k-1}, \chi_k, \mathbf{Y}^{k-1})$ of measurement $\mathbf{y}_k(i)$ given track component c_{k-1}. This is parametrized by:
  - the a priori model probabilities $\mu_{k|k-1}(c_{k-1}, \sigma)$ and measurement $\mathbf{y}_k(i)$ likelihood given track component c_{k-1} and model σ, $p(\mathbf{y}_k(i)|c_{k-1}, \sigma, \chi_k, \mathbf{Y}^{k-1})$.

In the case of non-parametric object tracking (without prior knowledge of the clutter measurement density) the outputs also include:
  - the values of the selection gate volumes for each trajectory model σ of each component c_{k-1}, $V_k(c_{k-1}, \sigma)$; and
  - the number of selected measurements by each model σ of each component c_{k-1}, $m_k(c_{k-1}, \sigma)$.

Each track has a set of components, and each component c_{k-1} is an IMM block of M models. Each model σ has a single Gaussian propagated object trajectory pdf $p(\mathbf{x}_k|c_{k-1}, \sigma, \mathbf{Y}^{k-1})$, as per (5.20)–(5.21). First calculate the mean and covariance of the measurement position, for each track component c_{k-1} and for each model σ (Section 2.2.1):

$$\left[\hat{\mathbf{y}}_{k|k-1}(c_{k-1}, \sigma), \mathbf{S}_k(c_{k-1}, \sigma)\right] = \mathrm{MP}\left[\hat{\mathbf{x}}_{k|k-1}(c_{k-1}, \sigma), \mathbf{P}_{k|k-1}(c_{k-1}, \sigma), \mathbf{H}, \mathbf{R}\right]. \tag{5.75}$$

Then form a set of measurements selected by trajectory model σ of component c_{k-1}, denoted by $\mathbf{y}_k(c_{k-1}, \sigma)$, as the elements $\mathbf{y}$ of $\mathbf{Y}_k$ satisfying

$$\begin{aligned} \mathbf{y}_k(c_{k-1}, \sigma) = \big\{ \mathbf{y} \in \mathbf{Y}_k : [\mathbf{y} - \hat{\mathbf{y}}_{k|k-1}(c_{k-1}, \sigma)]^T \mathbf{S}_k^{-1}(c_{k-1}, \sigma) \\ \times [\mathbf{y} - \hat{\mathbf{y}}_{k|k-1}(c_{k-1}, \sigma)] \le g \big\}, \end{aligned} \tag{5.76}$$

where $\sqrt{g}$ is termed the gate size, which depends on the gating probability P_G. Denote by $m_k(c_{k-1}, \sigma)$ the cardinality of set $\mathbf{y}_k(c_{k-1}, \sigma)$. The volume of the selection gate for the $\{c_{k-1}, \sigma\}$ component/trajectory model combination is given by

$$V_k(c_{k-1}, \sigma) = \frac{\pi^{n/2}}{\Gamma(n/2 + 1)} \sqrt{g|\mathbf{S}_k(c_{k-1}, \sigma)|} = C_n \sqrt{g|\mathbf{S}_k(c_{k-1}, \sigma)|}, \tag{5.77}$$

where $\Gamma(\cdot)$ denotes the gamma function, and n denotes the dimensionality of the surveillance area. C_n is the volume of n-dimensional hypersphere with unit radius; for $n = 2$, $C_2 = \pi$, and for $n = 3$, $C_3 = 4/3\pi$.

The pseudo-function comprising (5.75)–(5.77) is denoted by

$$[\mathbf{y}_k(c_{k-1}, \sigma), V_k(c_{k-1}, \sigma)] = \mathrm{MS}_1[\mathbf{Y}_k, \hat{\mathbf{x}}_{k|k-1}(c_{k-1}, \sigma), \mathbf{P}_{k|k-1}(c_{k-1}, \sigma), \mathbf{H}, \mathbf{R}]. \tag{5.78}$$

The set of measurements selected by the track is the union of all measurements selected by the individual components:

$$\mathbf{y}_k = \bigcup_{c_{k-1}} \bigcup_{\sigma} \mathbf{y}_k(c_{k-1}, \sigma),$$

with cardinality denoted by m_k. The ith element of $\mathbf{y}_k$ is denoted by $\mathbf{y}_k(i)$. For non-parametric object tracking we also need the volume of the union of the selection gates. This is more involved and an efficient approximation is presented in Section 9.3.

Measurement likelihood

Expressions for the measurement likelihoods, given that they are the track detection, are derived in Section 5.4.2. Here we skip the derivations and simply repeat the results.

The likelihoods of the selected measurements are equal to their prior pdfs, given that they are the (possible) object detection. If a measurement is not selected by a track, or a track component, or a track component model, then the corresponding likelihoods are zero.

The track likelihood of measurement $\mathbf{y}_k(i)$ is the mean of the likelihoods of measurement $\mathbf{y}_k(i)$ on the track component level:

$$p_k(i) \overset{\triangle}{=} p(\mathbf{y}_k(i) \mid \chi_k, \mathbf{Y}^{k-1}) = \sum_{c_{k-1}=1}^{C_{k-1}} p(c_{k-1}) p_k(i, c_{k-1}),$$

and the measurement likelihood at the track component level is the mean of the likelihoods at the level of the individual trajectory models:

$$p_k(i, c_{k-1}) \overset{\triangle}{=} p(\mathbf{y}_k(i) \mid c_{k-1}, \chi_k, \mathbf{Y}^{k-1}) = \sum_{\sigma=1}^{M} \mu_{k|k-1}(c_{k-1}, \sigma) p_k(i, c_{k-1}, \sigma).$$

Finally, the track likelihood of measurement $\mathbf{y}_k(i)$ given the track component c_{k-1} and the trajectory model σ equals

$$\left[\{p_k(i, c_{k-1}, \sigma)\}_i\right] = \mathrm{ML}_1\left[\{\mathbf{y}_k(i)\}_i, \hat{\mathbf{x}}_{k|k-1}(c_{k-1}, \sigma), \mathbf{P}_{k|k-1}(c_{k-1}, \sigma), \mathbf{H}, \mathbf{R}\right].$$

The pseudo-function ML_1 is defined in Section 5.4.2 by

$$p_k(i, c_{k-1}, \sigma)$$

$$\overset{\triangle}{=} p(\mathbf{y}_k(i) | c_{k-1}, \sigma, \chi_k, \mathbf{Y}^{k-1})$$

$$= \begin{cases} \frac{1}{P_G} \mathcal{N}\left(\mathbf{y}_k(i); \hat{\mathbf{y}}_{k|k-1}(c_{k-1}, \sigma), \mathbf{S}_k(c_{k-1}, \sigma)\right), & \mathbf{y}_k(i) \in V_k(c_{k-1}, \sigma), \\ 0, & \mathbf{y}_k(i) \notin V_k(c_{k-1}, \sigma). \end{cases}$$

The conditional object measurement mean and covariance are calculated by

$$\left[\hat{\mathbf{y}}_{k|k-1}(c_{k-1}, \sigma), \mathbf{S}_k(c_{k-1}, \sigma)\right] = \text{MP}\left[\hat{\mathbf{x}}_{k|k-1}(c_{k-1}, \sigma), \mathbf{P}_{k|k-1}(c_{k-1}, \sigma), \mathbf{H}, \mathbf{R}\right],$$

which is defined in Section 2.2.1 by

$$\hat{\mathbf{y}}_{k|k-1}(c_{k-1}, \sigma) = \mathbf{H}\hat{\mathbf{x}}_{k|k-1}(c_{k-1}, \sigma),$$

$$\mathbf{S}_k(c_{k-1}, \sigma) = \mathbf{H}\mathbf{P}_{k|k-1}(c_{k-1}, \sigma)\mathbf{H}^T + \mathbf{R}.$$

5.5.3 Single-object tracking data association

The single-object tracking data association operation calculates the a posteriori data association probabilities as well as the a posteriori probability of object existence using the measurement likelihoods. The derivations are presented in Section 5.4.4. Here we skip the derivations and simply repeat and formalize the results.

Single-object tracking ignores any possible detections of objects being followed by other tracks. Therefore, these operations are performed on each track independently, and may be parallelized.

The inputs to the data association operation at time k are:

- the a priori probability of object existence $p(\chi_k|\mathbf{Y}^{k-1})$; and
- the likelihood $p_k(i)$ of each selected measurement $\mathbf{y}_k(i)$.

The outputs of the data association operation at time k are:

- the a posteriori probability of object existence $p(\chi_k|\mathbf{Y}^k)$; and
- the a posteriori data association probabilities conditioned on object existence, $\beta_k(i) \overset{\triangle}{=} p(\theta_k(i)|\chi_k, \mathbf{Y}^k)$, $i = 0, \ldots, m_k$, where $\theta_k(0)$ denotes the event that none of the selected measurements is the object detection, and $\theta_k(i)$, $i > 0$, denotes the event that measurement $\mathbf{y}_k(i)$ is the object detection.

Denote by Λ_k the measurement likelihood ratio at time k, then

$$\Lambda_k = 1 - P_D P_G + P_D P_G \sum_{i=1}^{m_k} \frac{p_k(i)}{\rho_k(i)}. \tag{5.79}$$

The a posteriori probability of object existence is given by

$$p(\chi_k|\mathbf{Y}^k) = \frac{\Lambda_k p(\chi_k|\mathbf{Y}^{k-1})}{1 - (1 - \Lambda_k)p(\chi_k|\mathbf{Y}^{k-1})}, \tag{5.80}$$

and the a posteriori data association probabilities are

$$\beta_k(i) \triangleq p(\theta_k(i)|\chi_k, \mathbf{Y}^k) = \frac{1}{\Lambda_k} \cdot \begin{cases} 1 - P_D P_G, & i = 0, \\ P_D P_G \frac{p_k(i)}{\rho_k(i)}, & i > 0. \end{cases} \tag{5.81}$$

Equations (5.79)–(5.81) are denoted by the pseudo-function

$$\left[p(\chi_k|\mathbf{Y}^k), \{\beta_k(i)\}_{i=0}^{m_k}\right] = \text{ISTDA}\left[p(\chi_k|\mathbf{Y}^{k-1}), \{p_k(i)\}_{i=1}^{m_k}\right]. \tag{5.82}$$

Equations (5.79), (5.80) and (5.81) are universal in the sense that they do not depend on the shape of the prior object measurement pdf $p(\mathbf{y}|\chi_k, \mathbf{Y}^{k-1})$. These equations are valid for $p(\mathbf{y}|\chi_k, \mathbf{Y}^{k-1})$ being a single Gaussian pdf, Gaussian mixture, as well as for some non-linear pdf represented by a particle filter or other non-linear estimator.

5.5.4 Multi-object tracking data association

A multi-object situation arises when tracks share measurements, i.e., when some measurements are selected (gated) by more than one track. If the objects (tracks) are sufficiently separated so that they do not share measurements, the multi-object tracking reverts to the single-object tracking.

For computational reasons, in each scan k tracks are separated in clusters of tracks which share selected measurements. The tracks and their selected measurements belong to the cluster. Multi-object data association operations are performed on each cluster of tracks simultaneously, and can not be parallelized. The operations on separate clusters, however, are independent, and may be parallelized on this level. To distinguish between tracks, we use track superscripts τ and η.

In this section we present the data association operations on one cluster and all tracks and measurements are assumed to belong to the cluster. The multi-object data association is derived in Section 5.4.5, here we present the final results.

The inputs to the multi-object data association operation at time k are:

- the a priori probability of object τ existence, $p(\chi_k^\tau|\mathbf{Y}^{k-1})$; and

- the measurement likelihoods with respect to track τ, $p_k^\tau(i) \triangleq$ $p(\mathbf{y}_k(i)|\chi_k^\tau, \mathbf{Y}^{k-1})$ for each track τ and each selected measurement $\mathbf{y}_k(i)$.

The outputs of the data association operation at time k are, for each track τ:

- the a posteriori probability of object τ existence $p(\chi_k^\tau | \mathbf{Y}^k)$; and
- the a posteriori data association probabilities conditioned on object existence, $\boldsymbol{\beta}_k^\tau(i) \triangleq p(\theta_k^\tau(i) | \chi_k^\tau, \mathbf{Y}^k)$, $i = 0, \ldots, m_k$, where $\theta_k^\tau(0)$ denotes the event that none of the selected measurements is the detection of object τ at time k, and $\theta_k^\tau(i > 0)$ denotes the event that selected measurement $\mathbf{y}_k(i)$ is the detection of object τ at time k.

A joint event is an allocation of all measurements to all tracks. A feasible joint event is a joint event which satisfies the following:

- each track is assigned zero or one measurement; and
- each measurement is allocated to zero or one track.

All feasible joint measurement to track allocations are enumerated and their a posteriori probabilities evaluated. They are used to calculate the a posteriori data association probabilities and the a posteriori probability of object existence for each track individually.

Two feasible joint events are different if assignment of at least one measurement is different. Different feasible joint events are mutually exclusive. For each joint event ε define:

- $T_0(\varepsilon)$: set of tracks allocated no measurement;
- $T_1(\varepsilon)$: set of tracks allocated one measurement;
- $i(\tau, \varepsilon)$: index of measurement allocated to track τ.

The a posteriori probability of joint events is given by

$$
p(\varepsilon | \mathbf{Y}^k) = c_k^{-1} \prod_{\tau \in T_0(\varepsilon)} \left(1 - P_D^\tau P_G^\tau P\{\chi_k^\tau | \mathbf{Y}^{k-1}\}\right)
$$
$$
\times \prod_{\tau \in T_1(\varepsilon)} \left(P_D^\tau P_G^\tau P\{\chi_k^\tau | \mathbf{Y}^{k-1}\} \frac{p_k^\tau(i(\tau, \varepsilon))}{\rho_k(i(\tau, \varepsilon))}\right),
\tag{5.83}
$$

where the normalization constant c_k is calculated by utilizing the fact that feasible joint events are mutually exclusive and that they form an exhaustive set

$$
\sum_\varepsilon p(\varepsilon | \mathbf{Y}^k) = 1.
$$

Please note that product operation on an empty set equals one.

Track-based data association probabilities

Once the a posteriori probabilities of feasible joint events are known, they are used to obtain the a posteriori probability of object existence and data association probabilities for each track.

The first step is to calculate the a posteriori probabilities of allocating a measurement i, $i \geq 0$, to track τ. Denote by $\Xi(\tau, i)$ the set of feasible joint events which allocate measurement i to track τ. The event that no measurement in a cluster is track τ detection is the union of all (mutually exclusive) feasible joint events which allocate no measurement to track τ and, as the events ε are mutually exclusive,

$$p\big(\theta_k^{\tau}(0)|\mathbf{Y}^k\big) = \sum_{\varepsilon \in \Xi(\tau, 0)} p\big(\varepsilon|\mathbf{Y}^k\big). \tag{5.84}$$

In the same manner, the probability that measurement i, $i > 0$, in a cluster is object τ detection (object τ detection also implies object τ existence) is

$$p\big(\chi_k^{\tau}, \theta_k^{\tau}(i)|\mathbf{Y}^k\big) = \sum_{\varepsilon \in \Xi(\tau, i)} p\big(\varepsilon|\mathbf{Y}^k\big).$$

The probability that no measurement in cluster is track τ detection and that object τ exists is

$$p\big(\chi_k^{\tau}, \theta_k^{\tau}(0)|\mathbf{Y}^k\big) = \frac{(1 - P_D^{\tau} P_G^{\tau}) p(\chi_k^{\tau}|\mathbf{Y}^{k-1})}{1 - P_D^{\tau} P_G^{\tau}\, p(\chi_k^{\tau}|\mathbf{Y}^{k-1})} p\big(\theta_k^{\tau}(0)|\mathbf{Y}^k\big).$$

Events $\{\chi_k^{\tau}, \theta_k^{\tau}(i)\}$ are mutually exclusive for different $i \geq 0$, and their union is the object existence event. Therefore, the a posteriori probability of object existence is

$$p\big(\chi_k^{\tau}|\mathbf{Y}^k\big) = \sum_{i=0}^{m_k} p\big(\chi_k^{\tau}, \theta_k^{\tau}(i)|\mathbf{Y}^k\big). \tag{5.85}$$

The data association probabilities are then given by

$$\beta_k^{\tau}(i) \triangleq p\big(\theta_k^{\tau}(i)|\chi_k^{\tau}, \mathbf{Y}^k\big) = \frac{p\big(\chi_k^{\tau}, \theta_k^{\tau}(i)|\mathbf{Y}^k\big)}{p\big(\chi_k^{\tau}|\mathbf{Y}^k\big)}, \quad i \geq 0. \tag{5.86}$$

Equations (5.83)–(5.86) are denoted by the pseudo-function

$$\big[\{p(\chi_k^{\tau}|\mathbf{Y}^k), \{\beta_k^{\tau}(i)\}_{i \geq 0}\}_{\tau}\big] = \mathrm{JMTDA}\big[\{p(\chi_k^{\tau}|\mathbf{Y}^{k-1}), \{p_k^{\tau}(i)\}_{i>0}\}_{\tau}\big]. \tag{5.87}$$

As can be easily verified, for single-track clusters ($T = 1$), the multi-object data association reverts to the single-object data association.

5.5.5 *Track trajectory update*

The track trajectory update as presented here is identical to both single- and multi-object tracking. These operations are performed on each track independently, and may be parallelized. The derivations are presented in Section 5.4.6, and only the results are repeated here.

The track trajectory update operation has two stages. In the first stage, the a posteriori trajectory state pdf is found with the maximum precision available (under the assumptions). This stage is called here the "component update," and is the subject of this section. The component update stage increases the number of track components by a factor of $1 + m_k$. Track components created at this stage are termed the "tentative track components."

To prevent the saturation of computational resources, the set of tentative track components is subjected to the component control which produces a final set of a posteriori track components at time k. The component control procedures result in sub-optimal object tracking algorithms, and are detailed separately in Section 5.6.

The component update operation at time k is independently performed for each track, and in this section we describe the component update operation on one track and (again) omit the track superscripts.

The inputs to the component update operation at time k are:

- the set of measurements $\mathbf{y}_k$ used to update track at time k, and for each selected measurement $\mathbf{y}_k(i)$:
  - the a priori pdf of selected measurement $\mathbf{y}_k(i)$ – measurement likelihoods with respect to track $p_k(i) \overset{\triangle}{=} p(\mathbf{y}_k(i)|\chi_k, \mathbf{Y}^{k-1})$, with respect to track components $p_k(i, c_{k-1}) \overset{\triangle}{=} p(\mathbf{y}_k(i)|c_{k-1}, \chi_k, \mathbf{Y}^{k-1})$ and with respect to trajectory models of individual track components $p_k(i, c_{k-1}, \sigma) \overset{\triangle}{=} p(\mathbf{y}_k(i)|c_{k-1}, \sigma, \chi_k, \mathbf{Y}^{k-1})$;
  - the a posteriori data association probabilities conditioned on object existence, $\beta_k(i) \overset{\triangle}{=} p(\theta_k(i)|\chi_k, \mathbf{Y}^k)$, $i = 0, \ldots, m_k$, where $\theta_k(0)$ denotes the event that none of the selected measurements is the object detection, and $\theta_k(i)$, $i > 0$ denotes the event that measurement $\mathbf{y}_k(i)$ is the object detection; and
- the a priori pdf of object trajectory state at time k, $p(\mathbf{x}_k|\chi_k, \mathbf{Y}^{k-1})$, parameterized by:
  - the set of track components indexed by c_{k-1}; and for each component c_{k-1},
  - the relative component probability $p(c_{k-1})$;
  - the a priori IMM estimator pdf parameterized by probabilities $\mu_{k|k-1}(c_{k-1}, \sigma)$, mean values $\hat{\mathbf{x}}_{k|k-1}(c_{k-1}, \sigma)$ and covariances $\mathbf{P}_{k|k-1}(c_{k-1}, \sigma)$ for all object trajectory models σ.

The output of the component update operation at time k is:

- the a posteriori pdf of object trajectory state at time $k - 1$, $p(\mathbf{x}_k | \chi_k, \mathbf{Y}^k)$, which is parameterized by:
  - the set of tentative track components indexed by $\tilde{c}_k$; and for each $\tilde{c}_k$,
  - the relative probability $p(\tilde{c}_k) \stackrel{\triangle}{=} p(\tilde{\xi}_k(\tilde{c}_k)|\chi_k, \mathbf{Y}^k)$;
  - the a posteriori IMM estimator pdf parameterized by model probabilities $\mu_{k|k}(\tilde{c}_k, \sigma)$, mean values $\hat{\mathbf{x}}_{k|k}(\tilde{c}_k, \sigma)$ and covariances $\mathbf{P}_{k|k}(\tilde{c}_k, \sigma)$ for all models σ.

Each pair of one a priori track component c_{k-1} and one selected measurement $\mathbf{y}_k(i)$ (including the "null" measurement $i = 0$) creates a new tentative track component $\tilde{\xi}_k(\tilde{c}_k)$,

$$\tilde{\xi}_k(\tilde{c}_k) = \{\xi_{k-1}(c_{k-1}), \theta_k(i)\}. \tag{5.88}$$

For reasons of space, $\tilde{c}_k$ is usually used instead of $\tilde{\xi}_k(\tilde{c}_k)$. Thus $p(\tilde{c}_k) \stackrel{\triangle}{=} p(\tilde{\xi}_k(\tilde{c}_k)|\chi_k, \mathbf{Y}^k)$.

The number of tentative track components equals

$$\tilde{C}_k = C_{k-1}(1 + m_k).$$

The a posteriori trajectory state pdf at time k is a mixture of tentative track component a posteriori state pdfs,

$$p(\mathbf{x}_k | \chi_k, \mathbf{Y}^k) = \sum_{\tilde{c}_k=1}^{\tilde{C}_k} p(\tilde{c}_k) p(\mathbf{x}_k | \tilde{c}_k, \chi_k, \mathbf{Y}^k) \tag{5.89}$$

with the posterior probabilities of individual tentative components given by

$$p(\tilde{c}_k) \stackrel{\triangle}{=} p(\tilde{\xi}_k(\tilde{c}_k)|\chi_k, \mathbf{Y}^k) = \beta_k(i) p(c_{k-1}) \cdot \begin{cases} 1, & i = 0, \\ \dfrac{p_k(i, c_{k-1})}{p_k(i)}, & i > 0. \end{cases} \tag{5.90}$$

Each tentative track component $\tilde{c}_k$ a posteriori object trajectory state pdf is a mixture of the individual pdfs of the object trajectory states conditioned on the individual object trajectory model σ:

$$p(\mathbf{x}_k | \tilde{c}_k, \chi_k, \mathbf{Y}^k) = \sum_{\sigma=1}^{M} \mu_{k|k}(\tilde{c}_k, \sigma) p(\mathbf{x}_k | \tilde{c}_k, \sigma, \chi_k, \mathbf{Y}^k),$$

where the a posteriori object trajectory model probabilities of the individual tentative components are given by

$$
\mu_{k|k}(\tilde{c}_k, \sigma) = \mu_{k|k-1}(c_{k-1}, \sigma) \cdot
\begin{cases}
1, & i = 0, \\
\dfrac{p_k(i, c_{k-1}, \sigma)}{p_k(i, c_{k-1})}, & i > 0,
\end{cases}
\tag{5.91}
$$

and the a posteriori trajectory state pdf of the individual object trajectory models of the individual components at time k have Gaussian pdfs

$$
p(\mathbf{x}_k | \tilde{c}_k, \sigma, \chi_k, \mathbf{Y}^k) = \mathcal{N}(\mathbf{x}_k; \hat{\mathbf{x}}_{k|k}(\tilde{c}_k, \sigma), \mathbf{P}_{k|k}(\tilde{c}_k, \sigma)),
$$

defined by its mean and covariance

$$
\begin{aligned}
&\left[\hat{\mathbf{x}}_{k|k}(\tilde{c}_k, \sigma), \mathbf{P}_{k|k}(\tilde{c}_k, \sigma) \right] \\
&= \begin{cases}
\left[\hat{\mathbf{x}}_{k|k-1}(c_{k-1}, \sigma), \mathbf{P}_{k|k-1}(c_{k-1}, \sigma) \right], & i = 0, \\
\mathrm{KF}_{\mathrm{E}} \left[\mathbf{y}_k(i), \hat{\mathbf{x}}_{k|k-1}(c_{k-1}, \sigma), \mathbf{P}_{k|k-1}(c_{k-1}, \sigma), \mathbf{H}, \mathbf{R} \right], & i > 0,
\end{cases}
\end{aligned}
\tag{5.92}
$$

where KF_{E} is the Kalman filter estimate, as defined in Section 2.2.1. If the tentative track component is formed by combining an a priori track component c_{k-1} with a null measurement $i = 0$, this operation will not change the object trajectory model mean and covariance. If the tentative track component is formed by combining an a priori track component c_{k-1} with measurement $\mathbf{y}_k(i)$, the Kalman filter estimation operation is used to calculate the object trajectory model mean and covariance.

Given that

$$
\sum_{i=0}^{m_k} \beta_k(i) = \sum_{c_{k-1}=1}^{C_{k-1}} p(c_{k-1}) = \sum_{\sigma=1}^{M} \mu_{k|k-1}(c_{k-1}, \sigma) = 1,
$$

it is straightforward to verify that

$$
\sum_{\tilde{c}_k=1}^{\tilde{C}_k} p(\tilde{c}_k) = \sum_{\sigma=1}^{M} \mu_{k|k}(\tilde{c}_k, \sigma) = 1.
$$

5.5.6 *Track output*

Track output is (usually) needed for confirmed tracks only, and most often consists of the estimated object trajectory state position. Particularly for multi-scan (multi-component) object trackers, this usually boils down to a choice of whether this estimate is represented by the most likely track component (akin to the maximum a posteriori estimation (MAP) and usually used by MHT), or the mean estimate across all track components (akin to minimum mean square estimation errors (MMSEE)). The MAP technique sometimes has a tendency to "jump" significantly from measurement to measurement, thus the authors in most cases prefer the MMSEE method.

Track output is not used as the input in the next track state update cycle, but is used by the system operators, data fusion center or by some other higher-level information processing. These operations are performed on each track independently, and may be parallelized.

The input to the track output operation at time k is:

- the a posteriori pdf of the object trajectory state pdf at time k, $p(\mathbf{x}_k|\chi_k, \mathbf{Y}^k)$, which is parameterized by:
  - the set of C_k track components indexed by c_k; and for each component c_k,
  - the relative probabilities of track components $p(c_k) \stackrel{\triangle}{=} p(\xi_k(c_k)|\chi_k, \mathbf{Y}^k)$; and
  - the a posteriori IMM estimator pdf parameterized by probabilities $\mu_{k|k}(c_k, \sigma)$, mean values $\hat{\mathbf{x}}_{k|k}(c_k, \sigma)$ and covariances $\mathbf{P}_{k|k}(c_k, \sigma)$ for all models σ.

In optimal object tracking, the a posteriori set of track components is the set of tentative track components, as described in Section 5.5.5. In most applied object tracking, the track component control presented in Section 5.6 transforms the set of tentative track components into a final a posteriori set of track components.

The output of the track output operation at time k is:

- the mean and covariance of the object trajectory state a posteriori pdf at time k, $\hat{\mathbf{x}}_{k|k}$ and $\mathbf{P}_{k|k}$ respectively.

The track output operation may be performed on each track independently and in parallel, and is defined by

$$[\hat{\mathbf{x}}_{k|k}, \mathbf{P}_{k|k}] = \text{GMix}\left[\left\{\hat{\mathbf{x}}_{k|k}(c_k, \sigma), \mathbf{P}_{k|k}(c_k, \sigma), p(c_k)\mu_{k|k}(c_k, \sigma)\right\}_{c_k, \sigma}\right], \quad (5.93)$$

where operation GMix is defined in Section 3.3.1, and calculates the mean and covariance of a Gaussian mixture pdf.

5.6 Track component control

Without component control, the number of components increases exponentially in time. Denote the number of a posteriori track components at time $k - 1$ by C_{k-1}, and the number of selected measurements at time k by m_k. Then the number of tentative track components $\tilde{C}_k$ at time k equals

$$\tilde{C}_k = (1 + m_k) C_{k-1}.$$

Track component management processes the set of tentative track components to produce the final a posteriori set of track components within computational capabilities.

As the a posteriori set of track components differs from the Bayesian set of tentative track components, this operation contributes to the sub-optimality of object tracking filters presented in this chapter.

The input to the component control operation at time k is:

- the tentative pdf of object trajectory state at time k, $p(\mathbf{x}_k | \chi_k, \mathbf{Y}^k)$, which consists of a set of $\tilde{C}_k$ tentative track components parameterized by:
  - the relative probabilities of tentative track components $p(\tilde{c}_k) \overset{\triangle}{=} p(\tilde{\xi}_k(\tilde{c}_k) | \chi_k, \mathbf{Y}^k)$; and for each tentative component $\tilde{c}_k$,
  - the a posteriori IMM pdf parameterized by probabilities $\mu_{k|k}(\tilde{c}_k, \sigma)$, mean values $\hat{\mathbf{x}}_{k|k}(\tilde{c}_k, \sigma)$ and covariances $\mathbf{P}_{k|k}(\tilde{c}_k, \sigma)$ for all models σ.

The output of the component control operation at time k is:

- the a posteriori pdf of object trajectory state at time k, $p(\mathbf{x}_k | \chi_k, \mathbf{Y}^k)$, which consists of a set of C_k a posteriori track components parameterized by:
  - the relative probabilities of track components $p(c_k) \overset{\triangle}{=} p(\xi_k(c_k) | \chi_k, \mathbf{Y}^k)$; and for each track component c_k,
  - the a posteriori IMM pdf parameterized by probabilities $\mu_{k|k}(c_k, \sigma)$, mean values $\hat{\mathbf{x}}_{k|k}(c_k, \sigma)$ and covariances $\mathbf{P}_{k|k}(c_k, \sigma)$ for all models σ.

A number of techniques may be used for control management (Blackman, 1986; Salmond, 1990; Blackman and Popoli, 1999; Williams and Mayback, 2003; Bochardt *et al.*, 2006):

- component merging;
- leaf pruning;
- sub-tree pruning.

5.6.1 Track component merging

Component merging involves merging two or more tentative track components into one. If all tentative track components merge into one a posteriori track component, resulting in the a posteriori object trajectory state pdf approximation by one Gaussian probability density function per trajectory model, we obtain the PDA-based algorithms:

- IMM-JITS and IMM-ITS become IMM-JIPDA and IMM-IPDA respectively;
- JITS and ITS become JIPDA and IPDA respectively.

When track components merge, the track component interpretation as the object trajectory state pdf given a measurement sequence may become inappropriate. In that case, the track trajectory state component pdf is the object trajectory state probability density function approximation given a set of measurement sequences. Track components remain mutually exclusive and exhaustive, and the track trajectory state pdf remains a Gaussian mixture.

The track components merging procedure has two parts. The first is the track component merge criteria (the choice of track components to merge into one), and the other part is the actual track component merging.

There are various track components merging criteria, some of which are rather complex and computationally expensive. The recent focus (Salmond, 1990; Williams and Mayback, 2003; Bochardt *et al.*, 2006) seems to be to merge track components in a manner with least distortion of the a posteriori track trajectory state pdf. The authors have found the following criterion, which is some decades old (Singer *et al.*, 1974), to be very simple and effective:

> Merge all tentative track components with common measurement history in last N_m scans.

The reason for the effectiveness of this criterion is that the measurements lose their "effect" the older they get. As shown in Singer *et al.* (1974), even a relatively short retained track component history N_m seems to capture measurement information effectively. A fringe benefit of this criterion is that the track component definition is retained, albeit limited to the window of last N_m scans:

$$\xi_k(c_k) = \left\{ \theta_{k-N_m+1}(i_{k-N_m+1}) \cdots \theta_k(i_k) \right\}.$$

In the remainder of this section we describe the merging of one set of track components Θ (of cardinality C_Θ) at time k into one merged track component. This procedure is valid regardless of the method used to choose the set of track components Θ. Depending on the approach used for track component merging, this procedure may need to be repeated many times, sometimes merging the "merged" track component with additional track components. In the case of the PDA-based algorithms (PDA, IPDA, IMM-PDA, IMM-IPDA, JPDA, JIPDA, IMM-JPDA and IMM-JIPDA) all track components are merged into one and $C_\Theta = \tilde{C}_k$.

Let $\overline{\Theta}$ denote the set of track components (of cardinality $C_{\overline{\Theta}}$) complementary to Θ. After the merging operation, all track components belonging to set Θ will be replaced by the merged track component, denoted here by $c_k = i_\Theta$. The total number of track components is reduced by $C_\Theta - 1$.

As all track components before merging are mutually exclusive, the merged track component i_Θ is mutually exclusive to all track components from the complementary set $\overline{\Theta}$. Relative probability of the merged track component is the sum of relative probabilities of the "constituent" track components

$$p(i_\Theta) = \sum_{\tilde{c}_k \in \Theta} p(\tilde{c}_k) = 1 - \sum_{\tilde{c}_k \in \overline{\Theta}} p(\tilde{c}_k). \tag{5.94}$$

The object trajectory state probability density function of merged track component i_Θ is defined by its IMM parameters. The a posteriori probability of the object trajectory model σ, given the merged component i_Θ, is given by

$$\mu_{k|k}(i_\Theta, \sigma) \overset{\Delta}{=} p(r_k = \sigma | i_\Theta, \chi_k, \mathbf{Y}^k)$$

$$= \frac{p(r_k = \sigma, i_\Theta | \chi_k, \mathbf{Y}^k)}{p(i_\Theta | \chi_k, \mathbf{Y}^k)}$$

$$= \frac{\sum_{\tilde{c}_k \in \Theta} p(r_k = \sigma | \tilde{c}_k, \chi_k, \mathbf{Y}^k) p(\tilde{c}_k | \chi_k, \mathbf{Y}^k)}{p(i_\Theta)},$$

where the second line is the Bayes' equation, and the third line is obtained by applying the total probability theorem. Thus,

$$\mu_{k|k}(i_\Theta, \sigma) = \frac{\sum_{\tilde{c}_k \in \Theta} \mu_{k|k}(\tilde{c}_k, \sigma) p(\tilde{c}_k)}{p(i_\Theta)}. \tag{5.95}$$

Define by $\beta(\tilde{c}_k, \sigma, \Theta) \overset{\Delta}{=} p(\tilde{c}_k | \sigma, \Theta, \chi_k, \mathbf{Y}^k)$ the relative probability that the tentative track component $\tilde{c}_k$ is correct, given that the set Θ of tentative component is correct, and given that the object trajectory model σ is correct. Given $\tilde{c}_k \in \Theta$, then $\{\tilde{c}_k, \Theta\} = \{\tilde{c}_k\}$ and

$$\beta(\tilde{c}_k, \sigma, \Theta) \overset{\Delta}{=} p(\tilde{c}_k | \sigma, \Theta, \chi_k, \mathbf{Y}^k) = \frac{p(\tilde{c}_k, \sigma, \Theta | \chi_k, \mathbf{Y}^k)}{p(\Theta, \sigma | \chi_k, \mathbf{Y}^k)} = \frac{p(\tilde{c}_k, \sigma | \chi_k, \mathbf{Y}^k)}{p(\Theta, \sigma | \chi_k, \mathbf{Y}^k)},$$

thus

$$\beta(\tilde{c}_k, \sigma, \Theta) = \frac{p(\tilde{c}_k) p(\sigma | \tilde{c}_k, \chi_k, \mathbf{Y}^k)}{\sum_{\tilde{c}_k \in \Theta} p(\tilde{c}_k) p(\sigma | \tilde{c}_k, \chi_k, \mathbf{Y}^k)} = \frac{p(\tilde{c}_k) \mu_{k|k}(\tilde{c}_k, \sigma)}{\sum_{\tilde{c}_k \in \Theta} p(\tilde{c}_k) \mu_{k|k}(\tilde{c}_k, \sigma)}. \tag{5.96}$$

The a posteriori object trajectory state probability density function, given IMM model (object trajectory model) σ and merged component i_Θ, is given by

$$p(\mathbf{x}_k | i_\Theta, \sigma, \chi_k, \mathbf{Y}^k) = \sum_{\tilde{c}_k \in \Theta} p(\mathbf{x}_k | \tilde{c}_k, i_\Theta, \sigma, \chi_k, \mathbf{Y}^k) p(\tilde{c}_k | \sigma, i_\Theta, \chi_k, \mathbf{Y}^k)$$

$$= \sum_{\tilde{c}_k \in \Theta} p(\mathbf{x}_k | \tilde{c}_k, \sigma, \chi_k, \mathbf{Y}^k) \boldsymbol{\beta}\, (\tilde{c}_k, \sigma, \Theta)\,,$$

which becomes

$$p(\mathbf{x}_k | i_\Theta, \sigma, \chi_k, \mathbf{Y}^k) = \sum_{\tilde{c}_k \in \Theta} \boldsymbol{\beta}\,(\tilde{c}_k, \sigma, \Theta)\, \mathcal{N}\left(\mathbf{x}_k; \hat{\mathbf{x}}_{k|k}(\tilde{c}_k, \sigma), \mathbf{P}_{k|k}(\tilde{c}_k, \sigma)\right)$$

$$\approx \mathcal{N}\left(\mathbf{x}_k; \hat{\mathbf{x}}_{k|k}(i_\Theta, \sigma), \mathbf{P}_{k|k}(i_\Theta, \sigma)\right),$$

with the Gaussian mixture approximated by a single Gaussian with identical mean and covariance,

$$[\hat{\mathbf{x}}_{k|k}(i_\theta, \sigma), \mathbf{P}_{k|k}(i_\Theta, \sigma)] = \mathrm{GMix}\Big[\big\{\hat{\mathbf{x}}_{k|k}(\tilde{c}_k, \sigma), \mathbf{P}_{k|k}(\tilde{c}_k, \sigma), \boldsymbol{\beta}\,(\tilde{c}_k, \sigma, \Theta)\big\}_{\tilde{c}_k \in \Theta}\Big].$$
$$(5.97)$$

The merged component i_Θ is defined by its relative probability $p(i_\Theta)$ (5.94), and for each object trajectory model σ a posteriori probability $\mu_{k|k}(i_\Theta, \sigma)$ (5.95) and object trajectory state mean $\hat{\mathbf{x}}_{k|k}(i_\Theta, \sigma)$ and covariance $\mathbf{P}_{k|k}(i_\Theta, \sigma)$ (5.96)–(5.97).

It is straightforward to verify that after the track component merging operation,

$$\sum_{c_k=1}^{C_k} p(c_k) = \sum_{\sigma=1}^{M} \mu_{k|k}(c_k, \sigma) = 1$$

holds.

5.6.2 *Track component leaf and sub-tree pruning*

Track components up to and including time k may be graphically represented. Without loss of generality, assume here that the track was initiated at time $\ell = 1$ by one measurement. Form a graph with each node representing a track component, arranged in levels which correspond to scan times. Graph nodes at level $\ell \in 1, \ldots, k$ are a posteriori track components c_ℓ at time ℓ. Vertices of the graph from level $\ell - 1$ to level ℓ represent measurements (as well as "null" measurements) which are paired with track components at the upper level $\ell - 1$ to form track components at level ℓ. It is straightforward to see that this graph is a tree, as the vertices go only from one level to the next, and each node in the tree has only one direct antecedent. An example is depicted in Figure 5.5.

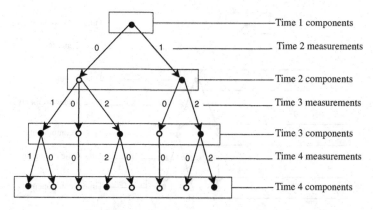

Figure 5.5 Track component tree.

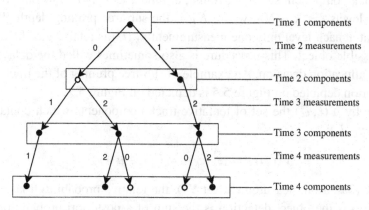

Figure 5.6 Track component tree – leaf pruning.

Leaf pruning at time k removes individual components with low relative probability $p(\tilde{c}_k)$. This technique may be used on its own, by removing enough track components to keep the total number of components below its limit. If this technique is used in conjunction with track component merging or with track component sub-tree pruning, it is used to quickly (with little computational requirements) remove track components with insignificant relative probabilities. Every track component $\tilde{c}_k$ with $p(\tilde{c}_k) < \tau_{c,t}$ is simply removed from memory, where $\tau_{c,t}$ is the track component termination threshold.

Due to the track component pruning operations, some nodes of the track component tree are removed. If a node on level $\ell < k$ has no descendants, it is removed from the tree. Thus, all leaves of the track component tree are on the level k and correspond to current track components. An example of tree pruning of the track component situation depicted on Figure 5.5 is depicted on Figure 5.6.

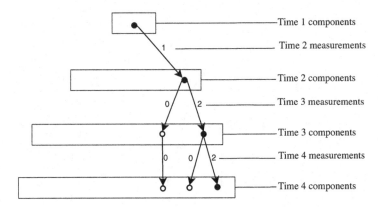

Figure 5.7 Track component tree – sub-tree pruning.

The track component sub-tree pruning at time k usually retains only one component at levels $\ell \leq k - N_P$, where N_P is the sub-tree pruning depth. This also means that at each level only one measurement $\mathbf{y}_\ell$ (i_ℓ) is retained as the detection of the possible object. This procedure is also sometimes called the delayed measurement allocation decision. An example of sub-tree pruning of the track component situation depicted on Figure 5.5 is depicted on Figure 5.7.

Denote by $\Upsilon(\ell, i_\ell)$ the set of tentative track components which contain measurement $\mathbf{y}_\ell$ (i_ℓ):

$$\Upsilon(\ell, i_\ell) = \{\tilde{\xi}_k(\tilde{c}_k) : \theta_\ell(i_\ell) \in \tilde{\xi}_k(\tilde{c}_k)\}.$$

At time k (based on measurement set $\mathbf{Y}^k$), the relative probability that measurement $\mathbf{y}_\ell$ (i_ℓ) is the object detection is the sum of a posteriori probabilities of all components which contain measurement $\mathbf{y}_\ell$ (i_ℓ):

$$p(\theta_\ell(i_\ell)|\chi_k, \mathbf{Y}^k) = \sum_{\tilde{c}_k \in \Upsilon(\ell, i_\ell)} p(\tilde{c}_k),$$

and the absolute probability of measurement $\mathbf{y}_\ell$ (i_ℓ) being the object detection at time k is

$$p(\theta_\ell(i_\ell), \chi_k|\mathbf{Y}^k) = p(\chi_k|\mathbf{Y}^k) \sum_{\tilde{c}_k \in \Upsilon(\ell, i_\ell)} p(\tilde{c}_k).$$

In single-object tracking usually only the measurement with the highest relative probability is retained at each level, while the other measurements are removed (with delay of N_P scans) by sub-tree pruning.

In multi-object tracking, the choice of the measurement to remain at level ℓ of track τ sub-tree is allocated using some optimization algorithm based on the absolute measurement probability. The obvious optimization constraint is that one measurement can be allocated to at most one track, with the exception of the "null" measurement. Often the auction optimization algorithm is used to allocate the measurement; the details of which are outside the scope of this book and may be found in Blackman and Popoli (1999).

In both track component leaf and track component sub-tree pruning algorithms, a subset Θ of components $\tilde{c}_k$ at time k is retained, and the complementary subset $\overline{\Theta}$ of components $\tilde{c}_k$ is removed:

$$\{c_k\} = \{\tilde{c}_k : \tilde{c}_k \in \Theta\}.$$

Denote by ΔP the relative probability of the set $\overline{\Theta}$,

$$\Delta P = \sum_{\tilde{c}_k \in \overline{\Theta}} p(\tilde{c}_k).$$

The relative probabilities of the remaining components have to be adjusted as

$$p(c_k) = p(\tilde{c}_k)/(1 - \Delta P),$$

and the probability of object existence is adjusted as

$$p(\chi_k|\mathbf{Y}^k) = p(\chi_k|\mathbf{Y}^k)(1 - \Delta P).$$

It is straightforward to verify that after the track component pruning operation

$$\sum_{c_k=1}^{C_k} p(c_k) = \sum_{\sigma=1}^{M} \mu_{k|k}(c_k, \sigma) = 1$$

holds.

5.7 Object-existence-based single-object tracking

Section 5.7 presents the object-existence-based single-object tracking algorithms, from the simplest (probabilistic data association or PDA) to the most complex (interacting multiple models – integrated track splitting or IMM-ITS).

Each algorithm assumes a track state which is a specialization of the general track model presented in Section 5.3. The specialization may be the existence of only one object trajectory model (non-maneuvering object tracking), merging of all track components into one (single scan object tracking), or assuming object existence (deterministic object existence).

These algorithms are recursive. Starting from the previously updated track state pdf (at time $k - 1$), measurements delivered by sensor(s) at time k are used to calculate the updated track state pdf (at time k). Only one update cycle for each algorithm is presented, as new track initialization is similar for all algorithms, and is presented in Section 9.4.

In some cases, PDA being one, the algorithms differ somewhat from the originally published version; the differences are duly noted.

There is a substantial amount of repetition between algorithms, as each subsequent algorithm is a generalization of the previous one. The intention of the authors was to describe each algorithm briefly, and to note the amount of specialization with respect to the optimal object tracking. Each algorithm should be read and understood by itself, without having to read and understand the other algorithms.

5.7.1 Probabilistic data association (PDA)

Probabilistic data association (PDA) is a single-scan estimator of trajectory of a single non-maneuvering object in clutter. Originally published in Bar-Shalom and Tse (1975), and presented in Chapter 4, PDA instantly became a popular choice as it is able to estimate object trajectory in significant clutter with very small computational requirements. PDA is the simplest algorithm presented in this chapter. Due to its simple structure, PDA is computationally the most efficient object trajectory state estimator presented in this chapter. It does not, however, provide the probability of object existence to be used as a track quality measure by the false track discrimination procedure. The implicit assumption is that the object exists, and the probability of object existence is, and remains, one. PDA propagates and updates each track in isolation (when updating one track, the other tracks are being ignored), and:

- assumes that the object must exist, i.e., the probability of object existence equals one, and is not updated,

$$p(\chi_{k-1}|\mathbf{Y}^{k-1}) = p(\chi_k|\mathbf{Y}^{k-1}) = p(\chi_k|\mathbf{Y}^k) = 1,$$

- assumes that the object does not maneuver, i.e., that the object follows a single trajectory model (the number of trajectory models, $M = 1$):

$$\mu_{k-1|k-1}(1) = \mu_{k|k-1}(1) = \mu_{k|k}(1) = 1,$$

- is a single-scan algorithm, i.e., all tentative track components are merged into one (the number of both prior and posterior track components equals one, $C_{k-1} = C_k = 1$), and both the a priori and a posteriori pdfs of the object trajectory state are approximated by a single Gaussian pdf:

$$p(c_{k-1}) = p(c_k) = 1.$$

Due to the single-object trajectory model, conditioning on the object trajectory model σ cancels out; ditto for the track state components:

$$p(\cdot|\sigma, \cdot) = p(\cdot|\cdot),$$

$$p(\cdot|c_{k-1}, \cdot) = p(\cdot|\cdot).$$

For example,

$$p(\mathbf{x}_k|\chi_k, \mathbf{Y}^{k-1}) = p(\mathbf{x}_k|c_{k-1}, \chi_k, \mathbf{Y}^{k-1}) = p(\mathbf{x}_k|c_{k-1}, \sigma, \chi_k, \mathbf{Y}^{k-1}),$$
$$p_k(i) = p_k(i, c_k - 1) = p_k(i, c_{k-1}, \sigma).$$

Track component control (Section 5.6) is an integral part of PDA track trajectory update. All tentative track components are merged into one, and the track update operation becomes the PDA estimation, which is detailed and presented in Section 4.3.

The original PDA (Bar-Shalom and Tse, 1975), in Section 4.3.1, assumes uniform clutter measurement density. In Section 5.5.3, PDA equations are somewhat rearranged to enable the use of non-uniform clutter measurement density. This is applicable in the case of parametric object tracking, when the clutter measurement density is available either theoretically, or by using some version of the multi-scan clutter mapping estimators.

On the other hand, if the clutter measurement density is a priori unknown (the non-parametric case), PDA estimates the clutter measurement density within the selection gate by replacing (9.3) by

$$\rho = m_k/V_k.$$

The single-scan PDA recursion is detailed in Algorithm 20. Only one track is implicitly assumed; however, if more than one tracks are initialized, this procedure is repeated for each existing track.

5.7.2 Interacting multiple models – probabilistic data association (IMM-PDA)

Interacting multiple models – probabilistic data association (IMM-PDA) is the maneuvering object extension of PDA, presented in Section 5.7.1. IMM-PDA is a single-object, single-scan, estimator of the maneuvering object trajectory in clutter. In common with PDA, IMM-PDA does not provide the probability of object existence to be used as a track quality measure by the false track discrimination procedure. The implicit assumption is that the object exists, and the probability of object existence is, and remains, one. IMM-PDA propagates and updates a single track in isolation (when updating one track, the other tracks are being ignored), and:

- assumes that the object must exist, i.e., the probability of object existence equals one, and is not updated,

$$p(\chi_{k-1}|\mathbf{Y}^{k-1}) = p(\chi_k|\mathbf{Y}^{k-1}) = p(\chi_k|\mathbf{Y}^k) = 1,$$

Algorithm 20 PDA filter recursion equations at time k

1: Time k inputs:
 - set $\mathbf{Y}_k$ of measurements delivered by the sensor; and
 - track trajectory state mean value $\hat{\mathbf{x}}_{k-1|k-1}$ and covariance $\mathbf{P}_{k-1|k-1}$.

2: Prediction {Sections 2.2.1, 5.5.1}:

$$\left[\hat{\mathbf{x}}_{k|k-1}, \mathbf{P}_{k|k-1}\right] = \mathrm{KF_P}\left[\hat{\mathbf{x}}_{k-1|k-1}, \mathbf{P}_{k-1|k-1}, \mathbf{F}, \mathbf{Q}\right].$$

3: Measurement selection {Section 5.5.2, (5.78)}:

$$[\mathbf{y}_k, V_k] = \mathrm{MS}_1\left[\mathbf{Y}_k, \hat{\mathbf{x}}_{k|k-1}, \mathbf{P}_{k|k-1}, \mathbf{H}, \mathbf{R}\right].$$

4: Likelihoods of all selected measurements i {(5.27)}:

$$\left[\{p_k(i)\}_i\right] = \mathrm{ML}_1\left[\{\mathbf{y}_k(i)\}_i, \hat{\mathbf{x}}_{k|k-1}, \mathbf{P}_{k|k-1}, \mathbf{H}, \mathbf{R}\right].$$

5: **if** non-parametric tracking **then**

6:     V_k calculated by (5.77) {Section 5.5.2}.

7:     Clutter measurement density estimation {Section 9.3}:

$$\rho = m_k / V_k.$$

8: **end if**

9: Single-object data association (sàns object existence) {Section 5.5.3}:

$$\left[-, \{\boldsymbol{\beta}_k(i)\}_{i=0}^{m_k}\right] = \mathrm{ISTDA}\left[-, \{p_k(i)\}_{i=1}^{m_k}\right].$$

10: Estimation/merging {Section 4.3}:

$$\left[\hat{\mathbf{x}}_{k|k}, \mathbf{P}_{k|k}\right] = \mathrm{PDA_E}\left[\hat{\mathbf{x}}_{k|k-1}, \mathbf{P}_{k|k-1}, \{\mathbf{y}_k(i)\}_{i=1}^{m_k}, \{\boldsymbol{\beta}_k(i)\}_{i=0}^{m_k}, \mathbf{H}, \mathbf{R}\right].$$

11: Output trajectory estimate:
 - track mean value $\hat{\mathbf{x}}_{k|k}$ and covariance $\mathbf{P}_{k|k}$.

- assumes that the object may maneuver, i.e., that the object may switch between more than one trajectory models (number of trajectory models $M \geq 1$),
- is a single-scan algorithm, i.e., all tentative track components are merged into one (the number of both prior and posterior track components equals one, $C_{k-1} = C_k = 1$),

$$p(c_{k-1}) = p(c_k) = 1.$$

Due to the single-track component, conditioning on the track component cancels out:

$$p(\cdot|c_{k-1}, \cdot) = p(\cdot|\cdot).$$

For example,

$$p(\mathbf{x}_k | \chi_k, \mathbf{Y}^{k-1}) = p(\mathbf{x}_k | c_{k-1}, \chi_k, \mathbf{Y}^{k-1}),$$

$$p_k(i, c_{k-1}, \sigma) = p_k(i, \sigma),$$

$$\boldsymbol{\mu}_{k|k}(\tilde{c}_k, \sigma) = \boldsymbol{\mu}_{k|k}(i, \sigma).$$

If the number of object trajectory models equals one, IMM-PDA reverts, both conceptually and algorithmically, to PDA.

Track component control, Section 5.6, is an integral part of the IMM-PDA track trajectory update. All tentative track components are merged into one and the track update operation becomes the PDA estimation, performed once for each object trajectory model σ.

The original IMM-PDA (Houles and Bar-Shalom, 1989), as well as the derivation in Chapter 4, assumes uniform clutter measurement density. In Section 5.5.3, the IMM-PDA equations are somewhat rearranged to enable the use of non-uniform clutter measurement density. This is applicable in the case of parametric object tracking, when the clutter measurement density is available either theoretically, or by using some version of the multi-scan clutter mapping estimators.

If the clutter measurement density is a priori unknown (the non-parametric case), IMM-PDA estimates the clutter measurement density within the selection gate by using the procedure defined in Section 9.3. This is also different from the original publications, due to an efficient approximation of the selection gate volume.

A single-scan IMM-PDA recursion is detailed in Algorithms 21 and 22. Only one track is implicitly assumed; however, if more than one tracks are initialized, this procedure is repeated for each existing track.

5.7.3 Integrated probabilistic data association (IPDA)

IPDA extends PDA by adding to it the recursive propagation and update of the probability of object existence, and, in the non-parametric case, a different and improved calculation of the clutter measurement density.

Integrated probabilistic data association (IPDA) is a single-object, single-scan estimator of non-maneuvering object trajectory in clutter and the probability of object existence. As the probability of object existence provides the tools for the false track discrimination procedure, IPDA is the simplest single-object tracking in clutter algorithm (presented in this chapter), which can be a basis for a complete object tracking solution. IPDA is also conceptually the simplest and computationally the most efficient random object existence estimator presented in this chapter.

Algorithm 21 IMM-PDA filter recursion equations at time k: part I

1: Time k inputs:
- set $\mathbf{Y}_k$ of measurements delivered by the sensor; and
- for each object trajectory model σ:
  - time $k-1$ a posteriori model probability $\mu_{k-1|k-1}(\sigma)$; and
  - trajectory estimate mean value $\hat{\mathbf{x}}_{k-1|k-1}(\sigma)$ and covariance $\mathbf{P}_{k-1|k-1}(\sigma)$.

2: IMM mixing and prediction {Sections 3.4.1, 5.5.1}:

$$\left[\left\{\mu_{k|k-1}(\sigma), \hat{\mathbf{x}}_{k|k-1}(\sigma), \mathbf{P}_{k|k-1}(\sigma)\right\}_\sigma\right]$$

$$= \text{IMM}_{\text{MP}}\left[\left\{\mu_{k-1|k-1}(\sigma), \hat{\mathbf{x}}_{k-1|k-1}(\sigma), \mathbf{P}_{k-1|k-1}(\sigma), \mathbf{F}_\sigma, \mathbf{Q}_\sigma\right\}_\sigma, \boldsymbol{\Gamma}\right].$$

3: Measurement selection {Section 5.5.2, (5.78)}:

$$[\mathbf{y}_k(\sigma), V_k(\sigma)] = \text{MS}_1\left[\mathbf{Y}_k, \hat{\mathbf{x}}_{k|k-1}(\sigma), \mathbf{P}_{k|k-1}(\sigma), \mathbf{H}, \mathbf{R}\right], \qquad \text{for each } \sigma,$$

$$\mathbf{y}_k = \bigcup_\sigma \mathbf{y}_k(\sigma).$$

4: Likelihoods of all selected measurements $i > 0$ {(5.27)}:

$$\left[\{p_k(i, \sigma)\}_i\right] = \text{ML}_1\left[\{\mathbf{y}_k(i)\}_i, \hat{\mathbf{x}}_{k|k-1}(\sigma), \mathbf{P}_{k|k-1}(\sigma), \mathbf{H}, \mathbf{R}\right], \qquad \text{for each } \sigma,$$

$$p_k(i) = \sum_\sigma \mu_{k|k-1}(\sigma) p_k(i, \sigma).$$

5: **if** non-parametric tracking **then**
6:     V_k calculated by (9.6) {Section 9.3}.
7:     Clutter measurement density estimation {Section 9.3}:

$$\rho = m_k / V_k.$$

8: **end if**
9: Single-object data association (sàns object existence) {Section 5.5.3}:

$$\left[-, \{\boldsymbol{\beta}_k(i)\}_{i=0}^{m_k}\right] = \text{ISTDA}\left[-, \{p_k(i)\}_{i=1}^{m_k}\right].$$

As such, it provides the object tracking in clutter solution for a number of actual systems.

IPDA propagates and updates each track in isolation (when updating one track, the other tracks are ignored), and:

- assumes random object existence and recursively calculates the probability of object existence;

Algorithm 22 IMM-PDA filter recursion equations at time k: part II

10: **for** each trajectory model σ **do**

11:     Posterior model probabilities $\mu_{k|k}(i, \sigma)$, given measurement $i \geq 0$ {(5.91)}:

$$\mu_{k|k}(i, \sigma) = \mu_{k|k-1}(\sigma) \begin{cases} 1, & i = 0, \\ \dfrac{p_k(i, \sigma)}{p_k(i)}, & i > 0. \end{cases}$$

12:     Posterior model probability $\mu_{k|k}(\sigma)$ {(5.95)}:

$$\mu_{k|k}(\sigma) = \sum_{i \geq 0} \beta_k(i) \mu_{k|k}(i, \sigma).$$

13:     Posterior data association probabilities $\beta(i, \sigma)$, $i \geq 0$ {(5.96)}:

$$\beta(i, \sigma) = \frac{\beta_k(i)}{\beta_k(0) + \sum_{j>0} \beta_k(j) p_k(j, \sigma)/p_k(j)} \begin{cases} 1, & i = 0, \\ \dfrac{p_k(i, \sigma)}{p_k(i)}, & i > 0. \end{cases}$$

14:     Estimation and merging {Section 4.3}:

$$\left[\hat{\mathbf{x}}_{k|k}(\sigma), \mathbf{P}_{k|k}(\sigma)\right]$$
$$= \mathrm{PDA_E}\left[\hat{\mathbf{x}}_{k|k-1}(\sigma), \mathbf{P}_{k|k-1}(\sigma), \{\mathbf{y}_k(i)\}_{i=1}^{m_k}, \{\beta_k(i, \sigma)\}_{i=0}^{m_k}, \mathbf{H}, \mathbf{R}\right].$$

15: **end for**

16: Output trajectory estimate:

$$[\hat{\mathbf{x}}_{k|k}, \mathbf{P}_{k|k}] = \mathrm{GMix}\left[\{\hat{\mathbf{x}}_{k|k}(\sigma), \mathbf{P}_{k|k}(\sigma), \mu_{k|k}(\sigma)\}_\sigma\right].$$

- assumes that the object does not maneuver, i.e., that the object follows a single trajectory model (the number of trajectory models, $M = 1$),

$$\mu_{k-1|k-1}(1) = \mu_{k|k-1}(1) = \mu_{k|k}(1) = 1;$$

- is a single-scan algorithm, i.e., all tentative track components are merged into one (the number of both prior and posterior track components equals one, $C_{k-1} = C_k = 1$),

$$p(c_{k-1}) = p(c_k) = 1.$$

Due to the single-object trajectory model, conditioning on the object trajectory model σ cancels out; ditto for the track state estimate components:

$$p(\cdot|\sigma, \cdot) = p(\cdot|\cdot),$$
$$p(\cdot|c_{k-1}, \cdot) = p(\cdot|\cdot).$$

For example,

$$p(\mathbf{x}_k|\chi_k, \mathbf{Y}^{k-1}) = p(\mathbf{x}_k|c_{k-1}, \chi_k, \mathbf{Y}^{k-1}) = p(\mathbf{x}_k|c_{k-1}, \sigma, \chi_k, \mathbf{Y}^{k-1}),$$

$$p_k(i) = p_k(i, c_k - 1) = p_k(i, c_{k-1}, \sigma).$$

Track component control (Section 5.6) is an integral part of the PDA track trajectory update. All tentative track components are merged into one, and the track update operation becomes the PDA estimation, which is detailed and presented in Section 4.3.

IPDA was originally proposed in Mušicki (1994) and Mušicki *et al.* (1994) for non-parametric object tracking. Use of non-uniform clutter measurement density in parametric object tracking was subsequently added in Mušicki and Evans (2004a).

In Section 5.5.3, IPDA equations are presented for the case of parametric object tracking, when the clutter measurement density is available either theoretically, or by using some version of the multi-scan clutter mapping estimators.

If the clutter measurement density is a priori unknown (the non-parametric case), IPDA estimates the clutter measurement density within the selection gate, as presented in Section 9.3.

A single-scan IPDA recursion is detailed in Algorithm 23. Only one track is implicitly assumed; however, if more than one tracks are initialized, this procedure is repeated for each existing track.

5.7.4 Interacting multiple models – integrated probabilistic data association (IMM-IPDA)

Interacting multiple models – integrated probabilistic data association (IMM-IPDA) is the maneuvering object extension of IPDA, presented in Section 5.7.3. It was introduced in Mušicki *et al.* (2004a) and Mušicki and Suvorova (2008).

IMM-IPDA is a single-object, single-scan, estimator of maneuvering object trajectory in clutter, as well as a recursive estimator of the probability of object existence. IMM-IPDA propagates and updates a single track in isolation (when updating one track, the other tracks are being ignored), and:

- assumes random object existence and recursively calculates the probability of object existence;
- assumes that the object may maneuver, i.e., that the object may switch between more than one trajectory models (the number of trajectory models, $M \geq 1$);
- is a single-scan algorithm, i.e., all tentative track components are merged into one (the number of both prior and posterior track components equals

---

Algorithm 23 IPDA filter recursion equations at time k

1: Time k inputs:
  - set $\mathbf{Y}_k$ of measurements delivered by the sensor;
  - probability of object existence $p(\chi_{k-1}|\mathbf{Y}^{k-1})$; and
  - object trajectory state mean value $\hat{\mathbf{x}}_{k-1|k-1}$ and covariance $\mathbf{P}_{k-1|k-1}$.

2: Track state propagation (Sections 2.2.1, 5.5.1):

$$p(\chi_k|\mathbf{Y}^{k-1}) = \gamma_{11} p(\chi_{k-1}|\mathbf{Y}^{k-1}),$$

$$\left[\hat{\mathbf{x}}_{k|k-1}, \mathbf{P}_{k|k-1}\right] = \text{KF}_\text{P}\left[\hat{\mathbf{x}}_{k-1|k-1}, \mathbf{P}_{k-1|k-1}, \mathbf{F}, \mathbf{Q}\right].$$

3: Measurement selection {Section 5.5.2, (5.78)}:

$$[\mathbf{y}_k, V_k] = \text{MS}_1\left[\mathbf{Y}_k, \hat{\mathbf{x}}_{k|k-1}, \mathbf{P}_{k|k-1}, \mathbf{H}, \mathbf{R}\right].$$

4: Selected measurements likelihood {(5.27)}: $\forall i$:

$$\left[\{p_k(i)\}_i\right] = \text{ML}_1\left[\{\mathbf{y}_k(i)\}_i, \hat{\mathbf{x}}_{k|k-1}, \mathbf{P}_{k|k-1}, \mathbf{H}, \mathbf{R}\right].$$

5: **if** non-parametric tracking **then**
6:   V_k calculated by (5.77) {Section 5.5.2}:
7:   Clutter measurement density estimation {Section 9.3}:

$$\rho = \hat{m}_k / V_k = (m_k - P_D P_G p(\chi_k|\mathbf{Y}^{k-1})) / V_k.$$

8: **end if**
9: Single-object data association {Section 5.5.3}:

$$\left[p(\chi_k|\mathbf{Y}^k), \{\boldsymbol{\beta}_k(i)\}_{i=0}^{m_k}\right] = \text{ISTDA}\left[p(\chi_k|\mathbf{Y}^{k-1}), \{p_k(i)\}_{i=1}^{m_k}\right].$$

10: Estimation/merging {Section 4.3}:

$$\left[\hat{\mathbf{x}}_{k|k}, \mathbf{P}_{k|k}\right] = \text{PDA}_\text{E}\left[\hat{\mathbf{x}}_{k|k-1}, \mathbf{P}_{k|k-1}, \{\mathbf{y}_k(i)\}_{i=1}^{m_k}, \{\boldsymbol{\beta}_k(i)\}_{i=0}^{m_k}, \mathbf{H}, \mathbf{R}\right].$$

11: Output object trajectory estimate:
  - mean value $\hat{\mathbf{x}}_{k|k}$ and covariance $\mathbf{P}_{k|k}$.

---

one, $C_{k-1} = C_k = 1$),

$$p(c_{k-1}) = p(c_k) = 1.$$

In other words, due to the single-track component, conditioning on the track component cancels out:

$$p(\cdot|c_{k-1}, \cdot) = p(\cdot|\cdot).$$

For example,

$$p(\mathbf{x}_k|\chi_k, \mathbf{Y}^{k-1}) = p(\mathbf{x}_k|c_{k-1}, \chi_k, \mathbf{Y}^{k-1}),$$

$$p_k(i, c_{k-1}, \sigma) = p_k(i, \sigma),$$

$$\mu_{k|k}(\tilde{c}_k, \sigma) = \mu_{k|k}(i, \sigma).$$

If the number of object trajectory models equals one, IMM-IPDA reverts, both conceptually and algorithmically, to IPDA.

Track component control (Section 5.6) is an integral part of the IMM-IPDA track trajectory update. All tentative track components are merged into one and the track trajectory update operation becomes the PDA estimation, performed once for each object trajectory model σ.

In Section 5.5.3, IMM-IPDA equations are presented for the case of parametric object tracking, when the clutter measurement density is available either theoretically, or by using some version of the multi-scan clutter mapping estimators.

On the other hand, if the clutter measurement density is a priori unknown (the non-parametric case), IMM-IPDA estimates the clutter measurement density within the selection gate, as presented in Section 9.3.

A single-scan IMM-IPDA recursion is detailed in Algorithm 24. Only one track is implicitly assumed; however, if more than one tracks are initialized, this procedure is repeated for each existing track. The algorithm shows that the trajectory estimate update, detailed in Algorithm 22, is identical in both IMM-PDA and IMM-IPDA. In other words, the trajectory estimate update does not depend on the probability of object existence, which is a desirable outcome. Algorithm 22 is also identical in IMM-JPDA and IMM-JIPDA, further indicating the seamless connections between algorithms in this chapter.

5.7.5 *Integrated track splitting (ITS)*

Integrated track splitting (ITS) was published in Mušicki *et al.* (2003, 2007). ITS extends IPDA by keeping multiple track components, or sufficient statistics of measurement scans for more than one measurement scan. The ITS object trajectory state estimate pdf approximates the true pdf better than IPDA. As a consequence, ITS is more successful in track retention, false track discrimination and object trajectory state estimation than IPDA, especially in difficult situations of low probability of detection and/or high clutter measurement density. As another, this time not so positive consequence, ITS requires significantly more computational resources than IPDA. This is tunable and the track component control provides a trade-off mechanism between computational requirements and object tracking performance.

Algorithm 24 IMM-IPDA filter recursion equations at time k

1: Time k inputs:
   - set $\mathbf{Y}_k$ of measurements delivered by the sensor; and
   - probability of object existence $p(\chi_{k-1}|\mathbf{Y}^{k-1})$; and
   - for each object trajectory model σ:
     - time $k-1$ a posteriori model probability $\mu_{k-1|k-1}(\sigma)$; and
     - trajectory estimate mean value $\hat{\mathbf{x}}_{k-1|k-1}(\sigma)$ and covariance $\mathbf{P}_{k-1|k-1}(\sigma)$.
2: Track state propagation: object existence and IMM mixing and prediction {Sections 3.4.1, 5.5.1}:

$$p(\chi_k|\mathbf{Y}^{k-1}) = \gamma_{11}\, p(\chi_{k-1}|\mathbf{Y}^{k-1}),$$

$$\left[\left\{\mu_{k|k-1}(\sigma), \hat{\mathbf{x}}_{k|k-1}(\sigma), \mathbf{P}_{k|k-1}(\sigma)\right\}_\sigma\right]$$
$$= \mathrm{IMM_{MP}}\left[\left\{\mu_{k-1|k-1}(\sigma), \hat{\mathbf{x}}_{k-1|k-1}(\sigma), \mathbf{P}_{k-1|k-1}(\sigma), \mathbf{F}_\sigma, \mathbf{Q}_\sigma\right\}_\sigma, \boldsymbol{\Gamma}\right].$$

3: Measurement selection {Section 5.5.2, (5.78)}:

$$[\mathbf{y}_k(\sigma), V_k(\sigma)] = \mathrm{MS}_1\left[\mathbf{Y}_k, \hat{\mathbf{x}}_{k|k-1}(\sigma), \mathbf{P}_{k|k-1}(\sigma), \mathbf{H}, \mathbf{R}\right], \qquad \text{for each } \sigma,$$

$$\mathbf{y}_k = \bigcup_\sigma \mathbf{y}_k(\sigma).$$

4: Likelihoods of all selected measurements $i > 0$ {(5.27)}:

$$\left[\{p_k(i, \sigma)\}_i\right] = \mathrm{ML}_1\left[\{\mathbf{y}_k(i)\}_i, \hat{\mathbf{x}}_{k|k-1}(\sigma), \mathbf{P}_{k|k-1}(\sigma), \mathbf{H}, \mathbf{R}\right], \qquad \text{for each } \sigma,$$

$$p_k(i) = \sum_\sigma \mu_{k|k-1}(\sigma) p_k(i, \sigma).$$

5: **if** non-parametric tracking **then**
6:     V_k calculated by (9.6) {Section 9.3}.
7:     Clutter measurement density estimation {Section 9.3}:

$$\rho = \hat{m}_k / V_k = (m_k - P_D P_G\, p(\chi_k|\mathbf{Y}^{k-1})) / V_k.$$

8: **end if**
9: Single-object data association {Section 5.5.3}:

$$\left[p(\chi_k|\mathbf{Y}^k), \{\boldsymbol{\beta}_k(i)\}_{i=0}^{m_k}\right] = \mathrm{ISTDA}\left[p(\chi_k|\mathbf{Y}^{k-1}), \{p_k(i)\}_{i=1}^{m_k}\right].$$

10: Algorithm 22 {IMM-PDA Part II – trajectory estimate update}

Integrated track splitting (ITS) is a single-object, multi-scan estimator of non-maneuvering object trajectory in clutter, as well as the probability of object existence.

ITS propagates and updates a single track in isolation; when updating one track, the other tracks are being ignored), and:

- assumes random object existence and recursively calculates the probability of object existence;
- assumes that the object does not maneuver, i.e., that the object follows a single trajectory model (the number of trajectory models, $M = 1$),

$$\mu_{k-1|k-1}(c_{k-1}, 1) = \mu_{k|k-1}(c_{k-1}, 1) = \mu_{k|k}(c_{k-1}, 1) = 1;$$

- is a multi-scan algorithm, with a non-negative number of retained track components.

Due to the single-object trajectory model, conditioning on the object trajectory model σ cancels out:

$$p(\cdot|\sigma, \cdot) = p(\cdot|\cdot).$$

For example,

$$p(\mathbf{x}_k|\chi_k, c_{k-1}, \mathbf{Y}^{k-1}) = p(\mathbf{x}_k|\chi_k, c_{k-1}, \sigma, \mathbf{Y}^{k-1}),$$

$$p_k(i, c_{k-1}) = p_k(i, c_{k-1}, \sigma).$$

Track component control (Section 5.6) is an integral part of practical ITS applications (although not necessary from a strictly theoretical point of view). However, as opposed to IPDA, the system designer has absolute freedom (limited by the computational resources, of course) of the track component control method and the number of retained track components. If merging component control is applied to the limit, and all track components are merged into one, ITS reverts to IPDA.

In Section 5.5.3, ITS equations are presented for the case of parametric object tracking, when the clutter measurement density is available either theoretically, or by using some version of the multi-scan clutter mapping estimators.

If the clutter measurement density is a priori unknown (the non-parametric case), ITS estimates the clutter measurement density within the selection gate, as presented in Section 9.3.

A single-scan ITS recursion is detailed in Algorithms 25 and 26. Only one track is implicitly assumed; however, if more than one tracks are initialized, this procedure is repeated for each existing track.

5.7.6 Interacting multiple models – integrated track splitting (IMM-ITS)

IMM-ITS, published in Mušicki *et al.* (2004b, 2007), extends ITS by including multiple-object trajectory models. Thus, the IMM-ITS object trajectory state

---

Algorithm 25 ITS filter recursion equations at time k: part I

1: Time k inputs:
  - set $\mathbf{Y}_k$ of measurements delivered by the sensor;
  - probability of object existence $p(\chi_{k-1}|\mathbf{Y}^{k-1})$; and
  - for each track component c_{k-1}:
    - relative probability $p(c_{k-1})$; and
    - mean $\hat{\mathbf{x}}_{k-1|k-1}(c_{k-1})$ and covariance $\mathbf{P}_{k-1|k-1}(c_{k-1})$.
2: Track state propagation {Sections 2.2.1, 5.5.1}:

$$p(\chi_k|\mathbf{Y}^{k-1}) = \gamma_{11} p(\chi_{k-1}|\mathbf{Y}^{k-1}).$$

3: **for** each track component c_{k-1} **do**

$$[\hat{\mathbf{x}}_{k|k-1}(c_{k-1}), \mathbf{P}_{k|k-1}(c_{k-1})] = \mathrm{KF_P}[\hat{\mathbf{x}}_{k-1|k-1}(c_{k-1}), \mathbf{P}_{k-1|k-1}(c_{k-1}), \mathbf{F}, \mathbf{Q}].$$

4: **end for**
5: Measurement selection {Section 5.5.2, (5.78)}.
6: **for** each track component c_{k-1} **do**

$$[\mathbf{y}_k(c_{k-1}), V_k(c_{k-1})] = \mathrm{MS}_1[\mathbf{Y}_k, \hat{\mathbf{x}}_{k|k-1}(c_{k-1}), \mathbf{P}_{k|k-1}(c_{k-1}), \mathbf{H}, \mathbf{R}].$$

7: **end for**

$$\mathbf{y}_k = \bigcup_{c_{k-1}} \mathbf{y}_k(c_{k-1}).$$

8: **for** each selected measurement $\mathbf{y}_k(i)$ **do**
9:   **for** each track component c_{k-1} **do**
10:     Selected measurement likelihoods {(5.27)}:

$$[\{p_k(i, c_{k-1})\}_i] = \mathrm{ML}_1[\{\mathbf{y}_k(i)\}_i, \hat{\mathbf{x}}_{k|k-1}(c_{k-1}), \mathbf{P}_{k|k-1}(c_{k-1}), \mathbf{H}, \mathbf{R}].$$

11:   **end for**
12:   Measurement likelihoods:

$$p_k(i) = \sum_{c_{k-1}} p(c_{k-1}) p_k(i, c_{k-1}).$$

13: **end for**
14: **if** non-parametric tracking **then**
15:   V_k calculated by (9.6) {Section 9.3}.
16:   Clutter measurement density estimation {Section 9.3}:

$$\rho = \hat{m}_k/V_k = (m_k - P_D P_G p(\chi_k|\mathbf{Y}^{k-1}))/V_k.$$

17: **end if**
18: Single-object data association {Section 5.5.3}:

$$\left[p(\chi_k|\mathbf{Y}^k), \{\boldsymbol{\beta}_k(i)\}_{i=0}^{m_k}\right] = \mathrm{ISTDA}\left[p(\chi_k|\mathbf{Y}^{k-1}), \{p_k(i)\}_{i=1}^{m_k}\right].$$

Algorithm 26 ITS filter recursion equations at time k: part II

19: Form tentative components $\tilde{c}_k = \{i, c_{k-1}\}, i = 0, \ldots, m_k$.

20: **for** each $\tilde{c}_k$ **do** {(5.90) and (5.92)}:

$$p(\tilde{c}_k) = \beta_k(i) p(c_{k-1}) \cdot \begin{cases} 1, & i = 0, \\ \dfrac{p_k(i, c_{k-1})}{p_k(i)}, & i > 0, \end{cases}$$

$$[\hat{\mathbf{x}}_{k|k}(\tilde{c}_k), \mathbf{P}_{k|k}(\tilde{c}_k)] = \begin{cases} [\hat{\mathbf{x}}_{k|k-1}(c_{k-1}), \mathbf{P}_{k|k-1}(c_{k-1})], & i = 0, \\ \mathrm{KF_E}[\mathbf{y}_k(i), \hat{\mathbf{x}}_{k|k-1}(c_{k-1}), \mathbf{P}_{k|k-1}(c_{k-1}), \mathbf{H}, \mathbf{R}], & i > 0. \end{cases}$$

21: **end for**

22: Component control {Section 5.6}:

$$\{\tilde{c}_k, \{p(\tilde{c}_k), \hat{\mathbf{x}}_{k|k}(\tilde{c}_k), \mathbf{P}_{k|k}(\tilde{c}_k)\}_{\tilde{c}_k}\} \rightarrow \{c_k, \{p(c_k), \hat{\mathbf{x}}_{k|k}(c_k), \mathbf{P}_{k|k}(c_k)\}_{c_k}\}.$$

23: Output trajectory estimate:

$$[\hat{\mathbf{x}}_{k|k}, \mathbf{P}_{k|k}] = \mathrm{GMix}\big[\{\hat{\mathbf{x}}_{k|k}(c_k), \mathbf{P}_{k|k}(c_k), p(c_k)\}_{c_k}\big].$$

---

estimate pdf approximates the true pdf even better than ITS. As a consequence, IMM-ITS further improves on the ITS performance in track retention, false track discrimination and object trajectory state estimation in the case of maneuvering objects.

Interacting multiple models – integrated track splitting (IMM-ITS) is a single-object, multi-scan, estimator of maneuvering object trajectory in clutter, as well as a recursive estimator of the probability of object existence. IMM-ITS propagates and updates a single track in isolation (when updating one track, the other tracks are being ignored), and:

- assumes random object existence and recursively calculates the probability of object existence;
- assumes that the object may maneuver, i.e., that the object may switch between more than one trajectory models (the number of trajectory models, $M \geq 1$);
- is a multi-scan algorithm, with a non-negative number of retained track components.

If the number of object trajectory models equals one, IMM-ITS reverts, both conceptually and algorithmically, to ITS.

Track component control (Section 5.6) is an integral part of the IMM-ITS application (although not necessary from a strictly theoretical point of view). However, as opposed to IMM-IPDA, the system designer has absolute freedom (limited by the computational resources, of course) of the track component control method and the number of retained track components. In common with ITS, the track component control provides a trade-off mechanism between computational requirements and object tracking performance. If merging component control is applied to the limit, and all track components are merged into one, IMM-ITS reverts to IMM-IPDA.

In Section 5.5.3, IMM-ITS equations are presented for the case of parametric object tracking, when the clutter measurement density is available either theoretically, or by using some version of multi-scan clutter mapping estimators.

If the clutter measurement density is a priori unknown (the non-parametric case), ITS estimates the clutter measurement density within the selection gate, as presented in Section 9.3.

A single-scan IMM-ITS recursion is detailed in Algorithms 27 and 28. Only one track is implicitly assumed; however, if more than one tracks are initialized, this procedure is repeated for each existing track.

5.8 Object-existence-based multi-object tracking

This section presents object-existence-based multi-object tracking algorithms, from the simplest (joint probabilistic data association or JPDA) to the most complex (interacting multiple models – joint integrated track splitting or IMM-JITS).

Each algorithm assumes a track state which is a specialization of the general track model presented in Section 5.3. The specialization may be the existence of only one object trajectory model (non-maneuvering object tracking), merging of all track components into one (single-scan object tracking) or assuming object existence (deterministic object existence).

All algorithms presented in this section share optimal multi-object tracking data association. However, due to the necessity to limit the number of track components commensurate to the computational resources available, as described in Section 5.6, these algorithms are not optimal.

All algorithms are recursive. Starting from the previously updated track state estimate (at time $k - 1$), measurements delivered by sensor(s) at time k are used to calculate the updated track state estimate (at time k). Only one update cycle for each algorithm is presented, as the new track initialization is similar for all algorithms, and is presented in Section 9.4.

In some cases, JPDA being one, the algorithms differ somewhat from the originally published version; the differences are noted.

There is a substantial amount of repetition between algorithms, as each subsequent algorithm is a generalization of the previous one. The intention of the authors was to describe each algorithm briefly, and to note the amount of specialization

---

Algorithm 27 IMM-ITS filter recursion equations at time k: part I

1: Time k inputs:
- set $\mathbf{Y}_k$ of measurements delivered by the sensor;
- probability of object existence $p(\chi_{k-1}|\mathbf{Y}^{k-1})$; and
- for each track component c_{k-1}:
  - relative probability $p(c_{k-1})$, and for each object trajectory model σ:
    * time $k-1$ a posteriori model probability $\mu_{k-1|k-1}(c_{k-1}, \sigma)$; and
    * mean $\hat{\mathbf{x}}_{k-1|k-1}(c_{k-1}, \sigma)$ and covariance $\mathbf{P}_{k-1|k-1}(c_{k-1}, \sigma)$.

2: Track state propagation {Sections 3.4.1, 5.5.1}:

$$p(\chi_k|\mathbf{Y}^{k-1}) = \gamma_{11} p(\chi_{k-1}|\mathbf{Y}^{k-1}).$$

3: **for** each track component c_{k-1} **do** {IMM mixing and prediction}

$$\left[\{\mu_{k|k-1}(c_{k-1}, \sigma), \hat{\mathbf{x}}_{k|k-1}(c_{k-1}, \sigma), \mathbf{P}_{k|k-1}(c_{k-1}, \sigma)\}_\sigma\right]$$
$$= \text{IMM}_{\text{MP}}\left[\{\mu_{k-1|k-1}(c_{k-1}, \sigma), \hat{\mathbf{x}}_{k-1|k-1}(c_{k-1}, \sigma),\right.$$
$$\left.\mathbf{P}_{k-1|k-1}(c_{k-1}, \sigma), \mathbf{F}_\sigma, \mathbf{Q}_\sigma\}_\sigma, \mathbf{\Gamma}\right].$$

4: **end for**
5: Measurement selection {Section 5.5.2, (5.78)}.
6: **for** (each track component c_{k-1}) and (each trajectory model σ) **do**

$$[\mathbf{y}_k(c_{k-1}, \sigma), V_k(c_{k-1}, \sigma)] = \text{MS}_1[\mathbf{Y}_k, \hat{\mathbf{x}}_{k|k-1}(c_{k-1}, \sigma), \mathbf{P}_{k|k-1}(c_{k-1}, \sigma), \mathbf{H}, \mathbf{R}].$$

7: **end for**

$$\mathbf{y}_k = \bigcup_{c_{k-1}} \bigcup_\sigma \mathbf{y}_k(c_{k-1}, \sigma).$$

8: **for** each selected measurement $\mathbf{y}_k(i)$ **do** {measurement likelihood}
9:    **for** each track component c_{k-1} **do**
10:        **for** each trajectory model σ **do** {(5.27)}

$$[\{p_k(i, c_{k-1}, \sigma)\}_i] = \text{ML}_1[\{\mathbf{y}_k(i)\}_i, \hat{\mathbf{x}}_{k|k-1}(c_{k-1}, \sigma), \mathbf{P}_{k|k-1}(c_{k-1}, \sigma), \mathbf{H}, \mathbf{R}].$$

11:        **end for**

$$p_k(i, c_{k-1}) = \sum_\sigma \mu_{k|k-1}(c_{k-1}, \sigma) p_k(i, c_{k-1}, \sigma).$$

12:    **end for**

$$p_k(i) = \sum_{c_{k-1}} p(c_{k-1}) p_k(i, c_{k-1}).$$

13: **end for**
14: **if** non-parametric tracking **then**

15:     V_k calculated by (9.6) {Section 9.3}.

16:     Clutter measurement density estimation {Section 9.3}:

$$\rho = \hat{m}_k / V_k = (m_k - P_D P_G p(\chi_k | \mathbf{Y}^{k-1})) / V_k.$$

17: **end if**

18: Single-object data association {Section 5.5.3}:

$$\left[p(\chi_k | \mathbf{Y}^k), \{\boldsymbol{\beta}_k(i)\}_{i=0}^{m_k} \right] = \text{ISTDA} \left[p(\chi_k | \mathbf{Y}^{k-1}), \{p_k(i)\}_{i=1}^{m_k} \right].$$

Algorithm 28 IMM-ITS filter recursion equations at time k: part II

19: Form tentative components $\tilde{c}_k = \{i, c_{k-1}\}, i = 0, \ldots, m_k$.

20: **for** each $\tilde{c}_k$ **do** {apply (5.90), (5.91), (5.92)}

$$p(\tilde{c}_k) = \boldsymbol{\beta}_k(i) p(c_{k-1}) \cdot \begin{cases} 1, & i = 0, \\ \dfrac{p_k(i, c_{k-1})}{p_k(i)}, & i > 0. \end{cases}$$

21:     **for** each trajectory model σ **do**

$$\mu_{k|k}(\tilde{c}_k, \sigma) = \mu_{k|k-1}(c_{k-1}, \sigma) \cdot \begin{cases} 1, & i = 0, \\ \dfrac{p_k(i, c_{k-1}, \sigma)}{p_k(i, c_{k-1})}, & i > 0, \end{cases}$$

$$\left[\hat{\mathbf{x}}_{k|k}(\tilde{c}_k, \sigma), \mathbf{P}_{k|k}(\tilde{c}_k, \sigma) \right]$$
$$= \begin{cases} \left[\hat{\mathbf{x}}_{k|k-1}(c_{k-1}, \sigma), \mathbf{P}_{k|k-1}(c_{k-1}, \sigma) \right], & i = 0, \\ \text{KF}_{\text{E}} \left[\mathbf{y}_k(i), \hat{\mathbf{x}}_{k|k-1}(c_{k-1}, \sigma), \mathbf{P}_{k|k-1}(c_{k-1}, \sigma), \mathbf{H}, \mathbf{R} \right], & i > 0. \end{cases}$$

22:     **end for**

23: **end for**

24: Component control {Section 5.6}:

$$\{\tilde{c}_k, \{p(\tilde{c}_k), \{\mu_{k|k}(\tilde{c}_k, \sigma), \hat{\mathbf{x}}_{k|k}(\tilde{c}_k, \sigma), \mathbf{P}_{k|k}(\tilde{c}_k, \sigma)\}_\sigma\}_{\tilde{c}_k}\}$$
$$\rightarrow \{c_k, \{p(c_k), \{\mu_{k|k}(c_k, \sigma), \hat{\mathbf{x}}_{k|k}(c_k, \sigma), \mathbf{P}_{k|k}(c_k, \sigma)\}_\sigma\}_{c_k}\}.$$

25: Output trajectory estimate:

$$[\hat{\mathbf{x}}_{k|k}, \mathbf{P}_{k|k}] = \text{GMix}\left[\{\hat{\mathbf{x}}_{k|k}(c_k, \sigma), \mathbf{P}_{k|k}(c_k, \sigma), p(c_k) \mu_{k|k}(c_k, \sigma)\}_{c_k, \sigma}\right].$$

with respect to the optimal object tracking. Each algorithm should be read and understood by itself, without having to read and understand the other algorithms.

5.8.1 *Joint probabilistic data association (JPDA)*

Joint probabilistic data association (JPDA) is a single-scan estimator of non-maneuvering multi-object trajectories in clutter. JPDA is the simplest, and also computationally most efficient, multi-object tracking algorithm presented in this chapter. It does not, however, provide the probability of object existence to be used as a track quality measure by the false track discrimination procedure.

- JPDA assumes that each track under consideration follows an existing unique object, i.e., the probability of object existence for each track equals one, and is not updated,

$$p(\chi_{k-1}|\mathbf{Y}^{k-1}) = p(\chi_k|\mathbf{Y}^{k-1}) = p(\chi_k|\mathbf{Y}^k) = 1.$$

- JPDA assumes that none of the objects being tracked maneuvers, i.e., that the objects follow a single trajectory model (the number of trajectory models, $M = 1$),

$$\mu_{k-1|k-1}(1) = \mu_{k|k-1}(1) = \mu_{k|k}(1) = 1.$$

Different objects (tracks) may follow different trajectory models, however.
- JPDA is a single-scan algorithm, i.e., all tentative track components are merged into one (the number of both prior and posterior track components equals one, $C_{k-1} = C_k = 1$), and both a priori and a posteriori pdf of object trajectory estimates are approximated by a single Gaussian pdf for each track:

$$p(c_{k-1}) = p(c_k) = 1.$$

Due to the single-object trajectory model, conditioning on the object trajectory model σ cancels out; ditto for the track state estimate components:

$$p(\cdot|\sigma, \cdot) = p(\cdot|\cdot),$$

$$p(\cdot|c_{k-1}, \cdot) = p(\cdot|\cdot).$$

For example,

$$p(\mathbf{x}_k|\chi_k, \mathbf{Y}^{k-1}) = p(\mathbf{x}_k|c_{k-1}, \chi_k, \mathbf{Y}^{k-1}) = p(\mathbf{x}_k|c_{k-1}, \sigma, \chi_k, \mathbf{Y}^{k-1}),$$

$$p_k(i) = p_k(i, c_k - 1) = p_k(i, c_{k-1}, \sigma).$$

Track component control (Section 5.6) is an integral part of the JPDA track trajectory update. All tentative track components are merged into one, and the track update operation becomes the PDA estimation, which is detailed and presented in Section 4.3.

The original JPDA (Bar-Shalom and Fortmann, 1988) assumes a uniform clutter measurement density not only per each cluster area, but across the whole of the surveillance area. In this section, the parametric JPDA is introduced, which uses prior values for the non-uniform clutter measurement density for each measurement individually.

Non-parametric JPDA equations are modified from the original ones, to calculate the clutter measurement density for each cluster separately. Details may be found in Section 9.3.

If the tracks are well separated, JPDA reverts to PDA. A single-scan JPDA recursion is detailed in Algorithm 29.

5.8.2 Interacting multiple models – joint probabilistic data association (IMM-JPDA)

Interacting multiple models – joint probabilistic data association (IMM-JPDA) is the maneuvering object extension of JPDA, presented in Section 5.8.1. IMM-JPDA is a single-scan estimator of maneuvering multi-object trajectories in clutter. In common with JPDA, IMM-JPDA does not provide the probability of object existence to be used as a track quality measure by the false track discrimination procedure.

- IMM-JPDA assumes that the object must exist for each track under consideration, i.e., the probability of object existence equals one, and is not updated,

$$p(\chi_{k-1}|\mathbf{Y}^{k-1}) = p(\chi_k|\mathbf{Y}^{k-1}) = p(\chi_k|\mathbf{Y}^k) = 1.$$

- IMM-JPDA assumes that the objects may maneuver, i.e., that each object may switch between more than one trajectory models (the number of trajectory models, $M \geq 1$). Each object may have its own set of trajectory models and model transition probabilities.
- IMM-JPDA is a single-scan algorithm, i.e., all tentative track components are merged into one (the number of both prior and posterior track components equals one, $C_{k-1} = C_k = 1$),

$$p(c_{k-1}) = p(c_k) = 1.$$

Due to the single-track component, conditioning on the track component cancels out:

$$p(\cdot|c_{k-1}, \cdot) = p(\cdot|\cdot).$$

Algorithm 29 JPDA filter recursion equations at time k

1: Time k inputs:
  - set $\mathbf{Y}_k$ of measurements delivered by the sensor; and
  - for each track τ trajectory estimate mean $\hat{\mathbf{x}}_{k-1|k-1}^{\tau}$ and covariance $\mathbf{P}_{k-1|k-1}^{\tau}$.

2: **for** each track τ **do**

3:   Track state prediction {Sections 2.2.1, 5.5.1}:

$$\left[\hat{\mathbf{x}}_{k|k-1}^{\tau}, \mathbf{P}_{k|k-1}^{\tau}\right] = \mathrm{KF}_{\mathrm{P}}\left[\hat{\mathbf{x}}_{k-1|k-1}^{\tau}, \mathbf{P}_{k-1|k-1}^{\tau}, \mathbf{F}, \mathbf{Q}\right].$$

4:   Measurement selection {Section 5.5.2, (5.78)}:

$$\left[\mathbf{y}_k^{\tau}, V_k^{\tau}\right] = \mathrm{MS}_1\left[\mathbf{Y}_k, \hat{\mathbf{x}}_{k|k-1}^{\tau}, \mathbf{P}_{k|k-1}^{\tau}, \mathbf{H}, \mathbf{R}\right].$$

5:   Measurement likelihood for each selected measurement i {(5.27)}:

$$\left[\{p_k^{\tau}(i)\}_i\right] = \mathrm{ML}_1\left[\{\mathbf{y}_k(i)\}_i, \hat{\mathbf{x}}_{k|k-1}^{\tau}, \mathbf{P}_{k|k-1}^{\tau}, \mathbf{H}, \mathbf{R}\right].$$

6: **end for**

7: **for** each cluster **do** {Section 5.5.4}:

$$\mathbf{y}_k = \bigcup_{\tau} \mathbf{y}_k^{\tau}.$$

8:   **if** non-parametric tracking **then**

9:     Cluster V_k calculated by (9.7) {Section 9.3}:

10:    Clutter measurement density estimation {Section 9.3, (9.11) and (9.7)}:

$$\rho = \mathrm{MTT}_{\mathrm{MK}}(\{1\}_{\tau}, \{p_k^{\tau}(i)\}_{\tau,i})/V_k.$$

11: **end if**

12: Multi-object data association (sàns object existence) {Section 5.5.4, (5.87)}:

$$\left[\{-, \{\boldsymbol{\beta}_k^{\tau}(i)\}_{i\geq 0}\}_{\tau}\right] = \mathrm{JMTDA}\left[\{1, \{p_k^{\tau}(i)\}_{i>0}\}_{\tau}\right].$$

13: **end for**

14: **for** each track τ **do**

15: Estimation/merging {Section 4.3}:

$$\left[\hat{\mathbf{x}}_{k|k}^{\tau}, \mathbf{P}_{k|k}^{\tau}\right] = \mathrm{PDA}_{\mathrm{E}}\left[\hat{\mathbf{x}}_{k|k-1}^{\tau}, \mathbf{P}_{k|k-1}^{\tau}, \{\mathbf{y}_k(i)\}_{i=1}^{m_k}, \{\boldsymbol{\beta}_k^{\tau}(i)\}_{i=0}^{m_k}, \mathbf{H}, \mathbf{R}\right].$$

16: Output trajectory estimate:
  - track mean value $\hat{\mathbf{x}}_{k|k}^{\tau}$ and covariance $\mathbf{P}_{k|k}^{\tau}$.

17: **end for**

For example,

$$p(\mathbf{x}_k|\chi_k, \mathbf{Y}^{k-1}) = p(\mathbf{x}_k|c_{k-1}, \chi_k, \mathbf{Y}^{k-1}),$$

$$p_k(i, c_{k-1}, \sigma) = p_k(i, \sigma),$$

$$\boldsymbol{\mu}_{k|k}(\tilde{c}_k, \sigma) = \boldsymbol{\mu}_{k|k}(i, \sigma).$$

If the number of object trajectory models equals one, IMM-JPDA reverts, both conceptually and algorithmically, to JPDA. If the tracks are well separated, IMM-JPDA reverts to IMM-PDA.

Track component control (Section 5.6) is an integral part of the IMM-JPDA track trajectory update. All tentative track components are merged into one and the track update operation becomes the PDA estimation, performed once for each object trajectory model σ.

The IMM-JPDA algorithm, as presented in this chapter, allows parametric object tracking in non-homogeneous clutter. If the clutter measurement density is a priori unknown (non-parametric case), IMM-JPDA estimates the clutter measurement density within each cluster by using the procedure defined in Section 9.3, which is somewhat different from the originally published approach (Blom and Bloem, 2002).

A single-scan IMM-JPDA recursion is detailed in Algorithm 30. As mentioned previously, the algorithm shows that the trajectory estimate update, detailed in Algorithm 22, is identical in both IMM-PDA and IMM-JPDA, as well as in both IMM-IPDA and IMM-JIPDA (below). In other words, the trajectory estimate update does not depend on the probability of object existence, which is a desirable outcome.

5.8.3 *Joint integrated probabilistic data association (JIPDA)*

Joint integrated probabilistic data association (JIPDA) was published in Mušicki and Evans (2002), and Mušicki *et al.* (2004b). JIPDA extends JPDA by adding the recursive propagation and update of the probability of object existence. From a different viewpoint, JIPDA extends IPDA by providing multi-object capabilities.

JIPDA is a single-scan estimator of non-maneuvering trajectories of multiple possible objects in clutter and their probabilities of object existence. As the probability of object existence provides the tools for the false track discrimination procedure, JIPDA is the simplest multi-object tracking in clutter algorithm presented in this chapter, which can be a basis for a complete object tracking solution. Due to its simple structure, JIPDA is computationally the most efficient multi-object random object existence estimator presented here. An application of linear multi-target tracking, described in Section 9.2, also enables simple and computationally efficient multi-object tracking.

Algorithm 30 IMM–JPDA filter recursion equations at time k

1: Time k inputs:
- set $\mathbf{Y}_k$ of measurements delivered by the sensor; and
- for each track τ and each object trajectory model σ:
  - time $k-1$ a posteriori model probability $\mu^\tau_{k-1|k-1}(\sigma)$; and
  - mean $\hat{\mathbf{x}}^\tau_{k-1|k-1}(\sigma)$ and covariance $\mathbf{P}^\tau_{k-1|k-1}(\sigma)$.

2: **for** each track τ **do**

3:    IMM mixing and prediction {Sections 3.4.1: 5.5.1}:

$$\left[\left\{\mu^\tau_{k|k-1}(\sigma), \hat{\mathbf{x}}^\tau_{k|k-1}(\sigma), \mathbf{P}^\tau_{k|k-1}(\sigma)\right\}_\sigma\right]$$
$$= \mathrm{IMM_{MP}}\left[\left\{\mu^\tau_{k-1|k-1}(\sigma), \hat{\mathbf{x}}^\tau_{k-1|k-1}(\sigma), \mathbf{P}^\tau_{k-1|k-1}(\sigma), \mathbf{F}_\sigma, \mathbf{Q}_\sigma\right\}_\sigma, \mathbf{\Gamma}\right].$$

4:    Measurement selection for each model σ {Section 5.5.2, (5.78)}:

$$\left[\mathbf{y}^\tau_k(\sigma), V^\tau_k(\sigma)\right] = \mathrm{MS}_1\left[\mathbf{Y}_k, \hat{\mathbf{x}}^\tau_{k|k-1}(\sigma), \mathbf{P}^\tau_{k|k-1}(\sigma), \mathbf{H}, \mathbf{R}\right].$$

5:    **for** each selected measurement i **do** {measurement likelihood, (5.27)}

$$\left[\left\{p^\tau_k(i, \sigma)\right\}_i\right] = \mathrm{ML}_1\left[\{\mathbf{y}_k(i)\}_i, \hat{\mathbf{x}}^\tau_{k|k-1}(\sigma), \mathbf{P}^\tau_{k|k-1}(\sigma), \mathbf{H}, \mathbf{R}\right], \qquad \text{for each } \sigma,$$
$$p^\tau_k(i) = \sum_\sigma \mu^\tau_{k|k-1}(\sigma) p^\tau_k(i, \sigma).$$

6:    **end for**

7: **end for**

8: **for** each cluster **do** {Section 5.5.4}

$$\mathbf{y}_k = \bigcup_\tau \bigcup_\sigma \mathbf{y}^\tau_k(\sigma).$$

9:    **if** non-parametric tracking **then**

10:       V_k calculated by (9.7) {Section 9.3}.

11:       Clutter measurement density estimation {Section 9.3, (9.11) and (9.7)}:

$$\rho = \mathrm{MTT_{MK}}(\{1\}_\tau, \{p^\tau_k(i)\}_{\tau,i}) / V_k.$$

12:    **end if**

13:    Multi-object data association (sàns object existence) {Section 5.5.4, (5.87)}:

$$\left[\left\{-, \{\beta^\tau_k(i)\}_{i\geq 0}\right\}_\tau\right] = \mathrm{JMTDA}\left[\{1, \{p^\tau_k(i)\}_{i>0}\}_\tau\right].$$

14: **end for**

15: For each track τ apply Algorithm 22 {IMM-PDA, part II: trajectory estimation update}.

- JIPDA assumes random object existence and recursively calculates the probability of object existence for each track under consideration.
- JIPDA assumes that each object (if it exists) does not maneuver, i.e., that the object follows a single trajectory model (the number of trajectory models, $M = 1$),

$$\mu_{k-1|k-1}(1) = \mu_{k|k-1}(1) = \mu_{k|k}(1) = 1.$$

Different potential objects (tracks) may follow different trajectory models.

- JIPDA is a single-scan algorithm, i.e., all tentative track components of each track are merged into one (the number of both prior and posterior track components equals one, $C_{k-1} = C_k = 1$),

$$p(c_{k-1}) = p(c_k) = 1.$$

Due to the single-object trajectory model, conditioning on the object trajectory model σ cancels out; ditto for the track state estimate components:

$$p(\cdot|\sigma, \cdot) = p(\cdot|\cdot)$$

$$p(\cdot|c_{k-1}, \cdot) = p(\cdot|\cdot).$$

For example,

$$p(\mathbf{x}_k|\chi_k, \mathbf{Y}^{k-1}) = p(\mathbf{x}_k|c_{k-1}, \chi_k, \mathbf{Y}^{k-1}) = p(\mathbf{x}_k|c_{k-1}, \sigma, \chi_k, \mathbf{Y}^{k-1}),$$

$$p_k(i) = p_k(i, c_{k-1}) = p_k(i, c_{k-1}, \sigma).$$

Track component control (Section 5.6) is an integral part of the JIPDA track trajectory update. All tentative track components are merged into one, and the track update operation becomes the PDA estimation, which is detailed and presented in Section 4.3.

In Section 5.5.4, JIPDA equations are presented for the case of parametric object tracking, when the clutter measurement density is available either theoretically, or by using some version of the multi-scan clutter mapping estimators.

If the clutter measurement density is a priori unknown (the non-parametric case), JIPDA estimates the clutter measurement density within each cluster area, as presented in Section 9.3.

In the case of well-separated tracks, JIPDA reverts to IPDA. The single-scan JIPDA recursion is detailed in Algorithm 31.

5.8.4 Interacting multiple models – joint integrated probabilistic data association (IMM-JIPDA)

Interacting multiple models – joint integrated probabilistic data association (IMM-JIPDA) was published in Mušicki and Suvorova (2008). IMM-JIPDA is the

---

Algorithm 31 JIPDA filter recursion equations at time k

1: Time k inputs:
  - set $\mathbf{Y}_k$ of measurements delivered by the sensor;
  - for each track τ, the probability of object existence $p(\chi_{k-1}^\tau | \mathbf{Y}^{k-1})$; and
  - trajectory estimate mean $\hat{\mathbf{x}}_{k-1|k-1}^\tau$ and covariance $\mathbf{P}_{k-1|k-1}^\tau$.

2: **for** each track τ **do**

3:   Track state propagation {Sections 2.2.1: 5.5.1}:

$$p(\chi_k^\tau | \mathbf{Y}^{k-1}) = \gamma_{11} p(\chi_{k-1}^\tau | \mathbf{Y}^{k-1}),$$

$$\left[\hat{\mathbf{x}}_{k|k-1}^\tau, \mathbf{P}_{k|k-1}^\tau\right] = \mathrm{KF_P}\left[\hat{\mathbf{x}}_{k-1|k-1}^\tau, \mathbf{P}_{k-1|k-1}^\tau, \mathbf{F}, \mathbf{Q}\right].$$

4:   Measurement selection {Section 5.5.2, (5.78)}:

$$[\mathbf{y}_k^\tau, V_k^\tau] = \mathrm{MS_1}\left[\mathbf{Y}_k, \hat{\mathbf{x}}_{k|k-1}^\tau, \mathbf{P}_{k|k-1}^\tau, \mathbf{H}, \mathbf{R}\right].$$

5:   Measurement likelihood for each selected measurement i {(5.27)}:

$$\left[\{p_k^\tau(i)\}_i\right] = \mathrm{ML_1}\left[\{\mathbf{y}_k(i)\}_i, \hat{\mathbf{x}}_{k|k-1}^\tau, \mathbf{P}_{k|k-1}^\tau, \mathbf{H}, \mathbf{R}\right].$$

6: **end for**

7: **for** each cluster **do** {Section 5.5.4}

$$\mathbf{y}_k = \bigcup_\tau \mathbf{y}_k^\tau.$$

8:   **if** non-parametric tracking **then**

9:     Cluster V_k calculated by (9.7) {Section 9.3}.

10:    Clutter measurement density estimation {Section 9.3, (9.11) and (9.7)}:

$$\rho = \mathrm{MTT_{MK}}(\{p(\chi_k^\tau | \mathbf{Y}^{k-1})\}_\tau, \{p_k^\tau(i)\}_{\tau,i})/V_k.$$

11:    **end if**

12:    Multi-object data association {Section 5.5.4, (5.87)}:

$$\left[\{p(\chi_k^\tau | \mathbf{Y}^k), \{\boldsymbol{\beta}_k^\tau(i)\}_{i \geq 0}\}_\tau\right] = \mathrm{JMTDA}\left[\{p(\chi_k^\tau | \mathbf{Y}^{k-1}), \{p_k^\tau(i)\}_{i>0}\}_\tau\right].$$

13: **end for**

14: **for** each track τ **do**

15:    Estimation/merging {Section 4.3}:

$$\left[\hat{\mathbf{x}}_{k|k}^\tau, \mathbf{P}_{k|k}^\tau\right] = \mathrm{PDA_E}\left[\hat{\mathbf{x}}_{k|k-1}^\tau, \mathbf{P}_{k|k-1}^\tau, \{\mathbf{y}_k(i)\}_{i=1}^{m_k}, \{\boldsymbol{\beta}_k^\tau(i)\}_{i=0}^{m_k}, \mathbf{H}, \mathbf{R}\right].$$

16:    Output trajectory estimate:
  - track mean value $\hat{\mathbf{x}}_{k|k}^\tau$ and covariance $\mathbf{P}_{k|k}^\tau$.

17: **end for**

---

maneuvering object extension of JIPDA presented in Section 5.8.3, as well as the multi-object extension of IMM-IPDA presented in Section 5.7.4.

IMM-JIPDA is a single-scan estimator of maneuvering trajectories of multiple possible objects in clutter and their probabilities of object existence.

- IMM-JIPDA assumes random object existence and recursively calculates the probability of object existence for each track under consideration.
- IMM-JIPDA assumes that each object (if it exists) may maneuver, i.e., that the object may switch between more than one trajectory models (the number of trajectory models $M \geq 1$). Each object may have an individual set of trajectory models and model transition probabilities.
- IMM-JIPDA is a single-scan algorithm, i.e., all tentative track components of each track are merged into one (the number of both prior and posterior track components equals one, $C_{k-1} = C_k = 1$),

$$p(c_{k-1}) = p(c_k) = 1.$$

In other words, due to the single-track component, conditioning on the track component cancels out:

$$p(\cdot|c_{k-1}, \cdot) = p(\cdot|\cdot).$$

For example,

$$p(\mathbf{x}_k|\chi_k, \mathbf{Y}^{k-1}) = p(\mathbf{x}_k|c_{k-1}, \chi_k, \mathbf{Y}^{k-1}),$$

$$p_k(i, c_{k-1}, \sigma) = p_k(i, \sigma),$$

$$\mu_{k|k}(\tilde{c}_k, \sigma) = \mu_{k|k}(i, \sigma).$$

If the number of object trajectory models equals one, IMM-JIPDA reverts, both conceptually and algorithmically, to JIPDA. In the case of well-separated tracks, IMM-JIPDA reverts to IMM-IPDA.

Track component control (Section 5.6) is an integral part of the IMM-JIPDA track trajectory update. All tentative track components are merged into one and the track trajectory update operation becomes the PDA estimation, performed once for each object trajectory model σ.

In Section 5.5.4, IMM-JIPDA equations are presented for the case of parametric object tracking, when the clutter measurement density is available either theoretically, or by using some version of the multi-scan clutter mapping estimators.

If the clutter measurement density is a priori unknown (the non-parametric case), IMM-JIPDA estimates the clutter measurement density within the cluster area, as presented in Section 9.3.

Algorithm 32 IMM-JIPDA filter recursion equations at time k

1: Time k inputs:
   - set $\mathbf{Y}_k$ of measurements delivered by the sensor; and
   - for each track τ, the probability of object existence $p(\chi_{k-1}^\tau | \mathbf{Y}^{k-1})$; and
   - for each object trajectory model σ:
     - time $k - 1$ a posteriori model probability $\mu_{k-1|k-1}^\tau(\sigma)$; and
     - mean $\hat{\mathbf{x}}_{k-1|k-1}^\tau(\sigma)$ and covariance $\mathbf{P}_{k-1|k-1}^\tau(\sigma)$.

2: **for** each track τ **do**

3:     Object existence propagation {Section 5.5.1}:

$$p(\chi_k^\tau | \mathbf{Y}^{k-1}) = \gamma_{11} p(\chi_{k-1}^\tau | \mathbf{Y}^{k-1}),$$

4:     IMM mixing and prediction {Section 3.4.1}:

$$\left[\left\{ \mu_{k|k-1}^\tau(\sigma), \hat{\mathbf{x}}_{k|k-1}^\tau(\sigma), \mathbf{P}_{k|k-1}^\tau(\sigma) \right\}_\sigma \right]$$
$$= \mathrm{IMM_{MP}} \left[\left\{ \mu_{k-1|k-1}^\tau(\sigma), \hat{\mathbf{x}}_{k-1|k-1}^\tau(\sigma), \mathbf{P}_{k-1|k-1}^\tau(\sigma), \mathbf{F}_\sigma, \mathbf{Q}_\sigma \right\}_\sigma, \boldsymbol{\Gamma} \right].$$

5:     Measurement selection for each model σ {Section 5.5.2, (5.78)}:

$$[\mathbf{y}_k^\tau(\sigma), V_k^\tau(\sigma)] = \mathrm{MS_1} \left[\mathbf{Y}_k, \hat{\mathbf{x}}_{k|k-1}^\tau(\sigma), \mathbf{P}_{k|k-1}^\tau(\sigma), \mathbf{H}, \mathbf{R} \right].$$

6:     **for** each selected measurement i **do** {measurement likelihood, (5.27)}

$$[\{p_k^\tau(i, \sigma)\}_i] = \mathrm{ML_1} \left[\{\mathbf{y}_k(i)\}_i, \hat{\mathbf{x}}_{k|k-1}^\tau(\sigma), \mathbf{P}_{k|k-1}^\tau(\sigma), \mathbf{H}, \mathbf{R} \right], \qquad \text{for each } \sigma,$$

$$p_k^\tau(i) = \sum_\sigma \mu_{k|k-1}^\tau(\sigma) p_k^\tau(i, \sigma).$$

7:     **end for**

8: **end for**

9: **for** each cluster **do** {Section 5.5.4}

$$\mathbf{y}_k = \bigcup_\tau \bigcup_\sigma \mathbf{y}_k^\tau(\sigma).$$

10:    **if** non-parametric tracking **then**

11:        V_k calculated by (9.7) {Section 9.3}.

12:        Clutter measurement density estimation {Section 9.3, (9.11)}:

$$\rho = \mathrm{MTT_{MK}}(\{p(\chi_k^\tau | \mathbf{Y}^{k-1})\}_\tau, \{p_k^\tau(i)\}_{\tau,i}) / V_k.$$

13:    **end if**

14:    Multi-object data association {Section 5.5.4, (5.87)}:

$$\left[\left\{ p(\chi_k^\tau | \mathbf{Y}^k), \{\beta_k^\tau(i)\}_{i \geq 0} \right\}_\tau \right] = \mathrm{JMTDA} \left[\left\{ p(\chi_k^\tau | \mathbf{Y}^{k-1}), \{p_k^\tau(i)\}_{i>0} \right\}_\tau \right].$$

15: **end for**

16: For each track τ apply Algorithm 22 {IMM-PDA, part II: trajectory estimation update}.

A single-scan IMM-JIPDA recursion is detailed in Algorithm 32. Note (again) that the trajectory estimate update, detailed in Algorithm 22, is identical for IMM-PDA, IMM-IPDA, IMM-JPDA and IMM-JIPDA.

5.8.5 *Joint integrated track splitting (JITS)*

Joint integrated track splitting (JITS) was published in Mušicki *et al.* (2003) and Mušicki and Evans (2008). JITS extends ITS by adding multi-object tracking capabilities. In a similar fashion to ITS, JITS provides a better approximation to object trajectory state estimate pdf than JIPDA. As a consequence, JITS is more successful in track retention, false track discrimination and object trajectory state estimation than JIPDA, especially in difficult situations of the low probability of detection and/or high clutter measurement density. As another not so positive consequence, JITS requires significantly more computational resources than JIPDA. This is tunable and the track component control provides a trade-off mechanism between computational requirements and object tracking performance.

Joint integrated track splitting (JITS) is a multi-scan estimator of non-maneuvering trajectories of multiple possible objects in clutter and their probabilities of object existence.

- JITS assumes random object existence and recursively calculates the probability of object existence for each track under consideration.
- JITS assumes that each object (if it exists) does not maneuver, i.e., that the object follows a single trajectory model (the number of trajectory models, $M = 1$),

$$\mu_{k-1|k-1}(1) = \mu_{k|k-1}(1) = \mu_{k|k}(1) = 1.$$

Each track may have an individual trajectory model.

- JITS is a multi-scan algorithm, with a non-negative number of retained track components.

Due to the single-object trajectory model, conditioning on the object trajectory model σ cancels out:

$$p(\cdot|\sigma, \cdot) = p(\cdot|\cdot).$$

For example,

$$p(\mathbf{x}_k|c_{k-1}, \chi_k, \mathbf{Y}^{k-1}) = p(\mathbf{x}_k|c_{k-1}, \sigma, \chi_k, \mathbf{Y}^{k-1}),$$

$$p_k(i, c_{k-1}) = p_k(i, c_{k-1}, \sigma).$$

Track component control (Section 5.6) is an integral part of JITS applications (although not necessary from a strictly theoretical point of view). However, as

---

Algorithm 33 JITS filter recursion equations at time k

1: Time k inputs:
   - set $\mathbf{Y}_k$ of measurements delivered by the sensor;
   - for each track τ, the probability of object existence $p(\chi_{k-1}^\tau | \mathbf{Y}^{k-1})$; and
   - for each track component c_{k-1}:
     - relative component probability $p(c_{k-1}^\tau)$; and
     - estimate mean $\hat{\mathbf{x}}_{k-1|k-1}^\tau (c_{k-1})$ and covariance $\mathbf{P}_{k-1|k-1}^\tau (c_{k-1})$.

2: **for** each track τ **do**

3:     $p(\chi_k^\tau | \mathbf{Y}^{k-1}) = \gamma_{11} p(\chi_{k-1}^\tau | \mathbf{Y}^{k-1})$ {track state propagation, Sections 2.2.1, 5.5.1}.

4:     Trajectory state propagation for each track component c_{k-1}:

$$\left[\hat{\mathbf{x}}_{k|k-1}^\tau (c_{k-1}), \mathbf{P}_{k|k-1}^\tau (c_{k-1}) \right] = \mathrm{KF}_{\mathrm{P}} \left[\hat{\mathbf{x}}_{k-1|k-1}^\tau (c_{k-1}), \mathbf{P}_{k-1|k-1}^\tau (c_{k-1}), \mathbf{F}, \mathbf{Q} \right].$$

5:     Measurement selection for each track component c_{k-1} {Section 5.5.2, (5.78)}:

$$\left[\mathbf{y}_k^\tau (c_{k-1}), V_k^\tau (c_{k-1}) \right] = \mathrm{MS}_1 \left[\mathbf{Y}_k, \hat{\mathbf{x}}_{k|k-1}^\tau (c_{k-1}), \mathbf{P}_{k|k-1}^\tau (c_{k-1}), \mathbf{H}, \mathbf{R} \right].$$

6:     **for** each selected measurement i **do** {measurement likelihood, (5.27)}

$$\left[\left\{ p_k^\tau (i, c_{k-1}) \right\}_i \right] = \mathrm{ML}_1 \left[\left\{ \mathbf{y}_k (i) \right\}_i, \hat{\mathbf{x}}_{k|k-1}^\tau (c_{k-1}), \mathbf{P}_{k|k-1}^\tau (c_{k-1}), \mathbf{H}, \mathbf{R} \right],$$

$$\text{for each } c_{k-1},$$

$$p_k^\tau (i) = \sum_{c_{k-1}} p(c_{k-1}^\tau) p_k^\tau (i, c_{k-1}).$$

7:     **end for** {each selected measurement i}

8: **end for** {each track τ}

9: **for** each cluster **do** {Section 5.5.4}

$$\mathbf{y}_k = \bigcup_\tau \bigcup_{c_{k-1}} \mathbf{y}_k^\tau (c_{k-1}).$$

10:     **if** non-parametric tracking **then**

11:       V_k calculated by (9.7) {Section 9.3}.

12:       Clutter measurement density estimation {Section 9.3, (9.11)}:

$$\rho = \mathrm{MTT}_{\mathrm{MK}} (\{ p(\chi_k^\tau | \mathbf{Y}^{k-1}) \}_\tau, \{ p_k^\tau (i) \}_{\tau, i}) / V_k.$$

13:     **end if**

14:     Multi-object data association {Section 5.5.4, (5.87)}:

$$\left[\left\{ p(\chi_k^\tau | \mathbf{Y}^k), \left\{ \beta_k^\tau (i) \right\}_{i \geq 0} \right\}_\tau \right] = \mathrm{JMTDA} \left[\left\{ p(\chi_k^\tau | \mathbf{Y}^{k-1}), \left\{ p_k^\tau (i) \right\}_{i > 0} \right\}_\tau \right].$$

15: **end for** {each cluster}

16: For each track τ apply Algorithm 26 {ITS filter, part II: trajectory estimation update}.

---

opposed to JIPDA, the system designer has absolute freedom (limited by the computational resources, of course) of the track component control method and the number of retained track components. If merging component control is applied to the limit, and all track components are merged into one, JITS reverts to JIPDA. If the tracks are well separated, JITS reverts to ITS.

In Section 5.5.4, JITS equations are presented for the case of parametric object tracking, when the (non-homogeneous) clutter measurement density is available either theoretically, or by using some version of the multi-scan clutter mapping estimators.

If the clutter measurement density is a priori unknown (the non-parametric case), JITS estimates the clutter measurement density within each cluster area, as presented in Section 9.3.

A single-scan JITS recursion is detailed in Algorithm 33. The algorithm shows that the trajectory estimate update, detailed in Algorithm 26, is identical for both JITS and ITS. This confirms (again) the seamless connections between algorithms in this chapter.

5.8.6 Interacting multiple models – joint integrated track splitting (IMM-JITS)

Interacting multiple models – joint integrated track splitting (IMM-JITS) was published in Mušicki and Evans (2008). IMM-JITS extends JITS by including multiple-object trajectory models. Thus, the IMM-JITS object trajectory state estimate pdf approximates the true pdf even better than JITS. As a consequence, IMM-JITS further improves on the JITS performance in track retention, false track discrimination and object trajectory state estimation in the case of maneuvering objects.

IMM-JITS is a multi-scan estimator of maneuvering trajectories of multiple possible objects in clutter and their probabilities of object existence.

- IMM-JITS assumes random object existence and recursively calculates the probability of object existence for each track under consideration.
- IMM-JITS assumes that the object (if it exists) may maneuver, i.e., that the object may switch between more than one trajectory models (the number of trajectory models $M \geq 1$). Each track may have an individual set of trajectory models and associated transition probability matrices.

  If the number of object trajectory models equals one, IMM-JITS reverts, both conceptually and algorithmically, to JITS.
- IMM-JITS is a multi-scan algorithm, with a non-negative number of retained track components.

Algorithm 34 IMM-JITS filter recursion equations at time k

1: Time k inputs:
- set $\mathbf{Y}_k$ of measurements delivered by the sensor;
- for each track τ, the probability of object existence $p(\chi_{k-1}^\tau|\mathbf{Y}^{k-1})$; and
- for each component c_{k-1} and object trajectory model σ:
  - relative component probability $p(c_{k-1}^\tau)$;
  - time $k-1$ a posteriori model probability $\mu_{k-1|k-1}^\tau(c_{k-1}, \sigma)$; and
  - estimate mean $\hat{\mathbf{x}}_{k-1|k-1}^\tau(c_{k-1}, \sigma)$ and covariance $\mathbf{P}_{k-1|k-1}^\tau(c_{k-1}, \sigma)$.

2: **for** each track τ state propagation {Sections 3.4.1, 5.5.1}:

$$p(\chi_k^\tau|\mathbf{Y}^{k-1}) = \gamma_{11} p(\chi_{k-1}^\tau|\mathbf{Y}^{k-1}),$$

for each c_{k-1}: $\quad \left[\left\{ \mu_{k|k-1}^\tau(c_{k-1}, \sigma), \hat{\mathbf{x}}_{k|k-1}^\tau(c_{k-1}, \sigma), \mathbf{P}_{k|k-1}^\tau(c_{k-1}, \sigma) \right\}_\sigma \right]$

$= \text{IMM}_{\text{MP}} \Big[\left\{ \mu_{k-1|k-1}^\tau(c_{k-1}, \sigma), \hat{\mathbf{x}}_{k-1|k-1}^\tau(c_{k-1}, \sigma), \mathbf{P}_{k-1|k-1}^\tau(c_{k-1}, \sigma), \right.$

$\left. \mathbf{F}_\sigma, \mathbf{Q}_\sigma \right\}_\sigma, \boldsymbol{\Gamma} \Big].$

{Measurement selection, Section 5.5.2, (5.78)}

3: **for** each track τ, component c_{k-1} and trajectory model σ **do**

$$[\mathbf{y}_k^\tau(c_{k-1}, \sigma), V_k^\tau(c_{k-1}, \sigma)]$$

$$= \text{MS}_1 \Big[\mathbf{Y}_k, \hat{\mathbf{x}}_{k|k-1}^\tau(c_{k-1}, \sigma), \mathbf{P}_{k|k-1}^\tau(c_{k-1}, \sigma), \mathbf{H}, \mathbf{R} \Big].$$

4: **end for**

{Measurement likelihoods, Section 5.5.2, (5.27)}

5: **for** each selected measurement $i > 0$, track τ, component c_{k-1} and trajectory model σ **do**

$$[\{p_k^\tau(i, c_{k-1}, \sigma)\}_i] = \text{ML}_1[\{\mathbf{y}_k(i)\}_i, \hat{\mathbf{x}}_{k|k-1}^\tau(c_{k-1}, \sigma), \mathbf{P}_{k|k-1}^\tau(c_{k-1}, \sigma), \mathbf{H}, \mathbf{R}],$$

$$p_k^\tau(i, c_{k-1}) = \sum_\sigma \mu_{k|k-1}^\tau(c_{k-1}, \sigma) p_k^\tau(i, c_{k-1}, \sigma),$$

$$p_k^\tau(i) = \sum_{c_{k-1}} p(c_{k-1}^\tau) p_k^\tau(i, c_{k-1}).$$

6: **end for**

7: **for** each cluster **do** {Section 5.5.4}

$$\mathbf{y}_k = \bigcup_\tau \bigcup_{c_{k-1}} \bigcup_\sigma \mathbf{y}_k^\tau(c_{k-1}, \sigma).$$

8:    **if** non-parametric tracking **then**

9:       V_k calculated by (9.7) {Section 9.3}.

10:      Clutter measurement density estimation {Section 9.3, (9.11)}:

$$\rho = \text{MTT}_{\text{MK}}(\{p(\chi_k^\tau | \mathbf{Y}^{k-1})\}_\tau, \{p_k^\tau(i)\}_{\tau,i}) / V_k.$$

11:    **end if**

12:    Multi-object data association {Section 5.5.4, (5.87)}:

$$\left[\{p(\chi_k^\tau | \mathbf{Y}^k), \{\boldsymbol{\beta}_k^\tau(i)\}_{i \geq 0}\}_\tau\right] = \text{JMTDA}\left[\{p(\chi_k^\tau | \mathbf{Y}^{k-1}), \{p_k^\tau(i)\}_{i > 0}\}_\tau\right].$$

13: **end for** {each cluster}

14: For each track τ apply Algorithm 28 {IMM-ITS filter, part II: trajectory update}.

Track component control (Section 5.6) is an integral part of the IMM-JITS application (although not necessary from a strictly theoretical point of view). However, as opposed to IMM-JIPDA, the system designer has absolute freedom (limited by the computational resources, of course) of the track component control method and the number of retained track components. If the merging component control is applied to the limit, and all track components are merged into one, IMM-JITS reverts to IMM-JIPDA. If the tracks are well separated, IMM-JITS reverts to IMM-ITS.

In Section 5.5.4, IMM-JITS equations are presented for the case of parametric object tracking, when the (non-homogeneous) clutter measurement density is available either theoretically, or by using some version of the multi-scan clutter mapping estimators.

If the clutter measurement density is a priori unknown (the non-parametric case), IMM-JITS estimates the clutter measurement density within each cluster area, as presented in Section 9.3.

A single-scan IMM-JITS recursion is detailed in Algorithm 34. The algorithm shows that the trajectory estimate update, detailed in Algorithm 28, is identical for both IMM-JITS and IMM-ITS. This confirms (again) the seamless connections between algorithms in this chapter.

5.9 Summary

Existence-based object tracking algorithms are introduced in this chapter. This class of object tracking algorithms is important in practical aspects of tracking. In reality, the set of detections contains false detections. Therefore the object tracker needs a method to distinguish between true and false tracks while catering for missed detections. Recursively calculated probability of object existence provides a basis for solving this problem. This chapter introduces IMM-ITS and IMM-JITS as the generalized Bayesian solution of the problem with the most relaxed set of constraints. Then other algorithms, i.e., ITS, IMM-IPDA, IPDA, IMM-PDA, PDA, JITS, IMM-JIPDA, JIPDA, IMM-JPDA and JPDA are derived for specific scenarios by adding scenario-specific constraints.

6

Multiple-object tracking in clutter: random-set-based approach

Typically, multiple-object tracking problems are handled by extending the single-object tracking algorithms where each object is tracked as an isolated entity. The challenge comes when the targets are close by and there is ambiguity about the origin of the measurement, i.e., which measurements are from which track (in general). Using similar techniques of data association, multiple measurements are assigned to multiple objects (in general). However, such an extension of single-object trackers to multiple-object trackers assumes that one knows the number of objects present in the surveillance space, which is not true.

This problem leads to some of the serious advances and methods of "data association" logic of these trackers. The data association step calculates the origin of the measurements in a probabilistic manner. It hypothesizes the measurement origin and calculates probabilities for each of the hypotheses. For example, a single-object tracking algorithm considers two hypotheses under measurement origin uncertainty – "the measurement is from an object of interest" or "the measurement is from clutter." Such algorithms ignore the possibility of the measurements originating from other objects. This problem is partially solved by introducing the hypothesis "the measurement is from the ith (out of N) objects." But setting the number of objects to a specific value is a limitation by itself. Moreover, this approach does not provide any measure for the validity of the number of objects. Multi-object trackers need to estimate the number of objects and their individual states jointly.

More than one measurement within the validation gate of a track can be considered as a new track hypothesis. A multiple hypothesis tracker (MHT) takes this approach. However, this leads to an unbounded solution set because the framework does not support the track management. Ad-hoc rules like "an object becomes invalid if it does not get measurements for n number of consecutive scans" are applied to limit the number of objects, but these rules (and the parameters of such rules) are arbitrarily set and usually set based on operator's "intuition" rather than any systematic approaches. The integrated PDA filter (IPDA, Chapter 4) and its variant, joint IPDA (JIPDA, Chapter 9), proposes a bounded, albeit approximate, solution where each track (hypothesis) is maintained based on a quality measure (object existence probability) and therefore results in a dynamically changing but limited number of tracks. Another bounded solution is based on IMM-PDA[1], which resolves the problem through "proof by contradiction." It assumes the existence of objects but limits the number of objects based on the probability of observability. These rules are more systematic approaches than MHT-based filtering with respect to track management, but are not optimal methods to estimate the number of targets.

Random finite set formalism (RFS)

In summary, all these multiple-object tracking algorithms are extensions of the corresponding single-object trackers. Even though these provide an idea of the number of objects present, they do not have explicit modeling to estimate the object number. A more rigorous approach is based on (RFS) (Mahler, 2004a, 2004b). Fundamentally it approaches the modeling of multi-object tracking by introducing sets that contain a random number of single-object states, each of which is a random vector. It allows the dynamics of each object (member of the object set) to vary according to some motion model while the the number of objects (members in the set) is allowed to vary according to some point process model. The tracking problem is to compute the posterior density, called the global posterior density, of this set-valued quantity. This set-based representation enables the framework to capture the uncertainty in the number of objects. Moreover, this set notation renders the ordering of each single-object state immaterial, i.e., the set $\{x_1, x_2\}$ is equivalent to $\{x_2, x_1\}$. In this manner it captures the essential elements of the multiple-object tracking problem – estimating the number of objects and estimating the state of each of the objects. RFS-based calculus then resolves the estimation problem of both the number of objects and each of their states from the observations (which are also modeled as a set).

The RFS-based filter approaches the estimation problem in the same way as the standard single-object tracking algorithms does. It utilizes Bayes' theorem to propagate the density in time. However, the density is defined on sets rather than on normal vectors or matrices. This requires the statistics of a randomly varying finite set captured by some measure analogous distribution function for random variable. "Belief mass" is defined to be the required function whose derivative

---

[1] IMM-PDA is usually used to track maneuvering targets in clutter. However, in Bar-Shalom *et al.* (1989), it has been adopted to use for establishing target presence by using one model, called an observable model, with $0 < P_D \leq 1$, and the other model, called an unobservable model, $P_D = 0$.

gives rise to multi-object set-based Markov density and likelihood function. These calculus (derivative and integral operations) are performed on sets according to finite set statistics (FISST) calculus (Appendix B, Mahler, 2004a).

This chapter introduces the optimal Bayes formulation of the RFS filter and its associated object motion and sensor measurement equations. When the complete density of a randomly varying set (any number of objects may appear or disappear) is propagated through the optimal Bayes recursion of the multi-object set, a combinatorial problem arises. This problem is well recognized in the literature (Mahler, 2003a; Zajic and Mahler, 2003; Vo *et al.*, 2005; Sidenbladh, 2003; Vo and Ma, 2006). A more practical solution is to propagate only the first-order moment of the multi-target density. This gives rise to the probability hypothesis density (PHD) filter. The PHD is defined as the density that gives the number of targets in a space when integrated over that space. Because PHD is concerned about the number of targets, the recursion of PHD follows a point process modeling and relevant calculus.

This chapter introduces the PHD and its various practical implementations. A simple single-object tracker based on RFS formalism will be developed as an example of the RFS filter as well as an exercise of the RFS calculus. Following Challa *et al.* (2002c) the integrated PDA (IPDA) filter will be derived as a practical implementation of the theoretical RFS-based filter equations. The well-established joint IPDA filter is also derived from the PHD filter equations (Challa and Chakravorthy, 2009). Both the IPDA-based filter and RFS-based filter were independently developed at roughly the same time. These derivations show the theoretical connections between these two approaches and establishes IPDA and joint IPDA as a practical implementation of near optimal multiple-object trackers.

6.1 The optimal Bayesian multi-object tracking filter

In the subsequent sections, models for object dynamics and sensor measurements will be presented according to the random set terminologies. The relevant Markov transitions and likelihood functions will be derived and used in the optimal Bayesian recursion. This chapter makes use of finite state statistics, which was introduced by Ron Mahler (2004a), and a brief overview is presented in Appendix B.

6.1.1 Target dynamics and sensor measurement models in RFS

Using random set notation, the object motion model is given by

$$\Gamma_{k+1} = \Phi_k(X_k, V_k) \cup B_k(X_k), \qquad (6.1)$$

where $\Phi(\cdot)$ represents the change of object dynamics from time $t = k$ to $t = k + 1$. $B(\cdot)$ caters for the object birth process in the multiple-target case.

Similarly, the sensor model in random set notation is given by

$$\Sigma = T(X) \cup C(X), \tag{6.2}$$

where $T(\cdot)$ defines the measurements originated from the true objects, and $C(\cdot)$ accounts for clutter measurements.

6.1.2 Markov transition density and likelihood function in RFS

The statistics of the finitely varying random state set Γ_k is described by its belief-mass function,

$$\beta_{\Gamma_{k|k-1}}(\mathbb{S}|\mathbf{X}_{k-1}) = \text{Prob}(\Gamma_k \subseteq \mathbb{S}) = p(\Gamma_k \subseteq \mathbb{S}).$$

This is the total probability of finding all objects in region $\mathbb{S}$ at time k, if at time $k - 1$ they had a multi-object state $\mathbf{X}_{k-1}$.

Similarly, the statistics of the finitely varying random state set Σ_k is described by its belief-mass function,

$$\beta_{\Sigma_k}(\mathbb{S}|\mathbf{X}_k) = \beta_{\Sigma'_k \cup \Lambda_k}(\mathbb{S}|\mathbf{X}_k) = p(\Sigma'_k \cup \Lambda_k \subseteq \mathbb{S}).$$

This is the total probability that all observations in a sensor (or multi-sensor) scan will be found in any given region $\mathbb{S}$, if the object has state $\mathbf{x}_k$.

By differentiating these belief-mass functions (using the generalized Radon–Nikodym theorem), one can obtain multi-object Markov densities and multi-object measurement likelihoods (Goodman *et al.*, 1997). As belief-mass functions are functions of sets, the derivatives are set derivatives. Using FISST one can obtain the derivatives of belief-mass functions and hence the multi-object Markov densities and multi-object likelihoods as follows:

- The multi-object Markov density is a set derivative of the belief-mass function $\beta_{\Gamma_{k|k-1}}(\mathbb{S}|\mathbf{X}_{k-1})$ and can be represented as follows:

$$f_{k|k-1}(\mathbf{x}_k|\mathbf{X}_{k-1}) = \frac{\delta \beta_{\Gamma_{k|k-1}}(\mathbb{S}|\mathbf{X}_{k-1})}{\delta \mathbf{x}_k}.$$

- The multi-object measurement likelihood is a set derivative of the belief-mass function $\beta_{\Sigma_k}(\mathbb{S}|\mathbf{X}_k)$ of the corresponding sensor model and is given by

$$f_{\Sigma_k}(\mathbf{y}_k|\mathbf{x}_k) = \frac{\delta \beta_{\Sigma_k}(\mathbb{S}|\mathbf{X}_k)}{\delta \mathbf{y}_k}.$$

6.1.3 Optimal Estimation in RFS

These multi-object Markov transition densities and likelihood functions can then be used in the standard Bayesian non-linear filtering equations to obtain a recursive method for the multi-object posterior density. The general form of the Bayes' recursion is the same as in the case of the single-sensor, single-object tracking problem, and is given by

$$f_{k|k}(\mathbf{x}_k|\mathbf{y}^k) = \frac{1}{\Delta} f_{\Sigma_k}(\mathbf{y}_k|\mathbf{x}_k) \int f_{k|k-1}(\mathbf{x}_k|\mathbf{X}_{k-1}) f_{k-1|k-1}(\mathbf{X}_{k-1}|\mathbf{y}^{k-1}) \delta \mathbf{X}_{k-1}.$$

(6.3)

However, note that the integration in the Bayesian recursion equation above is a set integral. Consider the single object tracking problem with object existence uncertainty. The following scenario is assumed: if there are no object present in the scene, then this will continue to be the case. However, if there is one object in the scene then this object either will persist (with probability p_v) or it will vanish (with probability $1 - p_v$). Specifically, the goal is to determine if a target exists and, if it does, its state. This problem, **with at most one target** assumption, was first addressed by Mušicki *et al.* (1994) almost in parallel to the introduction of FISST by Mahler (1997). Mušicki *et al.* provide an approximate solution to this problem under linear Gaussian assumptions and named their techniques IPDA. Techniques based on this are presented in Chapters 5 and 9. This simple object tracking is suitable for introducing random set formalism and it can be used to derive useful filters.

6.2 The probabilistic hypothesis density approximations

6.2.1 First moment of RFS: the PHD

In a single-object tracking problem, two mostly used statistics of the posterior density are – the first-order moment (a vector, posterior expectation) and the second-order moment (matrix). Higher-order moments are ignored and these first two order moments are assumed to be sufficient statistics for the posterior density. In such a case, recursion of these two statistics ensures the complete propagation of the density. If advancing the similar argument by proposing that even the second-order moment can be neglected, the posterior density is described by only the first-order moment and the recursion of first-order moment alone is sufficient for a filter algorithm (this approach is taken in the constant gain Kalman filter). However, representing a complete density by its first-order moment is justified if the density is unimodal and significantly concentrated around its mean (low variance).

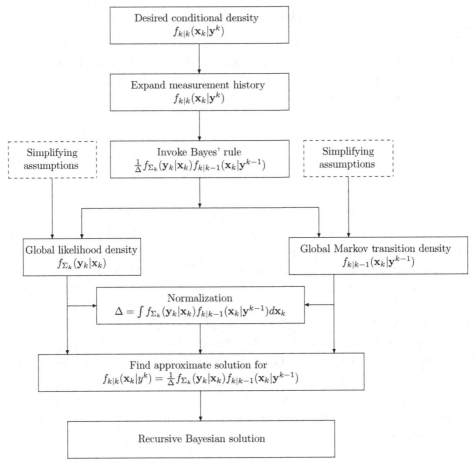

Summary of the recursive Bayesian framework for multiple-object tracking.

The same logic has been extended to derive implementable algorithms within the random set formalism. To propagate the higher-order moment (second order), one needs to resolve combinations of possibilities. Just to illustrate, let us assume a two-object case. Possible realizations for object state set X can be $\{\phi\}, \{\mathbf{x}_1\}, \{\mathbf{x}_2\}, \{\mathbf{x}_1, \mathbf{x}_2\}$. For a higher number of objects these combinations can become extremely large and therefore harder to manage. Therefore representing the density with only the first-order moment and propagating the first-order moment only is useful. This first-order moment is called "probabilistic hypothesis density" (PHD). It is defined below:

PHD is a density whose integration over a space results in the number of objects present within that space.

In this sense it is different from probability density that integrates to 1 when taken over the entire space. The PHD recursion filter is derived in the following sections.

6.2.2 PHD prediction

Let $f_{k|k}(X|Z^k)$ be the multi-object posterior density at k. Its prediction is given as

$$f_{k+1|k}(X|Z^k) = \int f_{k+1|k}(X|W) f_{k|k}(W|Z^k) \delta W,$$

where $f_{k+1|k}(X|W)$ is the multi-object Markov transition density. The probability generating functions (p.g.fl) of the predicted density is

$$G_{k+1|k}[h] = \int h^X f_{k+1|k}(X|Z^k) \delta X$$

$$= \int \left(\int h^X f_{k+1|k}(X|W) \delta X \right) f_{k|k}(W|Z^k) \delta W$$

$$= \int G_{k+1|k}[h|X] f_{k|k}(X|Z^k) \delta X, \tag{6.4}$$

where $G_{k+1|k}[h|X]$ is the p.g.fl of the $f_{k+1|k}(X|W)$. The PHD of the predicted density if the gradient derivative of p.g.fl evaluated at $h = 1$:

$$D_{k+1|k}(\mathbf{x}) = \frac{\delta G_{k+1|k}}{\delta \mathbf{x}}[1].$$

Now, the multi-object motion model consists of the dynamic transition of existing targets, the spawning of new objects from existing ones and the birth of new object. The model is given by

$$\Xi_{k+1|k} = T(X) \cup \Psi(X) \cup \Psi_0(X),$$

where:

- $T(X)$ is the set of surviving objects, $T(X) = T(\mathbf{x}_1) \cup T(\mathbf{x}_2) \cup \cdots \cup T(\mathbf{x}_\tau)$;
- $\Psi(X)$ is the set of spawned objects, $\Psi(X) = \Psi(\mathbf{x}_1) \cup \Psi(\mathbf{x}_2) \cup \cdots \cup \Psi(\mathbf{x}_\tau)$;
- $\Psi_0(X)$ is the set of new-born objects.

Then the p.g.fl $G_{k+1|k}[h|X]$ is given by

$$G_{k+1|k}[h|X] = (1 - p_s + p_s p_h)^X \times b_h^X \times e_h.$$

Therefore,

$$G_{k+1|k}[h] = \int (1 - p_s + p_s p_h)^X \times b_h^X \times e_h f_{k|k}(X|Z^k)\delta X$$

$$= e_h \int (1 - p_s + p_s p_h)^X \times b_h^X f_{k|k}(X|Z^k)\delta X$$

$$= e_h G_{k|k}[(1 - p_s + p_s p_h)b_h], \tag{6.5}$$

where:

- $b_h(\mathbf{x}) = \int h^X b_{k+1|k}(X|\mathbf{x})\delta X$ is the p.g.fl of $b_{k+1|k}(X|\mathbf{x})$;
- $e_h = \int h^X b_{k+1|k}(X)\delta X$ is the p.g.fl of $b_{k+1|k}(X)$;
- $p_h(\mathbf{x}) = \int h(\mathbf{y}) f_{k+1|k}(\mathbf{y}|\mathbf{x})$.

Substituting $\Phi[h] = (1 - p_s + p_s p_h)b_h$,

$$\frac{\delta G_{k+1|k}}{\delta \mathbf{x}}[h] = \frac{\delta G_{k+1|k}}{\delta \mathbf{x}}[h](e_h G_{k|k}[\Phi[h]])$$

$$= \left(\frac{\delta}{\delta \mathbf{x}} e_h\right) G_{k|k}[\Phi[h]] + e_h \left(\frac{\delta}{\delta \mathbf{x}} G_{k|k}[\Phi[h]]\right). \tag{6.6}$$

The PHD of $G_{k+1|k}\delta \mathbf{x}[h]$, $D_{k+1|k}(\mathbf{x})$ is the functional derivative evaluated at $h = 1$:

$$D_{k+1|k}(\mathbf{x}) = \frac{\delta G_{k+1|k}}{\delta \mathbf{x}}[1]$$

$$= \left(\frac{\delta}{\delta \mathbf{x}} e_h\right)\Bigg|_{h=1} G_{k|k}[\Phi[1]] + e_1 \left(\frac{\delta}{\delta \mathbf{x}} G_{k|k}[\Phi[h]]\right)\Bigg|_{h=1}$$

$$= b_{k+1|k}(\mathbf{x}) \cdot 1 + 1 \cdot \left(\frac{\delta}{\delta \mathbf{x}} G_{k|k}[\Phi[h]]\right)\Bigg|_{h=1},$$

where $b_{k+1|k}(\mathbf{x})$ is the PHD of the $k + 1|k(X)$. Now the PHD of $G_{k|k}[\Phi[h]]$ is

$$D_{k+1|k}(\mathbf{x}) = \int \frac{\delta \Phi_{\mathbf{w}}}{\delta \mathbf{x}}[1] \cdot D_{k|k}(\mathbf{w})d\mathbf{w},$$

where

$$\frac{\delta \Phi_{\mathbf{w}}}{\delta \mathbf{x}}[h] = \frac{\delta}{\delta \mathbf{x}}(1 - p_s(\mathbf{w}) + p_s(\mathbf{w})p_h(\mathbf{w}))b_h(\mathbf{w})$$

$$= p_s(\mathbf{w})\left(\frac{\delta}{\delta \mathbf{x}} p_h(\mathbf{w})\right) b_h(\mathbf{w}) + (1 - p_s(\mathbf{w}) + p_s(\mathbf{w})p_h(\mathbf{w}))\frac{\delta}{\delta \mathbf{x}} b_h(\mathbf{w})$$

$$= p_s(\mathbf{w}) f_{k+1|k}(\mathbf{x}|\mathbf{w})b_h(\mathbf{w}) + (1 - p_s(\mathbf{w}) + p_s(\mathbf{w})p_h(\mathbf{w}))b_{k+1|k}(\mathbf{x}|\mathbf{w}).$$

Therefore, setting $h = 1$, we get

$$\frac{\delta \Phi_{\mathbf{w}}}{\delta \mathbf{x}}[1] = p_s(\mathbf{w}) f_{k+1|k}(\mathbf{x}|\mathbf{w}) + b_{k+1|k}(\mathbf{x}|\mathbf{w}).$$

The time update of PHD is given by

$$D_{k+1|k}(\mathbf{x}) = b_{k+1|k}(\mathbf{x}) + \int p_s(\mathbf{w}) f_{k+1|k}(\mathbf{x}|\mathbf{w}) b_h(\mathbf{w}) + b_{k+1|k}(\mathbf{x}|\mathbf{w}) d\mathbf{w}.$$

6.2.3 PHD update

Let us denote $f_{k+1|k}(X|Z^k)$ as the time-predicted multi-object posterior and $Z_{k+1} = \{\mathbf{z}_1, \ldots, \mathbf{z}_m\}$ as the set of newly collected measurements. The data updated posterior is then given by Bayes' recursion formula,

$$f_{k+1|k+1}(X|Z^{k+1}) = K^{-1} f_{k+1}(Z|X, \mathbf{x}^*) f_{k+1|k}(X|Z^k),$$

where the normalization factor is

$$K = f_{k+1}(Z|Z^k) = \int f_{k+1}(Z|X, \mathbf{x}^*) f_{k+1|k}(X|Z^k) \delta X.$$

To resolve the PHD formulation, let a two-variable p.g.fl be defined as

$$F[g, h] = \int \int h^X g^Z f_{k+1}(Z|X, \mathbf{x}^*) f_{k+1|k}(X|Z^k) \delta X \delta Z$$

$$= \int \int h^X G_{k+1}[g|X, \mathbf{x}^*] f_{k+1|k}(X|Z^k) \delta X, \qquad (6.7)$$

where the multi-object measurement density is

$$G_{k+1}[g|X, \mathbf{x}^*] = \int g^Z f_{k+1}(Z|X, \mathbf{x}^*) \delta Z.$$

The normalization factor K can be written as

$$K = f_{k+1}(Z|Z^k) = \frac{\delta^m F}{\delta \mathbf{z}_m \ldots \delta \mathbf{z}_1}[0, 1].$$

Similarly, the PHD of $f_{k+1|k+1}(X|Z^{k+1})$ can be written as

$$D_{k+1|k+1}(\mathbf{x}|Z^{k+1}) = \frac{1}{f_{k+1}(Z_{k+1}|Z^k)} \frac{\delta^{m+1} F}{\delta \mathbf{z}_m \ldots \delta \mathbf{z}_1 \delta \mathbf{x}}[0, 1].$$

Now, the observation model is as follows:

$$\Sigma = \Sigma(\mathbf{x}_1) \cup \cdots \cup \Sigma(\mathbf{x}_n) \cup \Theta$$

where $\Sigma(\mathbf{x}_i)$ is the observation (empty if missed) from ith object and Θ is the state-independent Poisson false alarm. It is also assumed that $\Sigma(\cdot)$ and Θ are all statistically independent. This independences implies that

$$G_{k+1}[g|X, \mathbf{x}^*] = G_{k+1}[g|\mathbf{x}_1] \cdots G_{k+1}[g|\mathbf{x}_n]G_\Theta[g]$$

where

$$G_{k+1}[g|\mathbf{x}_i] = 1 - p_d(\mathbf{x}_i) + p_d(\mathbf{x}_i)p_g(\mathbf{x}_i),$$

$$p_g(\mathbf{x}) = \int g(\mathbf{z})f(\mathbf{z}|\mathbf{x})d\mathbf{z},$$

$$G_\Theta[g] = e^{\lambda\kappa[g]-\lambda},$$

$$\kappa[g] = \int g(\mathbf{z})c(\mathbf{z})d\mathbf{z}.$$

Then the two-variable p.g.fl becomes

$$F[g, h] = \int \int h^X (1 - p_D + p_D p_g)^X e^{\lambda\kappa[g]-\lambda} f_{k+1|k}(X|Z^k)\delta X$$

$$= e^{\lambda\kappa[g]-\lambda} \int ((1 - p_D + p_D p_g)h)^X f_{k+1|k}(X|Z^k)\delta X$$

$$= e^{\lambda\kappa[g]-\lambda} G_{k+1|k}[(1 - p_D + p_D p_g)h].$$

In order to result in a closed-form solution of the data-updated PHD, let it be assumed that the time-updated multi-object posterior $f_{k+1|k}(X|Z^k)$ is approximately Poisson:

$$G_{k+1|k}[h] = e^{\mu\sigma[h]-\mu},$$

$$\sigma[h] = \int h(\mathbf{x})s(\mathbf{x})d\mathbf{x}.$$

Therefore, the PHD of the density is

$$D_{k+1|k} = \mu s(\mathbf{x}).$$

Then,

$$G_{k+1|k}[(1 - p_D + p_D p_g)h] = \exp(\mu\sigma[h(1 - p_D)] + \mu\sigma[hp_D p_g] - mu),$$

$$F(g, h) = \exp(\lambda\kappa[g] - \lambda + \mu\sigma[h(1 - p_D)] + \mu\sigma[hp_D p_g] - \mu).$$

Setting $h = 1$,

$$F(g, 1) = \exp(\lambda\kappa[g] - \lambda + \mu\sigma[p_D] + \mu\sigma[p_D p_g]).$$

In general, the denominator of the data-updated posterior is given by (taking the derivative of $F(\cdot)$ and setting $g = 0$)

$$f_{k+1}(Z|Z^k) = e^{-\lambda - \mu\sigma[p_D]} \prod_{\mathbf{z} \in Z} \lambda c(\mathbf{z}) + \mu\sigma[p_D L_{\mathbf{z}}].$$

Similarly, the numerator of the data-updated posterior is given by

$$\frac{\delta^{m+1} F}{\delta \mathbf{z}_m \cdots \delta \mathbf{z}_1 \delta \mathbf{x}}[0, 1] = e^{-\lambda - \mu\sigma[p_D]} \prod_{\mathbf{z} \in Z} (\lambda c(\mathbf{z}) + \mu\sigma[p_D L_{\mathbf{z}}]) \, \mu(1 - p_D)(\mathbf{x})s(\mathbf{x})$$

$$+ e^{-\lambda - \mu\sigma[p_D]} \prod_{\mathbf{z} \in Z} (\lambda c(\mathbf{z}) + \mu\sigma[p_D L_{\mathbf{z}}])$$

$$\times \sum_{\mathbf{z} \in Z} \frac{\mu p_D(\mathbf{x}) L_{\mathbf{z}}(\mathbf{x})s(\mathbf{x})}{\lambda c(\mathbf{z}) + \mu\sigma[p_D L_{\mathbf{z}}]}.$$

Substituting the numerator and denominator, the data-updated PHD can be obtained:

$$D_{k+1|k+1}(\mathbf{x}) = \mu(1 - p_D)(\mathbf{x})s(\mathbf{x}) + \sum_{\mathbf{z} \in Z} \frac{\mu p_D(\mathbf{x}) L_{\mathbf{z}}(\mathbf{x})s(\mathbf{x})}{\lambda c(\mathbf{z}) + \mu\sigma[p_D L_{\mathbf{z}}]}$$

$$= \mu(1 - p_D)(\mathbf{x})D_{k+1|k}(\mathbf{x}) + \sum_{\mathbf{z} \in Z} \frac{D_{k+1|k}[p_D L_{\mathbf{z}}]}{\lambda c(\mathbf{z}) + D_{k+1|k}[p_D L_{\mathbf{z}}]}$$

$$\times \frac{p_D(\mathbf{x}) L_{\mathbf{z}}(\mathbf{x}) D_{k+1|k}(\mathbf{x})}{D_{k+1|k}[p_D L_{\mathbf{z}}]}.$$

6.2.4 The cardinalized PHD approximation

The PHD filter successfully avoids the problem of the computational intractability of the complete Bayes' recursion of RFS by approximating the multi-object posterior density as a Poisson distribution and propagating the first-order moment only. However, it has been shown to have unstable performance in estimating the object number $N_{k|k}$, especially in the presence of missed detections and very high clutter (Erdinc *et al.*, 2005). This instability stems from two main reasons, which are discussed below:

- PHD gives the total number of objects by linear approximation. For example, for the single-object case, PHD gives the expected number of objects at k as

$$N_{k|k} = (1 - p_{D,\mathbf{x}_{k-1}})N_{k|k-1},$$

or the true formula

$$N_{k|k} = \frac{(1 - p_{D,\mathbf{x}_{k-1}})N_{k|k-1}}{1 - N_{k|k-1}p_{D,\mathbf{x}_{k-1}}}.$$

If $N_{k|k-1} = 1$, the PHD approximation yields $(1 - p_{D,\mathbf{x}_{k-1}})$, where the true formula correctly results in 1. Thereby, PHD loses the information due to linearization.

- Secondly, the PHD employs the "expected a posteriori" estimator for the number of objects:

$$N_{k|k} = \sum_{n \geq 0} n . p_{k|k}(n|\mathbf{y}^k).$$

However, minor modes in the distribution that are induced by false alarms may cause the expected value to shift erratically. This will also result in instability in deciding the number of objects.

To overcome this instability one needs to propagate the higher orders of the true multi-object distribution. But this is not practical because the result would run again into a combinatorial problem that gave rise to the original first moment approximation. Therefore another alternative is proposed in Mahler (2007). The resulting algorithm is aptly named the "cardinalized PHD (CPHD) filter" because it propagates the entire distribution of object numbers (cardinality of the multi-target set) along with the first-order moment of the multi-object distribution. In this way, the CPHD filter retains the advantage of propagating only the first-order moment of the multi-object distribution (and thereby avoids the problem of combination under a large number of objects) and avoids the problem of approximating the object number with a linear function. Moreover, from the cardinality distribution, the number of objects is estimated by "maximum a posteriori (MAP)" rather than "expected a posteriori (EAP)" to avoid instability caused by the erratic shift of expectation under the high-clutter scenario. However, in cardinalized PHD, the model for spawning (from existing targets) needed to be omitted in order to achieve a closed-form recursion formula.

6.2.5 Summary of the PHD

One iteration of the PHD filter algorithm is detailed in Algorithm 35:

Algorithm 35 PHD filter recursion equations at time k

1: Time k inputs:
   - set $\mathbf{Y}_k$ of measurements delivered by the sensor; and
   - density at time $k - 1$, $D_{k-1|k-1}(\mathbf{x}_{k-1}|\mathbf{y}^{k-1})$.

Algorithm 36 PHD filter recursion equations at time k

2: Prediction:

$$D_{k|k-1}(\mathbf{x}_k|\mathbf{y}^{k-1}) = \gamma_k(\mathbf{x}_k) + \int \left[p_{S,k}(\mathbf{x}_{k-1}) f_{k|k-1}(\mathbf{x}_k|\mathbf{x}_{k-1}) + b_{k|k-1}(\mathbf{x}_k|\mathbf{x}_{k-1}) \right]$$
$$\times D_{k-1|k-1}(\mathbf{x}_{k-1}|\mathbf{y}^{k-1}) d\mathbf{x}_{k-1},$$

where:
- $\gamma_k(\mathbf{x}_k)$ is the intensity for objects appearing at time k;
- $p_{S,k}(\mathbf{x}_{k-1})$ is the object survival;
- $f_{k|k-1}(\mathbf{x}_k|\mathbf{x}_{k-1})$ is the single-object Markov transition density;
- $b_{k|k-1}(\mathbf{x}_k|\mathbf{x}_{k-1})$ is the intensity of spawning of object from existing ones.

3: Update:

$$D_{k|k}(\mathbf{x}_k|\mathbf{y}^k) = \left[(1 - p_{D,k}(\mathbf{x}_k)) + \sum_{\mathbf{z}_k \in \mathbf{y}_k} \frac{p_{D,k}(\mathbf{x}_k) f_k(\mathbf{z}_k|\mathbf{x}_k)}{\lambda_k c_k(\mathbf{z}_k) + \psi(\mathbf{z}_k|\mathbf{y}^{k-1})} \right] \times D_{k|k-1},$$

where:
- $\psi(\mathbf{z}_k|\mathbf{y}^{k-1}) = \int p_{D,k} f(\mathbf{z}_k|\mathbf{x}_k) D_{k|k-1}(\mathbf{x}_k|\mathbf{y}^{k-1})$;
- $p_{D,k}(\mathbf{x}_k)$ is the probability of detection;
- $f_k(\mathbf{z}_k|\mathbf{x}_k)$ is the single target likelihood function;
- λ_k and $c_k(\mathbf{z}_k)$ are the false alarm (clutter) intensity and false alarm spatial density respectively.

4: Number of targets. The number of objects in the region is given by:

$$N_{k|k} = \int D_{k|k}(\mathbf{x}_k|\mathbf{y}^k) d\mathbf{x}_k.$$

6.2.6 Summary of CPHD recursion

Before presenting the CPHD iteration, we first introduce some notation for clarity.

- Binomial coefficient:

$$\binom{l}{j}_C = \frac{l!}{j!(l-j)!}.$$

- Permutation coefficient:

$$\binom{l}{j}_P = \frac{l!}{(l-j)!}.$$

- Inner product:

$$\langle \alpha \beta \rangle = \begin{cases} \int \alpha(x)\beta(x)dx, & \text{if } \alpha, \beta \text{ are real-valued functions,} \\ \sum_{l=0}^{\infty} \alpha(l)\beta(l), & \text{if } \alpha, \beta \text{ are real sequences.} \end{cases}$$

- $e_j(\cdot)$ is the elementary symmetric function of order j defined for a finite set $\mathbf{y}$ by

$$e_j(\mathbf{y}) = \sum_{S \subseteq \mathbf{y}, |S|=j} \prod_{\zeta \in S} \zeta,$$

and $e_0(\mathbf{y}) = 1$

The CPHD filter algorithm is presented in Algorithm 37.

Algorithm 37 CPHD filter recursion equations at time k

1: Input at time k:
- Cardinality distribution of $k - 1$: the distribution of object number:

$$\sum_{n=0}^{\infty} p_{k-1|k-1}(n).$$

- Intensity at $k - 1$, $D_{k-1|k-1}(\mathbf{x}_{k-1}|\mathbf{y}^{k-1})$.

2: Prediction:

$$p_{k|k-1}(n) = \sum_{j=0}^{n} p_{\gamma,k}(n-j) \prod_{k|k-1} [D_{k-1|k-1}(\mathbf{x}_{k-1}|\mathbf{y}^{k-1}), p_{k-1}](j)$$

$$D_{k|k-1}(\mathbf{x}_k|\mathbf{y}^{k-1}) = \int p_{S,k}(\mathbf{x}_{k-1}) f_{k|k-1}(\mathbf{x}_k|\mathbf{x}_{k-1})d\mathbf{x}_{k-1} + \gamma_{k,\mathbf{x}_k}$$

where:

$$\prod_{k|k-1} [D, p](j) = \sum_{l=j}^{\infty} \binom{l}{j}_C \frac{\langle p_{S,k}, D \rangle^j \left(1 - p_{S,k}, D\right)^{l-j}}{\langle 1, v \rangle} p(l);$$

- $f_{k|k-1}(.|\mathbf{x}_{k-1})$ is the single-object transition density at time given past state $\mathbf{x}_{k-1}$;
- γ is the intensity of spontaneous birth;
- $p_{\gamma,k}(n-j)$ is the cardinality distribution of spontaneous births.

3: Update:

$$p_{k|k} = \frac{W_k^0[D_{k|k-1}(\mathbf{x}_k|\mathbf{y}^{k-1}); \mathbf{y}_k](n)p_{k|k-1}(n)}{\langle W_k^0[D_{k|k-1}(\mathbf{x}_k|\mathbf{y}^{k-1}); \mathbf{y}_k](n), p_{k|k-1}(n) \rangle},$$

$$D_{k|k}(\mathbf{x}_k|\mathbf{y}^k)$$

$$= (1 - p_{D,k}(\mathbf{x}_k)) \frac{\left\langle W_k^1[D_{k|k-1}(\mathbf{x}_k|\mathbf{y}^{k-1}); \mathbf{y}_k], p_{k|k-1}\right\rangle}{\left\langle W_k^0[D_{k|k-1}(\mathbf{x}_k|\mathbf{y}^{k-1}); \mathbf{y}_k], p_{k|k-1}\right\rangle} D_{k|k-1}(\mathbf{x}_k|\mathbf{y}^{k-1}),$$

$$+ \sum_{\mathbf{z}_k \in \mathbf{y}_k} \psi_{k,\mathbf{z}_k} \frac{\left\langle W_k^1[D_{k|k-1}(\mathbf{x}_k|\mathbf{y}^{k-1}); \mathbf{y}_k \setminus \{\mathbf{z}_k\}], p_{k|k-1}\right\rangle}{\left\langle W_k^0[D_{k|k-1}(\mathbf{x}_k|\mathbf{y}^{k-1}); \mathbf{y}_k], p_{k|k-1}\right\rangle} D_{k|k-1}(\mathbf{x}_k|\mathbf{y}^{k-1}),$$

where

- $$W_k^u[D, \mathbf{y}](n) = \sum_{j=0}^{min(|\mathbf{y}|,n)} (|\mathbf{y}| - j)! \, p_{K,k}(|\mathbf{y}| - j) \binom{n}{j+u}_P$$

$$\times \frac{\left\langle 1 - p_{D,k}(\mathbf{x}_k), D\right\rangle^{n-(j+u)}}{\langle 1, D\rangle^n} e_j(\Xi_k(D, \mathbf{y}_k));$$

- $$\psi_k(\mathbf{x}, \mathbf{z}) = \frac{\langle 1, \lambda_k c_k\rangle}{\lambda_k c_k(\mathbf{z}_k)} f_k(\mathbf{z}|\mathbf{x}) p_{D,k}(\mathbf{x});$$

- $$\Xi_k(D, \mathbf{y}) = \{\langle D, \psi_{k,\mathbf{z}}\rangle : \mathbf{z} \in \mathbf{y}_k\};$$

- $f_k(.|\mathbf{x}_k)$ is the single-object measurement likelihood at k given the target state $\mathbf{x}_k$;
- $p_{D,k}(\mathbf{x}_k)$ is the probability of object detection at time k given the target state $\mathbf{x}_k$;
- $p_{K,k}(\cdot)$ is the cardinality distribution of clutter.

6.3 Approximate filters

The PHD filter recursion formula in (35) does not have a closed-form solution in general. Therefore, even though PHD provides an alternative to propagate the first-order moment of the RFS distribution, it is hard to implement it. However, the particle PHD filter (Sidenbladh, 2003; Zajic and Mahler, 2003; Vo *et al.*, 2005) and Gaussian mixture PHD filter (Vo and Ma, 2006) are proposed, which result in a closed-form solution of the general PHD recursion formula under various conditions and approximations.

6.3.1 Gaussian mixture PHD filter

Another approximate implementation of PHD filter is called the "Gaussian mixture PHD (GMPHD) filter." In order to obtain a closed-form solution under the GMPHD filter algorithm, there are certain conditions, which are summarized below:

1. Each object follows a linear Gaussian dynamical model given by

$$f_{k|k-1}(\mathbf{x}_k|\mathbf{x}_{k-1}) = \mathcal{N}(\mathbf{x}_k; \mathbf{F}_{k-1}\mathbf{x}_{k-1}, \mathbf{Q}_{k-1}),$$

and each sensor has a linear Gaussian likelihood model given by

$$f_k(\mathbf{z}_k|\mathbf{x}_k) = \mathcal{N}(\mathbf{z}_k; \mathbf{H}_k\mathbf{x}_k, \mathbf{R}_k),$$

where:
- $\mathcal{N}(a; b, c)$ denotes a Gaussian density of random variable a with mean b and covariance c;
- $\mathbf{F}_{k-1}$ is the state transition matrix;
- $\mathbf{Q}_{k-1}$ and $\mathbf{R}_{k-1}$ are the process and observation noise covariances respectively.

2. The object persistence (survival) and detection probabilities are state independent:

$$p_{D,k}(\mathbf{x}_k) = p_{D,k},$$

$$p_{S,k}(\mathbf{x}_{k-1}) = p_{S,k}.$$

3. The object birth and spawn intensities are in the form of a Gaussian mixture, given by

$$\gamma_k(\mathbf{x}_k) = \sum_{i=1}^{J_{\gamma,k}} w_{\gamma,k}^i \mathcal{N}(\mathbf{x}_k; \mathbf{m}_{\gamma,k}^i, \mathbf{P}_{\gamma,k}^i),$$

$$b_{k|k-1}(\mathbf{x}_k|\mathbf{x}_{k-1}) = \sum_{j=1}^{J_{b,k}} w_{b,k}^j \mathcal{N}(\mathbf{x}_k; \mathbf{F}_{b,k-1}^j\mathbf{x}_{k-1} + \mathbf{u}_{b,k-1}^j, \mathbf{Q}_{b,k-1}^j),$$

where:
- $w_{\gamma,k}^i, \mathbf{m}_{\gamma,k}^i, \mathbf{P}_{\gamma,k}^i$ are the weight, mean and covariance of the ith spawn component, $i = 1, 2, \ldots, J_{\gamma,k}$, and $J_{\gamma,k}$ is the total number of components;
- $w_{b,k}^j, \mathbf{F}_{b,k-1}^j\mathbf{x}_{k-1} + \mathbf{u}_{b,k-1}^j, \mathbf{Q}_{b,k-1}^j$ are the weight, mean and covariance of the jth birth component, $j = 1, 2, \ldots, J_{b,k}$, and $J_{b,k}$ is the total number of components.

It is also assumed that the density at $k - 1$ is a Gaussian mixture of the form

$$D_{k-1|k-1}(\mathbf{x}_{k-1}|\mathbf{y}^{k-1}) = \sum_{i=1}^{J_{k-1}} w_{k-1}^i \mathcal{N}(\mathbf{x}_{k-1}; \mathbf{m}_{k-1}^i, \mathbf{P}_{k-1}^i).$$

Under the stated conditions and assumptions, the PHD recursion is reduced to the GMPHD algorithm, which is detailed in Algorithm 38.

Algorithm 38 Gaussian mixture PHD filter recursion at time k

1: Input to the recursion at time k:

- Density at time $k-1$, $D_{k-1|k-1}(\mathbf{x}_{k-1}|\mathbf{y}^{k-1}) = \sum\limits_{i=1}^{J_{k-1}} w_{k-1}^i \mathcal{N}(\mathbf{x}_{k-1}; \mathbf{m}_{k-1}^i, \mathbf{P}_{k-1}^i)$.

2: Prediction:

$$D_{k|k-1}(\mathbf{x}_k|\mathbf{y}^{k-1}) = D_{S,k|k-1}(\mathbf{x}_k|\mathbf{y}^{k-1}) + D_{b,k|k-1}(\mathbf{x}_k|\mathbf{y}^{k-1}) + \gamma_k(\mathbf{x}_k),$$

where:

$$\gamma_k(\mathbf{x}_k) = \sum_{i=1}^{J_{\gamma,k}} w_{\gamma,k}^i \mathcal{N}(\mathbf{x}_k; _{\gamma,k}^i, \mathbf{P}_{\gamma,k}^i);$$

$$D_{S,k|k-1}(\mathbf{x}_k|\mathbf{y}^{k-1}) = \sum_{i=1}^{J_{k-1}} w_{S,k}^i \mathcal{N}(\mathbf{x}_k; _{S,k}^i, \mathbf{P}_{S,k|k-1}^i);$$

$$\left[\mathbf{m}_{S,k}^i, \mathbf{P}_{S,k|k-1}^i\right] = KF_P\left[\mathbf{m}_{k-1}^i, \mathbf{P}_{k-1}^i, \mathbf{F}_k, \mathbf{Q}_k\right];$$

$$w_{S,k}^i = p_{S,k} w_{k-1}^i;$$

$$D_{b,k|k-1}(\mathbf{x}_k|\mathbf{y}^{k-1}) = \sum_{i=1}^{J_{k-1}} w_{b,k} \sum_{j=1}^{J_{b,k}} \mathcal{N}(\mathbf{x}_k; _{b,k}^{i,j}, \mathbf{P}_{b,k|k-1}^{i,j});$$

$$\mathbf{m}_{b,k}^{i,j} = \mathbf{F}_{b,k-1}^i + \mathbf{u}_{b,k}^j;$$

$$\mathbf{P}_{b,k|k-1}^{i,j} = \mathbf{F}_{b,k-1}^j \mathbf{P}_{b,k-1}^i (\mathbf{F}_{b,k-1}^j)^T + \mathbf{Q}_{b,k-1}^j.$$

3: Update:

$$D_{k|k}(\mathbf{x}_k|\mathbf{y}^k) = (1 - p_{D,k}) D_{k|k-1}(\mathbf{x}_k|\mathbf{y}^{k-1}) + \sum_{\mathbf{z}_k \in Z_k} (\mathbf{z}_k) D_{z,k}(\mathbf{x}_k; \mathbf{z}_k),$$

where:

$$D_{z,k}(\mathbf{x}_k; \mathbf{z}_k) = \sum_{i=1}^{J_{k|k-1}} w_k^i(\mathbf{z}_k) \mathcal{N}(\mathbf{x}_k; \mathbf{m}_{k|k}^i(\mathbf{z}_k), \mathbf{P}_{k|k}^i);$$

$$w_k^i(\mathbf{z}_k) = \frac{p_{D,k} w_{k|k-1}^i q_k^i(\mathbf{z}_k)}{\lambda_k c_k(\mathbf{z}_k) + p_{D,k} \sum\limits_{i=1}^{J_{k|k-1}} w_{k|k-1}^j q_k^j(\mathbf{z}_k)};$$

$$q_k^i(\mathbf{z}_k) = \mathcal{N}(\mathbf{z}_k; \mathbf{H}_k \mathbf{m}_{k|k-1}^i, \mathbf{H}_k \mathbf{P}_{k|k-1}^i \mathbf{H}_k^T + \mathbf{R}_k);$$

$$\left[\mathbf{m}_{k|k}^i(\mathbf{z}_k), \mathbf{P}_{k|k}^i\right] = KF_E\left[\mathbf{z}_k, \mathbf{m}_{k|k-1}^i, \mathbf{P}_{k|k-1}, \mathbf{H}_k, ^i, \mathbf{R}_k\right].$$

6.3.2 Particle PHD filter

In a similar way to particle filter single-object tracking, the particle PHD filter approximates the density by "particle representation" and propagates the density according to the prediction and update step of the PHD filter. The algorithm is presented in Algorithm 39.

Looking at the resampling stage of the particle PHD filter, it must be noted that the new weights, $\{w_k^i\}_{i=1}^{L_k}$, do not add up to 1, but rather to $\hat{N}_{k|k}$. The updated particles, $\tilde{b}\tilde{x}_k^i$, are copied ζ_k^i times with the constraint

$$\sum_{i=1}^{L_{k-1}+J_k} \zeta_k^i = L_k.$$

6.3.3 Gaussian mixture CPHD algorithm

Similar to the GMPHD algorithm for the standard PHD filter, there exists a car-dinalized version called the Gaussian mixture CPHD (GMCPHD) algorithm. It was first proposed in Vo *et al.* (2006). The GMCPHD algorithm assumes the object dynamic evolution, sensor models, object survival/detection probabilities and object birth intensities in the same way that GMPHD does (described in Section 6.3.1 – note that there are no spawning intensities present in the cardinalized PHD formulation). Moreover, the intensity and cardinality distribution at $k - 1$ are assumed to be

$$D_{k-1|k-1}(\mathbf{x}_{k-1}|\mathbf{y}) = \sum_{i=1}^{J_{k-1}} w_{k-1}^i \mathcal{N}(\mathbf{x}_{k-1}; \mathbf{m}_{k-1}^i, \mathbf{P}_{k-1}^i),$$

and $\sum_{n=0}^{\infty} p_{k-1}(n)$ respectively. Under these assumptions, the CPHD algorithm reduces to the GMCPHD algorithm. The steps of one iteration of GMCPHD are detailed in Algorithm 40.

6.3.4 Track labelings

PHD based filter avoids the "track-to-measurement" association and thereby reduces the computational complexity (when compared to data-association-based multiple-object tracking algorithms. It recursively updates the density and state estimates. However, the update and state estimation process does not maintain track identification. To propagate the track id, the estimates states (or chosen peaks in the density) need to be associated with already existing objects. This process is called "track labeling" or "track to estimate association."

Algorithm 39 Particle PHD filter recursion at time k

1: Time k inputs:
   - set $\mathbf{Y}_k$ of measurements delivered by the sensor; and
   - weights and samples of time $k-1$ - w_{k-1}^i, $\mathbf{x}_{k-1}^i$ respectively, $i = 1, 2, \ldots, L_{k-1}$.

2: **for** each $i = 1, 2, \ldots, L_{k-1}$ **do**
   - sample $\tilde{\mathbf{x}}_k^i \simeq q_k(.|\mathbf{x}_{k-1}^i, \mathbf{y}_k)$;
   - $\tilde{w}_{k|k-1}^i = \frac{\phi_{k|k-1}(\tilde{\mathbf{x}}_k^i, \mathbf{x}_{k-1}^i)}{q_k(\tilde{\mathbf{x}}_k^i|\mathbf{x}_{k-1}^i)} w_{k-1}^i$.

3: **end for**

4: **for** each $i = L_{k-1}+1, L_{k-1}+2, \ldots, L_{k-1}+sJ_k$ **do**
   - sample $\tilde{\mathbf{x}}_k^i \simeq p_k(.|\mathbf{y}_k)$;
   - $\tilde{w}_{k|k-1}^i = \frac{1}{J_k} \frac{\gamma_k(\tilde{\mathbf{x}}_k^i)}{p_k(\tilde{\mathbf{x}}_k^i|\mathbf{y}_k)}$.

5: **end for**
   where:
   - $p(.|\mathbf{y}_k)$ is the importance density such that $\gamma_k(\mathbf{x}_k) > 0$ implies $p(.|\mathbf{y}_k) > 0$;
   - $q_k(.|\mathbf{x}_{k-1}, \mathbf{y}_k)$ is the importance density such that $\phi_{k|k-1}(.|\mathbf{x}_k, \mathbf{x}_{k-1}) > 0$ implies $q_k(.|\mathbf{x}_{k-1}, \mathbf{y}_k) > 0$;
   - $\phi_{k|k-1}(.|\mathbf{x}_k, \mathbf{x}_{k-1}) = p_{S,k}(\mathbf{x}_{k-1}) f_{k|k-1}(\mathbf{x}_k|\mathbf{x}_{k-1}) + b_{k|k-1}(\mathbf{x}_k|\mathbf{x}_{k-1})$;
   - J_k is the number of completely new-born object out of the birth process.

6: **for** each $\mathbf{z}_k \in \mathbf{y}_k$ **do** $C_k(\mathbf{z}_k) = \sum_{j=1}^{L_{k-1}+J_k} \psi_{k,\mathbf{z}_k}(\tilde{\mathbf{x}}_k^j) \tilde{w}_{k|k-1}^j$.

7: **end for**

8: **for** each $i = 1, 2, \ldots, L_{k-1}, J_k$ **do**
$$\tilde{w}_k^i = \left[1 - p_{D,k}(\tilde{\mathbf{x}}_k^i) + \frac{p_{D,k}(\tilde{\mathbf{x}}_k^i) f_k(z_k|\tilde{\mathbf{x}}_k^i)}{\lambda_k c_k(\mathbf{z}_k) + C_k(\mathbf{z}_k)} \right] \tilde{w}_{k|k-1}^i.$$

9: **end for**

10:
   - Compute the total mass $\hat{N}_{k|k} = \sum_{j=1}^{L_{k-1}+J_k} \tilde{w}_k^j$.
   - Resample $\left\{ \frac{\tilde{w}_k^i}{\hat{N}_{k|k}}, \tilde{\mathbf{x}}_k^i \right\}_{i=1}^{L_{k-1}+J_k}$ to get $\left\{ \frac{w_k^i}{\hat{N}_{k|k}}, \mathbf{x}_k^i \right\}_{i=1}^{L_k}$.
   - Rescale the weights by $\hat{N}_{k|k}$ to get $\{ w_k^i, \mathbf{x}_k^i \}_{i=1}^{L_k}$.

Let the rth active track at k be denoted by $\tau_{k_s:k}^r$, where r is the label for the track that started at some time in the past k_s. The track has state estimate and covariance $\mathbf{x}_{k|k}^r$ and $\mathbf{P}_{k|k}^r$ respectively:

$$\tau_{k_s:k}^r = \{ \mathbf{x}_{k|k}^r, \mathbf{P}_{k|k}^r \}.$$

Algorithm 40 Gaussian mixture CPHD filter recursion equations at time k

---

1: Input at k:

- Cardinality distribution at $k - 1$: $\sum\limits_{n=0}^{\infty} p_{k-1}(n)$.

- Intensity at $k - 1$: $D_{k-1|k-1}(\mathbf{x}_{k-1}|\mathbf{y}) = \sum\limits_{i=1}^{J_{k-1}} w_{k-1}^i \mathcal{N}(\mathbf{x}_{k-1}; \mathbf{m}_{k-1}^i, \mathbf{P}_{k-1}^i)$.

2: Prediction:

$$p_{k|k-1}(n) = \sum_{j=0}^{n} p_{\gamma,k}(n-j) \prod_{k|k-1} [D_{k-1|k-1}(\mathbf{x}_{k-1}|\mathbf{y}^{k-1}), p_{k-1}](j),$$

$$D_{k|k-1}(\mathbf{x}_k|\mathbf{y}^{k-1}) = D_{S,k|k-1}(\mathbf{x}_k|\mathbf{y}^{k-1}) + \gamma_{k,\mathbf{x}_k},$$

where:

-
$$\gamma_k(\mathbf{x}_k) = \sum_{i=1}^{J_{\gamma,k}} w_{\gamma,k}^i \mathcal{N}(\mathbf{x}_k; _{\gamma,k}^i, \mathbf{P}_{\gamma,k}^i); \tag{6.8}$$

-
$$D_{S,k|k-1}(\mathbf{x}_k|\mathbf{y}^{k-1}) = \sum_{i=1}^{J_{k-1}} w_{S,k}^i \mathcal{N}(\mathbf{x}_k; _{S,k}^i, \mathbf{P}_{S,k|k-1}^i);$$

$$[\mathbf{m}_{S,k}^i, \mathbf{P}_{S,k|k-1}^i] = KF_P[\mathbf{m}_{k-1}^i, \mathbf{P}_{k-1}^i, \mathbf{F}_k, \mathbf{Q}_k];$$

$$w_{S,k}^i = p_{S,k} w_{k-1}^i.$$

3: Update:

$$p_{k|k}(n) = \frac{W_k^0[D_{k|k-1}(\mathbf{x}_k|\mathbf{y}^{k-1}); \mathbf{y}_k](n) p_{k|k-1}(n)}{\langle W_k^0[D_{k|k-1}(\mathbf{x}_k|\mathbf{y}^{k-1}); \mathbf{y}_k], p_{k|k-1} \rangle},$$

$$D_{k|k}(\mathbf{x}_k|\mathbf{y}^k) = (1 - p_{D,k}(\mathbf{x}_k)) \frac{\langle W_k^1[w_{k|k-1}; \mathbf{y}_k], p_{k|k-1} \rangle}{\langle W_k^0[w_{k|k-1}; \mathbf{y}_k], p_{k|k-1} \rangle} D_{k|k-1}(\mathbf{x}_k|\mathbf{y}^{k-1})$$

$$+ \sum_{\mathbf{z}_k \in \mathbf{y}_k} D_{z,k}(\mathbf{x}_k; \mathbf{z}_k),$$

where:

- $D_{z,k}(\mathbf{x}_k; \mathbf{z}_k) = \sum\limits_{i=1}^{J_{k|k-1}} w_k^i(\mathbf{z}_k) \mathcal{N}(\mathbf{x}_k; \mathbf{m}_{k|k}^i(\mathbf{z}_k), \mathbf{P}_{k|k}^i);$

-
$$W_k^u[w, \mathbf{y}](n) = \sum_{j=0}^{min(|\mathbf{y}|,n)} (|\mathbf{y}| - j)! p_{K,k}(|\mathbf{y}| - j) \binom{n}{j+u}_P$$

$$\times \frac{\langle 1 - p_{D,k}(\mathbf{x}_k), D \rangle^{n-(j+u)}}{\langle 1, w \rangle^n} e_j(\Xi_k(w, \mathbf{y}_k));$$

- $$\psi_k(w, \mathbf{z}) = \left\{ \frac{\langle 1, \lambda_k c_k \rangle}{\lambda_k c_k(\mathbf{z}_k)} p_{D,k} w^T q_k(z) : \mathbf{z} \in \mathbf{y}_k \right\};$$

- $w_{k|k-1} = \left[w_{k|k-1}^1, \ldots, w_{k|k-1}^{J_{k|k-1}} \right]^T;$

- $q_k(\mathbf{z}) = \left[q_k^1(\mathbf{z}), \ldots, q_{k|k-1}^{J_{k|k-1}}(\mathbf{z}) \right]^T;$

- $q_k^i(\mathbf{z}) = p_{D,k} w_{k|k-1}^i q_k^i(\mathbf{z}_k) \mathcal{N}(\mathbf{z}; \mathbf{H}_k \mathbf{m}_{k|k-1}^i, \mathbf{H}_k \mathbf{P}_{k|k-1}^i \mathbf{H}_k^T + \mathbf{R}_k);$

$$w_k^i(\mathbf{z}_k) = \frac{\left\langle W_k^1[w_{k|k-1}; \mathbf{y}_k \setminus \{\mathbf{z}_k\}], p_{k|k-1} \right\rangle}{\left\langle W_k^0[w_{k|k-1}; \mathbf{y}_k], p_{k|k-1} \right\rangle};$$

- $$\left[\mathbf{m}_{k|k}^i(\mathbf{z}_k), \mathbf{P}_{k|k}^i \right] = KF_E \left[\mathbf{z}_k, \mathbf{m}_{k|k-1}^i, \mathbf{P}_{k|k-1}, \mathbf{H}_k, {}^i, \mathbf{R}_k \right].$$

Let the peaks in the updated density at k be a list $- \mathcal{P}_{k|k}$. The basic idea of the track labeling is to match peak(s) from the peak list with each of the tracks in the existing list.

One approach is "peak-to-track" association (Lin *et al.*, 2006). This approach validates the peaks against the predicted position of the existing objects. The validated peaks are then used to update the existing objects. In the case of more than one validated peak, new tracks are initiated with the history of the old objects copied and the updated state from the new validated peak is added to the history. Some constraints can be imposed in order to limit the candidate assignments (Lin *et al.*, 2006). Moreover, if a track is not associated with any peak continuously for some pre defined number of scans, the track is terminated.

The other approach uses the weights, $w_{k|k}^i$, of the peaks in the updated density to decide the "track-to-estimate" association (Panta *et al.*, 2005). If the weight is more than some suitably chosen threshold, the peak is a candidate for assignment. If more than one peak satisfies the criterion, new tracks are initiated with the history of the associated track copied over and the new peak is added to the list. Moreover, a decision to terminate the ith track is taken if the weight $w_{k|k}^i$ is below some predefined threshold or if the track cannot be associated with any peak for some pre defined number of scans, or a suitable combination of both of these criteria.

Both new tracks are initiated if a certain peak is not associated with any of the existing tracks. The initial weight, state estimate and covariances are chosen to suit the particular application, constraints and other prior knowledge.

6.3.5 State estimation

The PHD recursion formula propagates the density or first-order moment of the multi-object posterior. But it does not give the estimate of state. Object state estimation needs to be carried out separately after the density recursion is done. The general state estimation process of the PHD filter is described below.

The integration of density at k gives the expected number of tracks, $N_{k|k}$:

$$N_{k|k} = \sum_{i=0}^{J_{k|k}} w_{k|k}^i,$$

where $J_{k|k}$ is the number of peaks.

For the particle PHD filter, the next step is to group the existing particles into $N_{k|k}$ clusters, where the cluster centers are considered as the estimate of states. For the Gaussian mixture PHD filter, the state estimate can be directly obtained from the peaks of the $N_{k|k}$ surviving Gaussians.

Similar to the standard PHD filter recursion, CPHD recursion does not explicitly provide the state estimate of the objects. The cardinalized PHD filter propagates the entire cardinality distribution along with the first-order moment density. Moreover, to be neutral to spurious peaks caused by heavy clutter scenarios, CPHD prefers to estimate the object number by the "maximum a posteriori" method,

$$N_{k|k} = arg \max_n p_{k|k}(n).$$

The Gaussian mixture CPHD filter chooses the $N_{k|k}$ Gaussian with most weights $(w_{k|k}^i)$ and the peaks of the Gaussians are the desired state estimates.

6.4 Object-existence-based tracking filters

In the terminology of random sets, we consider a finite random set Γ_k of object states whose only instantiations are the single event $\mathbf{x}_k = \emptyset$ (i.e., no targets are present) or the infinitely numerous events $\mathbf{x}_k = \{\mathbf{x}_k\}$ (i.e., a single object is present and has state $\mathbf{x}_k$), where $\mathbf{x}_k$ is the random variable representing the state of the object. A sensor collects reports originating from the object with a probability of detection P_D and, in addition, collects reports from clutter objects that are uniformly distributed in space with probability of detection $\mathbf{P}_{FA}$ (i.e., the probability of detection of clutter sources). We denote the random measurements set as Σ_k and its instantiations as $\mathbf{y}_k$, where $|\mathbf{y}_k| = m_k$, i.e., at time k the sensor collects m_k measurements. The problem is to determine the best estimate of the object state if it exists, given all the measurements from the sensor.

We will apply this formalism to solve a simple object tracking in clutter problem with object existence uncertainty.

6.4.1 Random set model for object dynamics

Random set motion models are usually described using (6.1), where $B_k(\cdot)$ deals with the process of object birth. For the problem under consideration, $\mathbf{X}_{k-1} = \emptyset$ or $\mathbf{X}_{k-1} = \{\mathbf{x}_{k-1}\}$ (i.e., there can be up to one object in the scene and there is no birth process) and define the instantiations of Γ_k as $\mathbf{X}_k = \emptyset$ or $\mathbf{X}_k = \{\mathbf{x}_k\}$, with probability $p_{\overline{v}} = 1 - p_v$ and p_v, respectively. Since there is no birth process, the random set motion model could be reduced to $\Phi_k(\mathbf{X}_{k-1}, V_{k-1})$ without the birth process. This model encapsulates a random set evolution, where, if no objects are present in the scene, this will continue to be the case. However, if there is one object in the scene then this object either will persist (with probability p_v) or it will vanish (with probability $1 - p_v$). The dynamics described by the random set $\Phi_{k-1}(\mathbf{X}_{k-1})$ also covers object existence (persistence), as $\Phi_{k-1}(\mathbf{X}_{k-1})$ can take on the empty set as a value. The statistics of the finitely varying random state-set Γ_k is described by its belief-mass function

$$\beta_{\Gamma_{k|k-1}}(\mathbb{S}|\mathbf{X}_{k-1}) = p(\Gamma_k \subseteq \mathbb{S}).$$

This is the total probability of finding all objects in region $\mathbb{S}$ at time k. If there is at most one object, then it is the total probability of finding the object, if present, in any region $\mathbb{S}$.

If $\mathbf{X}_{k-1} = \emptyset$, Γ_k can have only one (set) value, i.e., $\emptyset$ (from the assumption that if the object does not exist at time $k - 1$, then it will continue to be the case), the belief measure is given by

$$\beta_{\Gamma_{k|k-1}}(\mathbb{S}|\emptyset) = p(\Gamma_k = \emptyset \subseteq \mathbb{S}) = 1.$$

Writing this in terms of the density function, we have

$$f_{k|k-1}(\emptyset|\emptyset) = \frac{\delta\beta_{\Gamma_{k|k-1}}}{\delta\{\mathbf{x}_{k-1}\}}(\emptyset) = 1.$$

On the other hand, if $\mathbf{X}_{k-1} = \{\mathbf{x}_{k-1}\}$, then Γ_k can take on two (set) values, $\{\mathbf{x}_k\}$ or $\emptyset$ with probabilities p_v and $1 - p_v$ respectively. In such a case, the belief measure can be calculated as follows:

$$\beta_{\Gamma_{k|k-1}}(\mathbb{S}|\{\mathbf{x}_{k-1}\}) = p(\Gamma_k = \emptyset \subseteq \mathbb{S}|\{\mathbf{x}_{k-1}\}) + p(\Gamma_k = \{\mathbf{x}_k\} \subseteq \mathbb{S}|\{\mathbf{x}_{k-1}\})$$

$$= 1 - p_v + p_v \int p(\mathbf{x}_k|\mathbf{x}_{k-1})d\mathbf{x}_k.$$

The complete Markov transition density for at most one object scenario can be written as

$$f(\emptyset|\emptyset) = 1,$$

$$f(\emptyset|\{\mathbf{x}_{k-1}\}) = 1 - p_v,$$

$$f(\{\mathbf{x}_{k-1}\}|\{\mathbf{x}_{k-1}\}) = p_v \, p(\mathbf{x}_k|\mathbf{x}_{k-1}).$$

6.4.2 Random set model for sensor measurements

Let Σ_k be the random set which models the observation set of a sensor which is a union of the random set corresponding to the target Σ'_k and the random set corresponding to clutter Λ_k. The case where there is no clutter is considered first. If there are no clutter objects and if the object exists, the possible realizations for Σ'_k are:

- $\Sigma'_k = \{\mathbf{y}_k\}$, where $\mathbf{y}_k$ is the measurement originating from the object and takes on any value $\mathbf{y}_k \in \mathcal{R}$, and
- $\Sigma'_k = \emptyset$,

due to the fact that the sensor has non-zero probability of no detection. Due to the random detection process, sometimes, even when the object exists, the sensor does not detect it. The statistics of this random set are described by the belief measure $\beta_{\Sigma'_k}(\mathbb{S}|X) = p(\Sigma'_k \subseteq \mathbb{S})$. The belief mass is the total probability that all observations in a sensor (or multi-sensor) scan will be found in any given region $\mathbb{S}$, if the object has state $\mathbf{x}_k = \{\mathbf{x}_k\}$:

$$\beta_{\Sigma'_k}(\mathbb{S}|\mathbf{X}_k) = p(\Sigma'_k \subseteq \mathbb{S}|\{\mathbf{x}_k\})$$

$$= p(\Sigma'_k = \emptyset|\{\mathbf{x}_k\}) + p(\Sigma'_k \neq \emptyset, \Sigma'_k \subseteq \mathbb{S}|\{\mathbf{x}_k\}).$$

Assuming that the detection process and the measurement process are independent and the probability of detection is P_D, then

$$\beta_{\Sigma'_k}(\mathbb{S}|\{\mathbf{X}_k\}) = p(\Sigma'_k = \emptyset|\{\mathbf{x}_k\}) + p(\Sigma'_k \neq \emptyset|\{\mathbf{x}_k\}) p(\Sigma'_k \subseteq \mathbb{S}|\{\mathbf{x}_k\})$$

$$= 1 - P_D + P_D \, p_{\Sigma'_k}(\mathbb{S}|\{\mathbf{x}_k\}).$$

Taking clutter and false alarms into account, the typical observation set Σ will have the general form

$$\Sigma_k = \Sigma'_k \cup \Lambda_k.$$

Here Σ'_k is the random set of measurements due to the object. It contains either one measurement or is the empty set. The subset Λ_k of observations, on the other hand, models the clutter process. The assumption is that clutter observations are generated by "clutter objects" and can be modeled in the same way as the object of actual interest, but whose noise statistics may differ from those of actual object objects.

For a clutter object, a zero probability of detection means that the clutter object is never observed by the sensor. Likewise, a unity probability of detection for a clutter object means that the clutter object always generates a spurious observation. Thus, for a clutter object, the probability of detection is $\mathbf{P}_{FA}$ (also known as the probability of false alarm). Hence the belief measure of the random set of observations generated by a clutter object is similar to the single object with missed detections,

$$\beta(\mathbb{S}) = 1 - \mathbf{P}_{FA} + \mathbf{P}_{FA} p_c(\mathbb{S}),$$

for some probability measure p_c, with spacial density c. It follows that the typical sensor observation set Σ_k will have the form $\Sigma_k = \Sigma'_k \cup \Lambda_k$, where Σ' models the object and $\Lambda_k = \Lambda_k(1) \cup \Lambda_k(2) \cup \cdots \Lambda_k(M)$, where each $\Lambda_k(j)$ models an individual clutter source at time k. Assuming that each $\Lambda_k(j)$ has the same probability measure p_c and probability of false alarm $\mathbf{P}_{FA}$ and that the object measurements are independent of clutter, we have

$$
\begin{aligned}
\beta_{\Sigma_k}(\mathbb{S}|\mathbf{X}k) &= p(\Sigma'_k, \Lambda_k \subseteq \mathbb{S}) \\
&= p(\Sigma'_k \subseteq \mathbb{S}) p(\Lambda_k \subseteq \mathbb{S}) \\
&= p(\Sigma'_k \subseteq \mathbb{S}) p(\Lambda_k(1) \subseteq \mathbb{S}, \ldots, \Lambda_k(M) \subseteq \mathbb{S}) \\
&= p(\Sigma'_k \subseteq \mathbb{S}) p(\Lambda_k(1) \subseteq \mathbb{S}) \cdots p(\Lambda_k(M) \subseteq \mathbb{S}) \\
&= \beta_{\Sigma'_k}(\mathbb{S}|\mathbf{X}k) \beta_c(\mathbb{S})^M.
\end{aligned}
$$

The derivative of this function gives the global likelihood $f_{\Sigma_k}(\mathbf{y}_k|\mathbf{x}_k)$. However, the belief function is composed of a product of two belief functions. Using the product rule of set derivatives, it can be shown that

$$
\begin{aligned}
f_{\Sigma_k}(\mathbf{y}_k|\mathbf{x}_k) &= f_{\Sigma'_k \cup \Lambda_k}(\mathbf{y}_k|\mathbf{x}_k) \\
&= \sum_{Z_k \subseteq \mathbf{y}_k} f_{\Sigma'_k}(Z_k|\mathbf{x}_k) f_{\Lambda_k}(\mathbf{y}_k - Z_k).
\end{aligned}
\tag{6.9}
$$

The global density of the clutter process is

$$
f_{\Lambda_k}\{\xi_1, \ldots, \xi_n\} = n! \binom{M}{n} \mathbf{P}_{FA}^n (1 - \mathbf{P}_{FA})^{M-n} c(\xi_1) \cdots c(\xi_n),
$$

where $\begin{pmatrix} M \\ n \end{pmatrix}$ gives the number of combinations of n out of M. If the clutter is assumed to be distributed uniformly in the surveillance volume V, then $c(\xi_i) = 1/V$. The global density reduces to

$$f_{\Lambda_k}\{\xi_1, \ldots, \xi_n\} = n! \frac{1}{V^n} \begin{pmatrix} M \\ n \end{pmatrix} \mathbf{P}_{FA}^n (1 - \mathbf{P}_{FA})^{M-n}.$$

If M is large and $\mathbf{P}_{FA}$ is small, then Λ may be approximated by a Poisson process. In that case, the global density can be further simplified to

$$f_{\Lambda_k}\{\xi_1, \ldots, \xi_n\} = n! \frac{1}{V^n} \frac{\lambda^n e^{-\lambda}}{n!}.$$

If there are m_k measurements from clutter, and denoting $f_{\Lambda_k}\{\xi_1, \ldots, \xi_n\}$ by $p_c(m_k)$, the clutter density becomes

$$p_c(m_k) = \frac{1}{V^{m_k}} \lambda^{m_k} e^{-\lambda}. \tag{6.10}$$

Note that $p_c(m_k) = p_c(m_k - 1)\lambda/V$.

The likelihood function from the object measurement process is given by

$$\beta_{\Sigma'_k}(\mathbb{S}|\mathbf{X}k) = p(\Sigma'_k = \emptyset|\{\mathbf{x}_k\}) + p(\Sigma'_k \neq \emptyset|\{\mathbf{x}_k\})p(\Sigma'_k \subseteq \mathbb{S}|\{\mathbf{x}_k\})$$

$$= 1 - P_D + P_D p_{\Sigma'_k}(\mathbb{S}|\{\mathbf{x}_k\})$$

$$= f(\emptyset|\mathbf{x}_k) + \int_{\mathbb{S}} f(\{\mathbf{y}_k\}|\mathbf{x}_k),$$

which implies that the likelihood function is given by

$$f(\emptyset|\mathbf{x}_k) = 1 - P_D, \tag{6.11}$$

$$f(\{\mathbf{y}_k\}|\mathbf{x}_k) = \frac{\delta}{\delta\{\mathbf{y}_k\}} P_D p_{\Sigma'_k}(\mathbb{S}|\{\mathbf{x}_k\})$$

$$= P_D p(\mathbf{y}_k|\mathbf{x}_k). \tag{6.12}$$

The overall likelihood function when the object exists, i.e., when $\mathbf{x}_k = \{\mathbf{x}_k\}$, is obtained by substituting (6.10), (6.11) and (6.12) into (6.9):

$$f_{\Sigma_k}(\mathbf{y}_k|\mathbf{x}_k) = f_{\Sigma'_k \cup \Lambda_k}(\mathbf{y}_k|\mathbf{x}_k)$$

$$= f_{\Sigma_k}(\emptyset|\mathbf{x}_k) f_{\Lambda_k}(\mathbf{y}_k(1), \ldots, \mathbf{y}_k(m_k))$$

$$+ \sum_i f_{\Sigma'_k}(\mathbf{y}_k(i)|\mathbf{x}_k) f_{\Lambda_k}(\mathbf{y}_k - \{\mathbf{y}_k(i)\}),$$

using $Z_k = \{\mathbf{y}_k(i)\} \equiv \{$measurement i originated from object and has value $\mathbf{y}_k(i)\}$,

$$f_{\Sigma_k}(\mathbf{y}_k|\mathbf{x}_k) = p_c(m_k)(1 - P_D) + p_c(m_{k-1})\left(P_D \sum_{i=1}^{m_k} p(\mathbf{y}_k(i)|\mathbf{x}_k)\right)$$

$$= p_c(m_k)\left(1 - P_D + \frac{P_D V}{\lambda} \sum_{i=1}^{m_k} p(\mathbf{y}_k(i)|\mathbf{x}_k)\right).$$

When the object does not exist, $\mathbf{x}_k = \emptyset$, the typical observation set Σ will have the general form

$$\Sigma_k = \Sigma'_k \cup \Lambda_k.$$

However, unlike the case where the object exists, Σ'_k takes only one value, i.e., $\Sigma'_k = \emptyset$. The likelihood has no contribution from the object and is equal to

$$f_{\Sigma_k}(\mathbf{y}_k|\emptyset) = f_{\Sigma'_k \cup \Lambda_k}(\mathbf{y}_k|\emptyset)$$

$$= f_{\Lambda_k}(\mathbf{y}_k|\emptyset) = p_c(m_k) = \frac{1}{V^{m_k}}\lambda^{m_k}e^{-\lambda}.$$

6.4.3 Bayes' update

The general form of the Bayesian recursion is given by (6.3):

$$f_{k|k}(\mathbf{x}_k|\mathbf{y}^k) = \frac{1}{\Delta} f_\Sigma(\mathbf{y}_k|\mathbf{x}_k) \int f_{k|k-1}(\mathbf{x}_k|Xk - 1) f_{k-1|k-1}(Xk - 1|\mathbf{y}^{k-1})\delta Xk - 1$$

$$= \frac{1}{\Delta} f_\Sigma(\mathbf{y}_k|\mathbf{x}_k)\{f_{k|k-1}(\mathbf{x}_k|\emptyset) f_{k|k-1}(\emptyset|\mathbf{y}^{k-1})$$

$$+ \int f_{k|k-1}(\mathbf{x}_k|\{\mathbf{x}_{k-1}\}) f_{k-1|k-1}(\{\mathbf{x}_{k-1}\}|\mathbf{y}^{k-1})\delta Xk - 1\}.$$

The posterior density $f_{k|k}(\mathbf{x}_k|\mathbf{y}^k)$ has the form in

$$f_{k|k}(\emptyset|\mathbf{y}^k) = \text{posterior probability that no objects are present,}$$

$$f_{k|k}(\{\mathbf{x}_k\}|\mathbf{y}^k) = \text{posterior probability of one object with state } \{\mathbf{x}_k\},$$

where $f_{k|k}(\mathbf{x}_k|\mathbf{y}^k)$ is a probability density in the sense that

$$\int f_{k|k}(\mathbf{x}_k|\mathbf{y}^k)\delta\mathbf{x}_k = f_{k|k}(\emptyset|\mathbf{y}^k) + \frac{1}{1!}\int f_{k|k}(\{\mathbf{x}_k\}|\mathbf{y}^k) = 1. \tag{6.13}$$

First,

$$f_{k|k}(\{\mathbf{x}_k\}|\mathbf{y}^k) = \frac{1}{\Delta} f_{\Sigma}(\mathbf{y}_k|\{\mathbf{x}_k\}) f_{k|k-1}(\{\mathbf{x}_k\}|\varnothing) f_{k|k-1}(\varnothing|\mathbf{y}^{k-1}) + \frac{1}{\Delta} f_{\Sigma}(\mathbf{y}_k|\{\mathbf{x}_k\})$$

$$\times \left(\int f_{k|k-1}\{\mathbf{x}_k\}|\{\mathbf{x}_{k-1}\} \right) f_{k-1|k-1}(\{\mathbf{x}_{k-1}\}|\mathbf{y}^{k-1})d\mathbf{X}k - 1$$

$$= \frac{p_c(m_k)}{\Delta} \left(1 - P_D + \frac{P_D V}{\lambda} \sum_{i=1}^{m_k} p(\mathbf{y}_k(i)|\mathbf{x}_k) \right)$$

$$\times \left(0 + p_v \int p(\mathbf{x}_k|\mathbf{x}_{k-1}) p(\mathbf{x}_{k-1}|\mathbf{y}^{k-1})d\mathbf{x}_{k-1} \right)$$

$$= \frac{p_c(m_k) p_v}{\Delta} \left(1 - P_D + \frac{P_D V}{\lambda} \sum_{i=1}^{m_k} p(\mathbf{y}_k(i)|\mathbf{x}_k) \right)$$

$$\times \int p(\mathbf{x}_k|\mathbf{x}_{k-1}) p(\mathbf{x}_{k-1}|\mathbf{y}^{k-1})d\mathbf{x}_{k-1}.$$

It is evident from (6.13) that

$$f_{k|k}(\varnothing|\mathbf{y}^k) = 1 - \int f_{k|k}(\{\mathbf{x}_k\}|\mathbf{y}^k).$$

Thus, by deriving the posterior for $f_{k|k}(\{\mathbf{x}_k\}|\mathbf{y}^k)$, we can simultaneously get the posterior for $f_{k|k}(\varnothing|\mathbf{y}^k)$. These last two equations form the Bayesian filtering equations for the problem of single-object tracking with object existence uncertainty. The posterior probability of object existence (persistence) is given by

$$p_v(\text{posterior}) = \int f_{k|k}(\{\mathbf{x}_k\}|\mathbf{y}^k).$$

Under Gaussian assumptions, one can derive the integrated probabilistic data association (IPDA) filter, as shown in the next section.

6.4.4 The integrated probabilistic data association filter

The integrated probabilistic data association (IPDA) filter, proposed in Mušicki *et al.* (1994), is derived based on the work of PDAF (Bar-Shalom and Fortmann, 1988) by introducing the concept of object existence. However, the same filter can be derived from the random set formalism by making approximations and introducing a Markov chain model for the evolution of object persistence. The random set object motion model in Mahler (2000) encapsulates a model of random set evolution, where, if no objects are present in the scene, this will continue to be the case. If, however, there is one object in the scene then this object either will persist (with probability p_v) or it will vanish (with probability $1 - p_v$). These probabilities are defined as probabilities of object persistence in Goodman *et al.* (1997). These parameters can be directly related to the object existence Markov chain used in IPDA. Both existence and persistence model the same thing, i.e., if the object is not present then it will continue to be the case and if the object is present then it will continue to be so with finite probability. From this point of view, there is no difference between the object existence probability defined in Mušicki *et al.* (1994) and the object persistence probability p_v, in Goodman *et al.* (1997). In IPDA, a Markov chain model is used to evolve the object existence (persistence) probability between sensor measurement times. In the original work of IPDA, two separate Markov chains were proposed – Markov chain 1 and Markov chain 2. Markov chain 2 deals with the observability issue along with the object existence (persistence). Here, the focus is only on Markov chain 1. Owing to this Markov model, all the equations containing target existence probability terms must replace the static probability of object existence p_v, to a dynamic probability of object existence $p_{k|k-1,v}$.

In the original derivation of IPDA, object existence (persistence) is defined as a random variable taking values $\mathfrak{N}_k$ and $\overline{\mathfrak{N}}_k$ modeling object existence (persistence) and object non-existence (non-persistence). The Markov chain that defines this transition between these states over time is given by

$$\begin{bmatrix} p(\mathfrak{N}_k) \\ p(\overline{\mathfrak{N}}_k) \end{bmatrix} = \begin{bmatrix} \gamma_{11} & \gamma_{12} \\ \gamma_{21} & \gamma_{22} \end{bmatrix} \begin{bmatrix} p(\mathfrak{N}_{k-1}) \\ p(\overline{\mathfrak{N}}_{k-1}) \end{bmatrix}.$$

One important fact to be noted is that, in all the works reported on IPDA, the Markov transition probability from non-existence to existence is zero, and non-existence to non-existence is 1. This in effect excludes the event that the object can come into existence from a non-existent state, enforcing the equivalence of existence and persistence. Thus, the probability of object persistence can be equated to the probability of object existence. As these probabilities evolve over time according to a Markov chain, we need to define three aspects of object existence:

- prior probability of object existence and non-existence, $p_{k-1|k-1,v}$ and $p_{k-1|k-1,\overline{v}}$;
- predicted probability of object existence and non-existence, $p_{k|k-1,v}$ and $p_{k|k-1,\overline{v}}$;
- posterior probability of object existence and non-existence, $p_{k|k,v}$ and $p_{k|k,\overline{v}}$.

Recognizing $p_{k|k-1,v} = p(\mathfrak{N}_k)$ and incorporating this Markov chain into the random set formalism, and involving linear Gaussian assumptions, the IPDA is derived. The Markov transition density is

$$\begin{bmatrix} p_{k|k-1,v} \\ p_{k|k-1,\bar{v}} \end{bmatrix} = \begin{bmatrix} \gamma_{11} & 0 \\ 1-\gamma_{11} & 1 \end{bmatrix} \begin{bmatrix} p_{k-1|k-1,v} \\ p_{k-1|k-1,\bar{v}} \end{bmatrix}.$$

Thus there is one free parameter γ_{11} and the predicted probability of object existence can be obtained as

$$p_{k|k-1,v} = \gamma_{11} p_{k-1|k-1,v} + 0 \times p_{k-1|k-1,\bar{v}}$$

$$= \gamma_{11} p_{k-1|k-1,v},$$

and the predicted probability of object non-existence is

$$p_{k|k-1,\bar{v}} = \gamma_{11} p_{k-1|k-1,v} + 1 \times p_{k-1|k-1,\bar{v}}$$

$$= (1-\gamma_{11}) p_{k-1|k-1,v} + p_{k-1|k-1,\bar{v}}.$$

As all practical algorithms use some form of gating and consider only the measurements that fall inside the chosen gate, the probability of detection P_D needs to be modified to include the probability that the target-originated measurement will fall within the gate, usually denoted by P_G. The overall probability then becomes $P_D P_G$ and is subsequently used in all equations in place of P_D.

Considering the simplifications that can be introduced into

$$f_{k|k}(\{\mathbf{x}_k\}|\mathbf{y}^k) = \frac{p_c(m_k) p_{k|k-1,v}}{\Delta} \left(1 - P_D P_G + \frac{P_D P_G V}{\lambda} \sum_{i=1}^{m_k} p(\mathbf{y}_k(i)|\mathbf{x}_k) \right)$$

$$\times \int p(\mathbf{x}_k|\mathbf{x}_{k-1}) p(\mathbf{x}_{k-1}|\mathbf{y}^{k-1}) d\mathbf{x}_{k-1},$$

the integral is identified as the Chapman–Kolmogorov integral, and solves to a Gaussian density if $p(\mathbf{x}_{k-1}|\mathbf{y}^{k-1})$ is Gaussian and the transition density $p(\mathbf{x}_{k-1}|x^{k-1})$ is also Gaussian. Under these Gaussian assumptions, the solution to the Chapman–Kolmogorov equation is equivalent to the Kalman predictor equation and results in a Gaussian density, usually represented by $N(\mathbf{x}_k; \hat{\mathbf{x}}_{k|k-1}, \mathbf{P}_{k|k-1})$.

The likelihood function

If the sensor is assumed to obtain target-originated measurements that are linear in the state of the target and is assumed to have been affected by additive white Gaussian noise, then the measurement likelihood,

$$p(\mathbf{y}_k(i)|\mathbf{x}_k) = N(\mathbf{y}_k(i); H\mathbf{x}_k, \mathbf{R}_k),$$

is also Gaussian, where H is the measurement matrix and $\mathbf{R}_k$ is the measurement noise covariance.

The conditional density

Under these assumptions, the posterior can be further simplified to:

$$f_{k|k}(\mathbf{x}_k|\mathbf{y}^k) = \frac{p_c(m_k)\,p_{k|k-1,v}}{\Delta}\left\{1 - P_D P_G + \frac{P_D P_G V}{\lambda}\sum_{i=1}^{m_k} N(\mathbf{y}_k^i; H\mathbf{x}_k, \mathbf{R}_k)\right\}$$

$$\times\, N(\mathbf{x}_k; \hat{\mathbf{x}}_{k|k-1}, \mathbf{P}_{k|k-1}).$$

If we define $\Lambda_k^i = N(\mathbf{y}_k(i); \hat{\mathbf{y}}_k, \mathbf{S}_k)$, where

$$\hat{\mathbf{y}}_k = H\hat{\mathbf{x}}_{k|k-1},$$

$$\mathbf{S}_k k = H P_{k|k-1} H^T + \mathbf{R}_k,$$

the posterior can be further simplified to

$$f_{k|k}(\mathbf{x}_k|\mathbf{y}^k)$$

$$= \frac{p_c(m_k)\,p_{k|k-1,v}}{\Delta}(1 - P_D P_G)N(\mathbf{x}_k; \hat{\mathbf{x}}_{k|k-1}\mathbf{P}_{k|k-1})$$

$$+ \frac{p_c(m_k)\,p_{k|k-1,v}}{\Delta}\frac{P_D P_G V}{\lambda}\sum_{i=1}^{m_k} N(\mathbf{y}_k(i); H\mathbf{x}_k, \mathbf{R}_k)N(\mathbf{x}_k; \hat{\mathbf{x}}_{k|k-1}, \mathbf{P}_{k|k-1}).$$

The updated density can now be written in a simplified form as

$$f_{k|k}(\mathbf{x}_k|\mathbf{y}^k) = \beta_{k,v}(0)N(\mathbf{x}_k; \hat{\mathbf{x}}_{k|k-1}, \mathbf{P}_{k|k-1}) + \sum_{i=1}^{m_k}\beta_{k,v}(i)N(\mathbf{x}_k; \hat{\mathbf{x}}_{k|k}^i, \mathbf{P}_{k|k}^i).$$

The updated density is the same as derived in Section 5.7.3.

The β_i are usually referred to as the data association probabilities and are given by

$$\beta_{k,\bar{v}}(0) = \frac{p_c(m_k)\,p_{k|k-1,\bar{v}}}{\Delta},$$

$$\beta_{k,v}(0) = \frac{p_c(m_k)\,p_{k|k-1,v}}{\Delta}(1 - P_D P_G),$$

$$\beta_{k,v}(i) = \frac{p_c(m_k)\,p_{k|k-1,v}}{\Delta}\frac{P_D P_G V}{\lambda}\Lambda_k^i,$$

$$\beta_{k,\bar{v}}(0) + \beta_{k,v}(0) + \sum_{i=1}^{m_k}\beta_{k,v}(i) = 1.$$

From this, the normalizing factor Δ can be obtained:

$$\Delta = p_c(m_k)\,p_{k|k-1,\bar{v}} + p_c(m_k)\,p_{k|k-1,v}(1 - P_D P_G) + p_c(m_k)\,p_{k|k-1,v}\frac{P_D P_G V}{\lambda}\Lambda_k^i.$$

Recognizing that $p_{k|k-1,\bar{v}} + p_{k|k-1,v} = 1$, and using $\delta_k = P_D P_G - \sum_{i=1}^{m_k} \frac{P_D P_G V}{\lambda} \Lambda_k^i$,

$$\Delta = p_c(m_k)(1 - \delta_k p_{k|k-1,v}).$$

Substituting,

$$\beta_{k,\bar{v}}(0) = \frac{p_{k|k-1,\bar{v}}}{(1 - \delta_k p_{k|k-1,v})},$$

$$\beta_{k,v}(0) = \frac{p_{k|k-1,v}}{(1 - \delta_k p_{k|k-1,v})}(1 - P_D P_G),$$

$$\beta_{k,v}(i) = \frac{p_{k|k-1,v}}{(1 - \delta_k p_{k|k-1,v})} \frac{P_D P_G V}{\lambda} \Lambda_k^i.$$

These data association probabilities are the same as derived in Section 5.7.3.

The posterior probability of target existence is

$$\int f_{k|k}(\mathbf{x}_k|\mathbf{y}^k) = \int \beta_{k,v}(0) N(\mathbf{x}_k; \hat{\mathbf{x}}_{k|k-1}, \mathbf{P}_{k|k-1}) + \sum_{i=1}^{m_k} \beta_{k,v}(i) N(\mathbf{x}_k; \hat{\mathbf{x}}_{k|k}^i, \mathbf{P}_{k|k}^i)$$

$$= \beta_{k,v}(0) + \sum_{i=1}^{m_k} \beta_{k,v}(i).$$

Thus,

$$p_{k|k,v} = \frac{(1 - \delta_k) p_{k|k-1,v}}{1 - \delta_k p_{k|k-1,v}}.$$

The posterior target existence probability is exactly the same as in Section 5.7.3.

The IPDAF has been derived using the formalism of random sets for single-object tracking in clutter, under linear Gaussian assumptions.

6.4.5 *Joint IPDA derivation from Gaussian mixture PHD filter*

The IPDA filter has already been derived from the random-set-based filter in Section 6.4.4, revealing the theoretical connection between the two frameworks. Similarly, in the multi-object tracking context it can be systematically shown that the joint IPDA (JIPDA) filtering algorithm can be derived from the Gaussian mixture PHD filter equations (Chakravorty and Challa, 2009). The following sections will go through the theoretical relation between the two algorithms and then follow the steps of the Gaussian mixture PHD that will result in the joint IPDA equations under the following conditions:

1. *Each object follows a linear Gaussian dynamical model.*
2. *Object survival and detection probabilities are independent of state.*
3. *There is no explicit object birth and spawning event.*

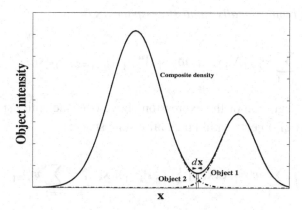

Figure 6.1 Object intensity in state space.

Joint IPDA and PHD filter

In the IPDA approach, the object state consists of dynamic parameters (position, velocity, etc.) and its existence. The IPDA algorithm provides a formulation to propagate the joint probability density given by

$$p(\mathbf{x}_{k-1}, \chi_{k-1} = 1 | Z_{1:k-1}) = p(\chi_{k-1} | Z_{1:k-1}) p(\mathbf{x}_{k-1} | Z_{1:k-1}), \qquad (6.14)$$

where $\mathbf{x}_{k-1}$ denotes the dynamic state of the object, $\chi_{k-1} = 1$ denotes the existence of the object and $\chi_{k-1} = 0$ denotes non-existence of the object. In IPDA, the object state density is assumed to be Gaussian. Therefore the expression in (6.14) becomes

$$p(\mathbf{x}_{k-1}, \chi_{k-1} = 1 | Z_{1:k-1}) = p(\chi_{k-1} = 1 | Z_{1:k-1}) p(\mathbf{x}_k | Z_{1:k-1})$$
$$= w_{k-1} \mathcal{N}(\mathbf{x}_{k-1}; \mathbf{m}_{k-1}, \mathbf{P}_{k-1}), \qquad (6.15)$$

where $w_{k-1} = p(\chi_{k-1} | Z_{1:k-1})$. In a multi-object environment, each object is represented by (6.15). The composite probability density for N such objects, where each has a density given by (6.15), is as follows:

$$\sum_{i=1}^{N} w_{k-1}^i \mathcal{N}(\mathbf{x}_{k-1}; \mathbf{m}_{k-1}^i, \mathbf{P}_{k-1}^i). \qquad (6.16)$$

The composite density in (6.16) can be intuitively interpreted as the object intensity. For simplicity (but without losing generality), let us assume that there are two targets, $N = 2$. The corresponding (assumed) component and composite density are shown as dotted and solid lines respectively in Figure 6.1.

In the figure, each dotted line corresponds to the component density function $w_{k-1}^i \mathcal{N}(\mathbf{x}_{k-1}; \mathbf{m}_{k-1}^i, \mathbf{P}_{k-1}^i)$ and hence does not integrate to one over the entire state space. The interpretation of the composite density is as follows: *the object* *intensity within infinitesimally small state-space d$\mathbf{x}$ is*

$$\sum_{i=1}^{N} w_{k-1}^{i} \mathcal{N}(\mathbf{x}_{k-1}; \mathbf{m}_{k-1}^{i}, \mathbf{P}_{k-1}^{i})|_{\mathbf{x}_{k-1}=d\mathbf{x}_{k-1}} d\mathbf{x}_{k-1}.$$

Therefore the integration of the expression above over the entire space will yield the total number of objects within the entire state space:

$$\int \sum_{i=1}^{N} w_{k-1}^{i} \mathcal{N}(\mathbf{x}_{k-1}; _{k-1}^{i}, \mathbf{P}_{k-1}^{i}) d\mathbf{x}_{k-1} = \sum_{i=1}^{N} w_{k-1}^{i}.$$

As a result, the composite density in (6.16), which is derived from the IPDA formalism of object tracking, conforms to the definition of the probability hypothesis density. The composite density can then be called the probability hypothesis density, following PHD terminology, and can be expressed as shown:

$$D_{k-1|k-1}(\mathbf{x}_{k-1}|Z_{1:k-1}) = \sum_{i=1}^{N} w_{k-1}^{i} \mathcal{N}(\mathbf{x}_{k-1}; _{k-1}^{i}, \mathbf{P}_{k-1}^{i}).$$

In IPDA terminology:

1. w_{k-1}^{i} is the prior object existence probability of the ith target;
2. $\mathcal{N}(\mathbf{x}_{k-1}; \mathbf{m}_{k-1}^{i}, \mathbf{P}_{k-1}^{i})$ is the prior state probability density of the ith object.

JIPDA derivation

Based on the equivalence established in Section 6.4.5 between IPDA-based joint object state (both dynamic and existence) density and the PHD, the recursion of GMPHD (as in Section 6.3.1) will be shown to result in the joint IPDA (JIPDA) algorithm under some required approximations. These approximations are listed here again for clarity:

1. *Each object follows a linear Gaussian dynamical model.*
2. *Object survival and detection probabilities are independent of state.*
3. *There is no explicit object birth and spawning event.*

These approximations are exactly the same as the necessary conditions for GMPHD with $D_{b,k|k-1}(\mathbf{x}_{k}|Z_{1:k-1}) = 0$ and $\gamma_{k}(\mathbf{x}_{k}) = 0$ (to cater for no object birth and spawning event). Therefore, the GMPHD recursion is applicable for the JIPDA derivation. Starting from the initial density, $D_{k-1|k-1}(\mathbf{x}_{k-1}|Z_{1:k-1})$, these recursion steps will be described next.

Prediction

$$D_{k|k-1}(\mathbf{x}_k|Z_{1:k-1}) = D_{S,k|k-1}(\mathbf{x}_k|Z_{1:k-1}) + D_{b,k|k-1}(\mathbf{x}_k|Z_{1:k-1}) + \gamma_k(\mathbf{x}_k). \tag{6.17}$$

The object birth and spawning are not explicitly modeled in the IPDA approach. Therefore, the corresponding intensities, $D_{b,k|k-1}(\mathbf{x}_k|Z_{1:k-1}) = 0$ and $\gamma_k(\mathbf{x}_k) = 0$. However, the existing objects survive with probability $p_{S,k}$, and the corresponding predicted intensity is given as follows:

$$D_{S,k|k-1}(\mathbf{x}_k|Z_{1:k-1}) = \sum_{i=1}^{N} w_{k|k-1}^i \mathcal{N}\left(\mathbf{x}_k; \mathbf{m}_{S,k}^i, \mathbf{P}_{S,k|k-1}^i\right), \tag{6.18}$$

$$\left[\mathbf{m}_{S,k}^i, \mathbf{P}_{S,k|k-1}^i\right] = K F_{\text{pred}}\left[\mathbf{F}_{k-1}, \mathbf{m}_{k-1}^i, -, \mathbf{P}_{k-1}^i, \mathbf{Q}_{k-1}\right], \tag{6.19}$$

$$w_{k|k-1}^i = p_{S,k} w_{k-1}^i. \tag{6.20}$$

According to (6.17), the predicted density is

$$D_{k|k-1}(\mathbf{x}_k|Z_{1:k-1}) = \sum_{i=1}^{N} w_{k|k-1}^i \mathcal{N}\left(\mathbf{x}_k; \mathbf{m}_{S,k}^i, \mathbf{P}_{S,k|k-1}^i\right). \tag{6.21}$$

Update

$$D_{k|k}(\mathbf{x}_k|Z_{1:k})$$

$$= (1 - p_{D,k})D_{k|k-1}(\mathbf{x}_k|Z_{1:k-1}) + \sum_{\mathbf{z}_k \in Z_k} D_{z,k}(\mathbf{x}_k; \mathbf{z}_k)$$

$$= \sum_{i=1}^{N}(1 - p_{D,k})w_{k|k-1}^i \mathcal{N}(\mathbf{x}_k; \mathbf{m}_{S,k}^i, \mathbf{P}_{S,k|k-1}^i)$$

$$+ \sum_{\mathbf{z}_k \in Z_k}\left[\sum_{i=1}^{N} \frac{p_{D,k} w_{k|k-1}^i q_k^i(\mathbf{z}_k)}{\lambda_k c_k(\mathbf{z}_k) + p_{D,k} \sum_{i=1}^{N} w_{k|k-1}^j q_k^j(\mathbf{z}_k)}\mathcal{N}(\mathbf{x}_k; \mathbf{m}_{k|k}^i(\mathbf{z}_k), \mathbf{P}_{k|k}^i(\mathbf{z}_k))\right]$$

$$= \sum_{i=1}^{N} w_{k|k-1}^i \left[(1 - p_{D,k})\mathcal{N}(\mathbf{x}_k; \mathbf{m}_{S,k}^i, \mathbf{P}_{S,k|k-1}^i)\right.$$

$$\left. + \sum_{\mathbf{z}_k \in Z_k} \frac{p_{D,k} q_k^i(\mathbf{z}_k)}{\lambda_k c_k(\mathbf{z}_k) + p_{D,k} \sum_{i=1}^{N} w_{k|k-1}^j q_k^j(\mathbf{z}_k)}\mathcal{N}(\mathbf{x}_k; \mathbf{m}_{k|k}^i(\mathbf{z}_k), \mathbf{P}_{k|k}^i(\mathbf{z}_k))\right]$$

$$= \sum_{i=1}^{N} \left[w_{k|k-1}^{i} \hat{\beta}_{k,0}^{i} \mathcal{N}(\mathbf{x}_k; \mathbf{m}_{S,k}^{i}, \mathbf{P}_{S,k|k-1}^{i}) \right.$$

$$\left. + \sum_{\mathbf{z}_k \in Z_k} w_{k|k-1}^{i} \hat{\beta}_{k,\mathbf{z}_k}^{i} \mathcal{N}(\mathbf{x}_k; \mathbf{m}_{k|k}^{i}(\mathbf{z}_k), \mathbf{P}_{k|k}^{i}(\mathbf{z}_k)) \right], \qquad (6.22)$$

where:

$$\hat{\beta}_{k,0}^{i} = (1 - p_{D,k}),$$

$$\hat{\beta}_{k,\mathbf{z}_k}^{i} = \frac{p_{D,k} q_k^{i}(\mathbf{z}_k)}{\lambda_k c_k(\mathbf{z}_k) + p_{D,k} \sum_{i=1}^{N} w_{k|k-1}^{j} q_k^{j}(\mathbf{z}_k)},$$

$$\left[\mathbf{m}_{k|k}^{i}(\mathbf{z}_k), \mathbf{P}_{k|k}^{i}(\mathbf{z}_k) \right] = KF_{\text{update}} \left[\mathbf{m}_{k|k-1}^{i}, \mathbf{z}_k, \mathbf{H}_k, \mathbf{P}_{k|k-1}^{i} \right].$$

JIPDA approximations

The GMPHD filter results in a Gaussian mixture for the density expression as in (6.22). The next step for GMPHD is to employ either "track to peak" assignment (Lin *et al.*, 2006) or "data-association" (Panta *et al.*, 2005) scheme. However, IPDA approximates the sum of Gaussians for each i (each term within the outermost summation sign in (6.22)) by a single Gaussian. For the ith object, the updated state density is the Gaussian mixture approximation of component densities. This approximation can be obtained from (6.22) and is shown below:

$$\left[w_{k|k-1}^{i} \hat{\beta}_{k,0}^{i} \mathcal{N}(\mathbf{x}_k; \mathbf{m}_{S,k}^{i}, \mathbf{P}_{S,k|k-1}^{i}) + \sum_{\mathbf{z}_k \in Z_k} w_{k|k-1}^{i} \hat{\beta}_{k,\mathbf{z}_k}^{i} \mathcal{N}(\mathbf{x}_k; \mathbf{m}_{k|k}^{i}(\mathbf{z}_k), \mathbf{P}_{k|k}^{i}) \right]$$

$$= \left[\hat{\beta}_{k,0}^{i} + \sum_{\mathbf{z}_k \in Z_k} \hat{\beta}_{k,\mathbf{z}_k}^{i} \right] w_{k|k-1}^{i}$$

$$\times \left[\beta_{k,0}^{i} \mathcal{N}(\mathbf{x}_k; \mathbf{m}_{S,k}^{i}, \mathbf{P}_{S,k|k-1}^{i}) \sum_{\mathbf{z}_k \in Z_k} \beta_{k,\mathbf{z}_k}^{i} \mathcal{N}(\mathbf{x}_k; \mathbf{m}_{k|k}^{i}(\mathbf{z}_k), \mathbf{P}_{k|k}^{i}) \right]$$

$$= w_{k|k}^{i} \mathcal{N}(\mathbf{x}_k; \mathbf{m}_{k|k}^{i}, \mathbf{P}_{k|k}^{i}),$$

where:

$$w_{k|k}^i = \left[\hat{\beta}_{k,0}^i + \sum_{\mathbf{z}_k \in Z_k} \hat{\beta}_{k,\mathbf{z}_k}^i \right] w_{k|k-1}^i, \tag{6.23}$$

$$\beta_{k,0}^i = \frac{\hat{\beta}_{k,0}^i}{\hat{\beta}_{k,0}^i + \sum_{\mathbf{z}_k \in Z_k} \hat{\beta}_{k,\mathbf{z}_k}^i}, \tag{6.24}$$

$$\beta_{k,\mathbf{z}_k}^i = \frac{\hat{\beta}_{k,\mathbf{z}_k}^i}{\hat{\beta}_{k,0}^i + \sum_{\mathbf{z}_k \in Z_k} \hat{\beta}_{k,\mathbf{z}_k}^i}, \tag{6.25}$$

$$\mathbf{m}_{k|k}^i = \beta_{k,0}^i \mathbf{m}_{S,k}^i + \sum_{\mathbf{z}_k \in Z_k} \beta_{k,\mathbf{z}_k}^i \mathbf{m}_{k|k}^i(\mathbf{z}_k), \tag{6.26}$$

$$\mathbf{P}_{k|k}^i = \beta_{k,0}^i \left(\mathbf{P}_{S,k|k-1}^i + (\mathbf{m}_{S,k}^i - \mathbf{m}_{k|k}^i)(\mathbf{m}_{S,k}^i - \mathbf{m}_{k|k}^i)' \right)$$
$$+ \sum_{\mathbf{z}_k \in Z_k} \beta_{k,\mathbf{z}_k}^i \left(\mathbf{P}_{k|k}^i + (\mathbf{m}_{k|k}^i(\mathbf{z}_k) - \mathbf{m}_{k|k}^i)(\mathbf{m}_{k|k}^i(\mathbf{z}_k) - \mathbf{m}_{k|k}^i)' \right). \tag{6.27}$$

The update density in (6.22) can now be expressed by substituting the approximate updated Gaussian obtained in (6.23):

$$D_{k|k}(\mathbf{x}_k|Z_{1:k}) = \sum_{i=1}^{N} w_{k|k}^i \mathcal{N}(\mathbf{x}_k; \mathbf{m}_{k|k}^i, \mathbf{P}_{k|k}^i). \tag{6.28}$$

In joint IPDA terminology:

1. $w_{k|k}^i$ is the posterior existence probability of the ith object, $w_{k|k}^i = p(\chi_k^i = 1|Z_{1:k})$;
2. $\beta_{k,0}^i$, $\beta_{k,\mathbf{z}_k}^i$ are the joint data association probabilities of the ith object (in the multiple-object tracking context).

Track management in JIPDA

The states, as well as the existence probabilities, of the existing N objects are updated according to (6.28). In order to provide a measure of track quality, JIPDA divides the track list based on the existence probabilities $w_{k|k}^i$. The general scheme for such division is summarized below:

1. "Confirmed track": ith track is confirmed if $w_{k|k}^i \geq \gamma_{\text{conf}}$.
2. "Terminated track": ith track is terminated if $w_{k|k}^i < \gamma_{\text{term}}$.
3. "Tentative track": ith track is tentative if $\gamma_1 \leq w_{k|k}^i \leq \gamma_2$,

(where the thresholds, $\gamma_{(.)}$, are chosen suitably). Moreover, several tracks may also be merged if they share a common measurement history over the past few (suitably chosen number of) scans and/or if they are close (suitably defined). Once the track management scheme is applied, all or some of the N objects may survive to continue into the next scan. Accordingly the object density of (6.28) is readjusted to

$$D_{k|k}(\mathbf{x}_k|Z_{1:k}) = \sum_{i=1}^{M} w_{k|k}^i \mathcal{N}(\mathbf{x}_k; \mathbf{m}_{k|k}^k, \mathbf{P}_{k|k}^i), \qquad (6.29)$$

where $M \leq N$. Moreover, the joint IPDA does not explicitly model the object birth or spawning. However, it initiates a new object at k based on a suitable initiation method (discussed in Section 9.4). Each new object is assigned with a certain existence probability and initial state density (of Gaussian form). These initiated object are then added to the overall object density, and the final expression becomes

$$D_{k|k}(\mathbf{x}_k|Z_{1:k}) = \sum_{i=1}^{M} w_{k|k}^i \mathcal{N}(\mathbf{x}_k; \mathbf{m}_{k|k}^k, \mathbf{P}_{k|k}^i) + \sum_{j=1}^{M_{\text{new}}} w_{k|k}^j \mathcal{N}(\mathbf{x}_k; \mathbf{m}_{k|k}^j, \mathbf{P}_{k|k}^j)$$

$$= \sum_{i=1}^{M+M_{\text{new}}} w_{k|k}^i \mathcal{N}(\mathbf{x}_k; \mathbf{m}_{k|k}^k, \mathbf{P}_{k|k}^i), \qquad (6.30)$$

where:

- M_{new} is the number of new tracks initiated at k;
- $w_{k|k}^j$, $\mathbf{m}_{k|k}^j$ and $\mathbf{P}_{k|k}^j$ are the initial object existence probability, mean and covariance for the jth initiated object respectively.

Terminating the existing objects and adding new objects in the density does not pose any problems with the definition of density because the integration of the adjusted density in (6.30) over the entire space still provides the total number of objects within the space. This expression of density in (6.30) becomes the input density for $k + 1$ and thus completes the proof of the joint IPDA recursion from the GMPHD formulation.

6.5 Performance bounds

In standard single-object tracking filters or their extensions into the multi-target tracking problem, the performance of the algorithm is captured by the (time averaged) RMS error. While RMS captures the performance in terms of state error, it is not well suited for evaluating multi-object tracking algorithms because it does not include the error in its estimation of object numbers. In a true multi-object tracking algorithm, the number of objects is a random variable, and therefore any multi-object algorithm needs to be awarded (or punished) for its correct (or incorrect) evaluation of the number of objects along with the state estimate for each of those objects. The "Wasserstein distance" is proposed as the correct measure of performance evaluation in Hoffman and Mahler (2002) for the multi-object tracking context. Here a summary of the evaluation process is presented.

The performance evaluation of any tracking filter is done by calculating the "distance" between the truth and the result from the filter. In RFS terminology, this distance is between two sets X and Y, where $X = \{\mathbf{x}_1, \ldots, \mathbf{x}_n\}$, $Y = \{\mathbf{y}_1, \ldots, \mathbf{y}_m\}$.

Let $\mathbf{d}(\mathbf{x}, \mathbf{y})$ be the distance metric between elements $\mathbf{x}$ and $\mathbf{y}$ on Euclidean N space. From any finite subset, an empirical distribution is constructed as

$$\delta_X(\mathbf{x}) = \frac{1}{n} \sum_{i=1}^{n} \delta_{\mathbf{x}_i}(\mathbf{x}),$$

$$\delta_Y(\mathbf{y}) = \frac{1}{m} \sum_{i=1}^{m} \delta_{\mathbf{y}_i}(\mathbf{y}).$$

The Wasserstein distance between X and Y is defined as

$$d_p^W(X, Y) = \inf_{c} \sqrt[p]{\sum_{i=1}^{n} \sum_{j=1}^{m} C_{i,j} d(\mathbf{x}_i, \mathbf{y}_j)^p},$$

$$d_\infty^W(X, Y) = \inf_{c} \max_{1 \le i \le n, 1 \le j \le m} \sqrt[p]{\tilde{C}_{i,j} d(\mathbf{x}_i, \mathbf{y}_j)^p},$$

where:

- the infimum is taken over all $n \times m$ "transportation matrices" $C = C_{i,j}$;

- $$\tilde{C}_{i,j} = \begin{cases} 1, & \text{if } C_{i,j} \neq 0, \\ 0, & \text{otherwise}; \end{cases}$$

- C is an $n \times m$ transportation matrix if $C_{i,j} \geq 0$ for all $i = 1, \ldots, n$, $j = 1, \ldots, m$, and if for all $i = 1, \ldots, n$, $j = 1, \ldots, m$,

$$\sum_{i=1}^{n} C_{i,j} = \frac{1}{m}, \qquad \sum_{j=1}^{m} C_{i,j} = \frac{1}{n},$$

(which implies $\sum_{i=1}^{n} \sum_{j=1}^{m} C_{i,j} = 1$).

In general $p \geq 2$ is used. The distance can be explained intuitively as follows. Suppose there is only one ground truth, $G = \{\mathbf{g}\}$, while the tracker output consists of several objects, $X = \{x_1, x_2, \ldots, x_n\}$. Among all the outputs, $\{x_1, \ldots, x_{n-1}\}$ are assumed to be close to the ground truth while $\{x_n\}$ is the outlier. Setting $p = 1$ in the distance equation will almost ignore the remaining "outlier" objects. However, setting $p \geq 2$ will put more and more weight on the "outliers" and therefore will increasingly punish the tracker for an incorrect estimation of the number of objects (or the targets away from the truth). This was exactly the objective of the performance evaluation of a true multi-object tracker. Therefore, "Wasserstein distance" is appropriate for the evaluation purpose and is now widely used in the literature (Mahler, 2003b; Panta *et al.*, 2005; Vo *et al.*, 2005, 2006).

6.6 Illustrative example

For a simulation study, a two-dimensional surveillance region of size 500×500 is considered. Each object state consists of position and velocity in each dimension given by $\mathbf{x}_k = [x_k \quad \dot{x}_k \quad y_k \quad \dot{y}_k]^T$. The object survival probability $p_{S,k} = 0.98$. Each object follows a linear dynamic model given by

$$\mathbf{x}_k = \mathbf{F}_{k-1}\mathbf{x}_{k-1} + \mathbf{v}_{k-1},$$

where

$$\mathbf{F}_{k-1} = \begin{bmatrix} 1 & T & 0 & 0 \\ 0 & 1 & 0 & 0 \\ 0 & 0 & 1 & T \\ 0 & 0 & 0 & 1 \end{bmatrix},$$

and $\mathbf{v}_{k-1}$ has a Gaussian distribution with mean zero and covariance $\mathbf{Q}_{k-1}$, which is given by

$$\mathbf{Q}_{k-1} = q^2 \cdot \begin{bmatrix} T^3/3 & T^2/2 & 0 & 0 \\ T^2/2 & T & 0 & 0 \\ 0 & 0 & T^3/3 & T^2/2 \\ 0 & 0 & T^2/2 & T \end{bmatrix}.$$

In the expressions above, T is the sampling period and is assumed to be 1 s. The process noise standard deviation q is assumed to be 2. Object birth is assumed to have Gaussian mixture intensity as follows:

$$\gamma_k(\mathbf{x}_k) = 0.25\mathcal{N}(\mathbf{x}_k; \mathbf{m}_\gamma^1, P_\gamma) + 0.25\mathcal{N}(\mathbf{x}_k; \mathbf{m}_\gamma^2, P_\gamma),$$

where $\frac{1}{\gamma} = [100 \quad 10 \quad 30 \quad -10]^T$, $\frac{2}{\gamma} = [400 \quad -10 \quad 30 \quad -10]^T$ and $P_\gamma = \text{diag}([25 \quad 0 \quad 25 \quad 0])$.

The sensor collects the positions of the objects in each scan. The observation model is as follows:

$$\mathbf{z}_k = \mathbf{H}_k\mathbf{x}_k + \mathbf{w}_k,$$

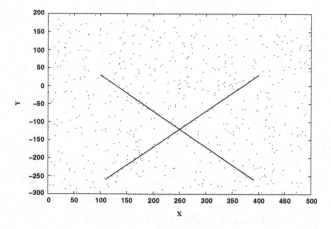

Figure 6.2 Object trajectories (solid line) and collected observations (dots) in one sample simulation run.

where

$$\mathbf{H}_k = \begin{bmatrix} 1 & 0 & 0 & 1 \\ 0 & 0 & 1 & 0 \end{bmatrix},$$

and $\mathbf{w}_k$ is the measurement noise that has Gaussian distribution with mean zero and covariance $\mathbf{R}$, given by

$$\mathbf{R} = \begin{bmatrix} 5 & 0 \\ 0 & 5 \end{bmatrix}. \tag{6.31}$$

The sensor detects each object with a probability of detection $p_{D,k} = 0.9$. The sensor also collects false detections (clutter) that are modeled as a Poisson RFS with intensity with density $\lambda_k = 2 \times 10^{-4} \, \mathrm{m}^{-2}$ or roughly 50 clutter returns over the whole surveillance region per scan. The simulation runs for 30 scans.

The object trajectory superimposed on all observations collected over the entire simulation time is shown in Figure 6.2. Figure 6.3 shows the tracker result for one sample run. The curve indicates that, when the objects were well separated, the estimated probability distribution of the number of objects and their states were converging to the true distributions indicated by the downward trend of the curve. However, once the objects came close and crossed, due to errors in association and/or estimation, the estimated distribution diverges from the true.

For evaluation purposes the tracker is executed for 200 Monte Carlo runs and the Wasserstein distance for each run is averaged at each time point. The resulting performance curve is shown in Figure 6.4.

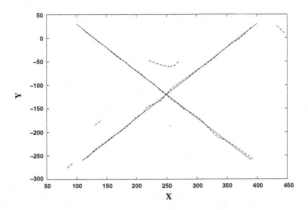

Figure 6.3 GM-PHD tracking result.

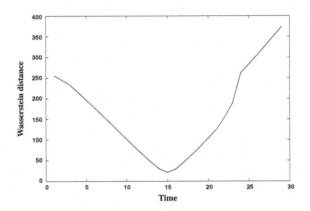

Figure 6.4 GM-PHD tracker performance.

6.7 Summary

In this chapter, a generalized model for multiple-object tracking is introduced. The random finite-set-based model captures the essential problem faced by any tracker algorithm under the multiple-object scenario: estimating the number of objects along with the estimation of dynamic states of each of those individual objects. The objects are modeled as a member of finite sets where the number of members can be randomly (but finitely) varying. Once the model is in place, Bayesian recursion resolves the problem of estimation. In this chapter, the optimal Bayesian recursion is discussed for random finite set models. The optimal recursion is also shown to reduce to the standard IPDA filter under the single-object scenario (with object existence uncertainty). Moreover, several approximated filters, PHD and CPHD, are discussed as a practical random-set-based filter because the optimal version causes combinatorial problems for a large density of objects.

7

Bayesian smoothing algorithms for object tracking

Estimation of an object state at a particular time based on measurements collected beyond that time is generally termed as smoothing or retrodiction. Smoothing improves the estimates compared to the ones obtained by filters owning to the use of more observations (or information). This comes at the cost of a certain time delay. However, these improvements are highly effective in applications like "situation awareness" or "threat assessment." These higher level applications improve operator efficiency if a more accurate picture of the actual field scenario is provided to them, even if it is with a time delay. For these applications, besides object state, parameters representing the overall scenario, like number of targets, their initiation/termination instants and locations, may prove to be very useful ones. A smoothing algorithm can result in a better estimation of the overall situational picture and thus help increase the effectiveness of the critical applications like situation/threat awareness. This chapter will introduce the Bayesian formulation of smoothing and derive the established smoothing algorithms under different tracking scenarios: non-maneuvering, maneuvering, clutter and in the presence of object existence uncertainty.

7.1 Introduction to smoothing

Filters, introduced in previous chapters, produce the "best estimate" of the object state at a particular time based on the measurements collected up to that time. Smoothers, on the other hand, produce an estimate of the state at a time based on measurements collected beyond the time in question (the predictor is another estimator where the estimation at a certain time is carried out based on measurements collected until a point before that time). The operations of these three types of estimators are schematically shown in Figure 7.1.

Based on how the smoother uses the collected measurements to estimate a past state, there are three types of smoothers:

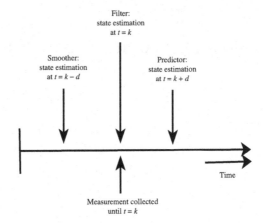

Figure 7.1 Estimation: filtering, smoothing, prediction.

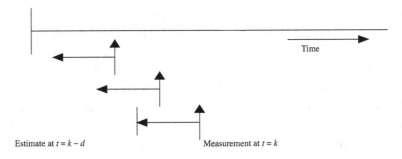

Figure 7.2 Fixed lag smoothing.

- *Fixed lag smoothing*: estimates the past state which lags by a fixed amount (Figure 7.2).
- *Fixed interval smoothing*: the states within time interval from $t = k - N$ to $t = k$ are estimated using the measurements within the window of the same interval from $t = k - N$ to $t = k$. The continuous smoothing is carried out by sliding the window while keeping the interval N fixed (Figure 7.3).
- *Fixed point smoothing*: estimates a state at a particular point of time in the past. This is carried out generally to improve the system initial conditions (Figure 7.4).

In the next section, the optimal Bayesian smoothing recursion will be presented.

7.2 Optimal Bayesian smoothing

One can build the state at a time $k - N$ from measurements received until after that particular time. In the context of Bayes' theorem, if the composite object state at time $t = k - N$ is denoted by $\mathbf{X}_{k-N}$ and the set of observations until time $t = k$

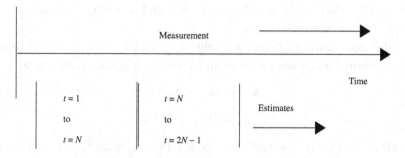

Figure 7.3 Fixed interval smoothing.

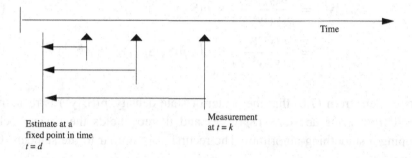

Figure 7.4 Fixed point smoothing.

is denoted by $\mathbf{y}^k = \{\mathbf{y}_k, \mathbf{y}_{k-1}, \ldots, \mathbf{y}_0\}$, the object state estimation is given by

$$p(\mathbf{s}_k|\mathbf{y}^k) = \frac{p(\mathbf{y}_k|\mathbf{s}_k)}{p(\mathbf{y}_k|\mathbf{y}^{k-1})} p(\mathbf{s}_k|\mathbf{y}^{k-1}). \tag{7.1}$$

Modeling the state vector $\mathbf{s}_k$ in (7.1) as the augmentation of past states from time $t = k - N$ to $t = k$ provides the basic smoothing framework to develop a smoothing algorithm. This approach of augmenting the state vector is an easy approach to design a Bayesian smoothing recursion. First the augmented state and observation models will be discussed in the next section. Often this augmented state approach is criticized for increasing the computational complexity of the smoothing algorithms. However, with the rate at which computing speeds are increasing, this criticism is no longer of any practical significance. This augmented state modeling allows the smoothing recursion to follow the same densities as the standard Bayesian filter does in Section 2.1.

7.2.1 Augmented models for smoothing

Using the same formalism introduced in Chapter 1, in the augmented state approach, the object dynamic state is re-modeled as

$$\mathbf{S}_k = \begin{bmatrix} \mathbf{s}_k^T & \mathbf{s}_{k-1}^T & \cdots & \mathbf{s}_{k-N}^T \end{bmatrix}^T \tag{7.2}$$

where $s_{(.)}$ is the state vector at a particular time and hence S_k is an augmentation of state vectors of past time instants.

With the redefinition of object dynamic state vector, both the object dynamic model and sensor model need to be modified (Section 1.3) respectively as

$$S_k = g(S_{k-1}, v_k) \tag{7.3}$$

$$y_k = l(S_k, w_k). \tag{7.4}$$

Therefore, for the augmented state model, the Bayesian formulation of (7.1) results in

$$p(S_k|y^k) = \frac{p(y_k|S_k)}{p(y_k|y^{k-1})} \times p(S_k|y^{k-1}) \tag{7.5}$$

$$= \frac{p(y_k|S_k)}{p(y_k|y^{k-1})} \times \int_{S_{k-1}} p(S_k|S_{k-1}) p(S_k|y^{k-1}). \tag{7.6}$$

It is evident from (7.6) that the posterior state density $p(S_k|y^k)$ is recursively obtained from prior density $p(S_k|y^{k-1})$ and thereby holds the basic block of developing a smoothing algorithm. The recursion is similar to the standard filter equation.

Therefore by introducing the augmented state vector, it is possible to utilize the same Bayesian formulation (stated in previous chapters) for finding out posterior state density. But in the process, the states $s_k, s_{k-1}, \ldots, s_{k-N}$ also get updated through the measurement at time $t = k$. Therefore, the same framework provides a filtered estimate for state s_k, the rest of the state vector elements, $s_{k-1}, s_{k-2}, \ldots, s_{k-N}$ are smoothed. For illustration purposes, a linear model for object dynamics and sensor model will be assumed for the derivation of the Kalman Smoother, PDA Smoother, IMM Smoother and IPDA Smoother.

7.3 Augmented state Kalman smoothing

7.3.1 Object dynamic model

Let us assume an object is moving in one dimension at a fairly constant speed. If we denote x_k as the position in one dimension and $\dot{x}_k$ as the velocity in that dimension at time $t = k$, the composite object state is $x_k = [x_k \quad \dot{x}_k]^T$. The object dynamic is then given by

$$x_k = F_k x_{k-1} + v_k,$$

where

$$F_k = \begin{bmatrix} 1 & T \\ 0 & 1 \end{bmatrix},$$

(T is the sampling interval) and v_k is the additive process noise to allow small deviation from constant speed constraint. The process noise is a zero mean, white Gaussian with variance $\mathbf{Q}_k$. For smoothing, the state vectors are augmented to give rise to $\mathbf{X}_k = [\mathbf{x}_k^T \quad \mathbf{x}_{k-1}^T \quad \cdots \quad \mathbf{x}_{k-N}^T]^T$. The past states are governed by the following transition equation:

$$\mathbf{X}_{k-m} = \mathbf{I}\mathbf{X}_{k-m},$$

where $\mathbf{I}$ is the identity matrix with appropriate dimension; the past states remain the same.

The augmented state dynamic model is then given by

$$\mathbf{X}_k = \mathbf{F}_k\mathbf{X}_{k-1} + \mathbf{V}_k, \tag{7.7}$$

where:

- $\mathbf{F_k}$ is the state transition matrix given by

$$\mathbf{F_k} = \begin{bmatrix} F_k & 0 & 0 & 0 \\ I & 0 & 0 & 0 \\ 0 & I & 0 & 0 \\ \vdots & \ddots & \cdots & \vdots \\ 0 & \cdots & I & 0 \end{bmatrix};$$

- $\mathbf{V}_k$ is the process noise vector given by

$$\mathbf{V}_k = \begin{bmatrix} v_k & 0 & \cdots & 0 \end{bmatrix}^T;$$

- $\mathbf{Q}_k$ is the process noise covariance matrix given by

$$\mathbf{Q}_k = \begin{bmatrix} Q_k & 0 & \cdots & 0 \\ 0 & 0 & \cdots & 0 \\ 0 & \vdots & \ddots & \vdots \\ 0 & \cdots & 0 & 0 \end{bmatrix}. \tag{7.8}$$

7.3.2 Sensor measurement model

We shall assume that, at any particular time $t = k$, the sensor is receiving the measurements of the current position of the object. In that case, the sensor measurement equation is given by

$$y_k = \mathbf{H}_k\mathbf{x}_k + \mathbf{w}_k,$$

with $\mathbf{H}_k = \begin{bmatrix} 1 & 0 \end{bmatrix}$ and $\mathbf{w}_k$ is zero mean, white Gaussian noise with variance $\mathbf{R}_k$. For the smoother the state vector is augmented. Hence, the smoother sensor equation is

$$y_k = \mathcal{H_k}X_k + W_k, \tag{7.9}$$

where

- $$H_k = \begin{bmatrix} H & 0 & \cdots & 0 \end{bmatrix}$$

(it is noted that with H_k defined as above, the measurement vector remains the same, $Y_k = y_k$);

- $W_k = w_k$ is the sensor noise vector. The sensor noises are drawn from a normal distribution with mean zero and variance $\mathcal{R}_k = R_k$ (because the sensor measurement vector is not changing).

7.3.3 The state estimate

The conditional mean of the posterior density $p(X_k|y^k)$ is the best MMSE estimate for the state X_k. The recursion for $p(X_k|y^k)$ is given by (7.6) and stated here for clarity:

$$p(X_k|y^k) = \frac{p(y_k|X_k)}{p(Y_k|Y^{k-1})} \times \int_{X_{k-1}} p(X_k|X_{k-1})p(X_{k-1}|y^{k-1}), \tag{7.10}$$

where:

- $p(y_k|y^{k-1})$ is the *normalization factor*;
- $p(y_k|X_k)$ is the *likelihood function*;
- $p(X_k|y^{k-1}) = \int_{X_{k-1}} p(X_k|X_{k-1})p(X_{k-1}|y^{k-1})$ is the *prediction density*.

The recursion formula along with different densities defined above is exactly the same as the ones for the standard Kalman filter in Section 2.2. The Gaussian and linear assumptions also hold for the smoother. Therefore, Theorem 2.1 is applicable and so are the derivation steps for the *Normalization factor, Likelihood function and Prediction densities*. Therefore we can use the same results as in Section 2.2 with the state vectors and matrices replaced by the augmented models given by (7.7) and (7.9). The resulting solution is the augmented counterpart of the standard Kalman filter and is termed as augmented state Kalman smoother.

7.3.4 Augmented state Kalman smoother equations

Starting from a prior knowledge of the state estimate $\hat{X}_{k_1|k-1}$ and associated error covariance $P_{k-1|k-1}$ at time $t = k - 1$, the augmented state Kalman smoother

(AS-KS) will calculate the object state estimate $\hat{\mathbf{X}}_{k|k}$ through the following steps:

$$\left[\hat{\mathbf{X}}_{k|k-1}, \mathbf{P}_{k|k-1}\right] = \text{KF}_{\text{P}}\left[\hat{\mathbf{X}}_{k_1|k-1}, \mathbf{P}_{k-1|k-1}, \mathcal{F}_k, \mathbf{Q}_k\right]$$

$$\left[\hat{\mathbf{X}}_{k|k}, \mathbf{P}_{k|k}\right] = \text{KF}_{\text{E}}\left[\mathbf{y}, \hat{\mathbf{X}}_{k|k-1}, \mathbf{P}_{k|k-1}, \mathcal{H}_{\mathbf{k}}, \mathcal{R}_k\right].$$

7.4 Smoothing for maneuvering object tracking

In Chapter 3, the maneuvering object tracking problem was discussed in detail and the Interacting Multiple Model (IMM) algorithm was shown to be most effective under the filter maneuvering scenario. The IMM-based smoothing formulation for maneuvering object tracking was carried out by Helmick *et al.* (1993, 1995, 1996). In Challa *et al.* (2002a, 2002b), the augmented model of Bayesian smoothing is utilized to extend the IMM algorithm into smoothing. The algorithm is called the "augmented state IMM" (AS-IMM) algorithm. The AS-IMM algorithm has the advantage of following the standard IMM algorithm steps (Section 3.4) just by augmenting the past states as described in (7.2).

7.4.1 Optimal Bayes' estimation of AS-IMM

For a simple multiple model approach for maneuvering object tracking, the object dynamic model (7.7) is modified as

$$\mathbf{X}_k = \mathcal{F}_k^j \mathbf{X}_{k-1} + \mathbf{U}_k^j \mathbf{V}^j(k), \tag{7.11}$$

The system may at any time switch between any two models $r_k = i$ and $r_k = j$, and the model transition is governed by a Markov chain with known transition probabilities

$$\pi_{ij} = P(r_k = j|r_k = i),$$

where $i, j = 1, 2, \ldots, d$.

The sensor model is also modified to cater for different models and is given by

$$\mathbf{y}_k = \mathbf{H}^j \mathbf{X}_k + \mathbf{W}_k. \tag{7.12}$$

Then, the underlying Bayesian formulation to obtain the posterior density of the augmented state based on the collected measurement set is given by

$$p(\mathbf{X}_k|\mathbf{y}^k) = \sum_{j=1}^d p(\mathbf{X}_k|\mathbf{M}_k^j, \mathbf{y}^k) P(\mathbf{M}_k^j|\mathbf{y}^k), \tag{7.13}$$

where the following notation is used:

$$\mathbf{M}_k^j = \left[M_k^{0,j} \quad M_k^{1,j} \quad \cdots \quad M_k^{N,j} \right]^T. \tag{7.14}$$

$M_k^{i,j}$ signifies that the model M^j that corresponds to the ith lag of the state augmented system is in effect at time k and where $i = 0, 1, 2, \ldots, N$.

The posterior density calculation in (7.13) is same as given in (3.45). The major challenge is to find out the model probabilities for all lags of the augmented state. These probabilities are discussed below.

7.4.2 The model probabilities of the AS-IMM

Provided that measurements with different time indices are independent of each other, the model probability update for model $M_k^{i,j}$ at time k using the current measurement $\mathbf{y}_k$ is given by

$$
\begin{aligned}
P(M_k^{i,j}|\mathbf{y}^k) &= \frac{1}{\delta} p(\mathbf{y}_k|M_k^{i,j}, \mathbf{y}^{k-1}) P(M_k^{i,j}|\mathbf{y}^{k-1}) \\
&= \frac{1}{\delta} p(\mathbf{y}_k|M_k^{i,j}, \mathbf{y}^{k-1}) P(M_k^{i,j}|\mathbf{y}^{k-1}),
\end{aligned}
\tag{7.15}
$$

where the normalization factor $\delta = \sum_{j=1}^{d} P(M_k^{i,j}|\mathbf{y}^k)$ is available once the right-hand side of (7.15) for each individual model $M_k^{i,j}$, $j = 1, \ldots, d$, is evaluated. Each term in the numerator of (7.15) will be derived here:

1. $p(\mathbf{y}_k|M_k^{i,j}, \mathbf{y}^{k-1})$ – the model-based measurement likelihood for the ith lag:

$$
\begin{aligned}
p(\mathbf{y}_k|M_k^{i,j}, \mathbf{y}^{k-1}) &= \sum_{h=1}^{d} p(\mathbf{y}_k|M_k^{i+1,h}, M_k^{i,j}, \mathbf{y}^{k-1}) P(M_k^{i+1,h}|M_k^{i,j}) \\
&\quad \times \sum_{l=1}^{d} P(M_k^{i,j}|M_k^{i-1,l}) \\
&= \sum_{h=1}^{d} \cdots \sum_{s=1}^{d} \sum_{w=1}^{d} \\
&\quad \times p(\mathbf{y}_k|M_k^{0,w}, M_k^{1,s}, \ldots, M_k^{i+1,h}, M_k^{i,j}, \mathbf{y}^{k-1}) \\
&\quad \times P(M_k^{0,w}|M_k^{1,s}), \ldots, P(M_k^{i+1,h}|M_k^{i,j}) \\
&\quad \times \sum_{l=1}^{d} P(M_k^{i,j}|M_k^{i-1,l}) \\
&= \frac{1}{\delta} \sum_{h=1}^{d} \cdots \sum_{s=1}^{d} \sum_{w=1}^{d} \\
&\quad \times \mathcal{N}\left(\mathbf{y}_k; \hat{\mathbf{y}}_{k|k-1}^{w}, \mathbf{S}_k^{w}\right) \pi_{sw} \cdots \pi_{jh} \sum_{l=1}^{d} \pi_{lj}, \quad (7.16)
\end{aligned}
$$

where the following relations are identified according to the structure of the augmented state filter:

$$p(\mathbf{y}_k|M_k^{0,w}, M_k^{1,s}, \ldots, M_k^{i+1,h}, M_k^{i,j}, \mathbf{y}^{k-1}) = p(\mathbf{y}_k|M_k^{0,w}, \mathbf{y}^{k-1})$$

$$= \mathcal{N}\left(\mathbf{y}_k; \hat{\mathbf{y}}_{k|k-1}^w, S_k^w\right)$$

$$P(M_k^{i+1,h}|M_k^{i,j}) = P(M_{k-i+1}^{0,h}|M_{k-i}^{0,j}) = \pi_{jh}$$

$$P(M_k^{i-1,l}|\mathbf{y}^{k-1}) = P(M_{k-1}^{i,l}|\mathbf{y}^{k-1}) = \mu_l^i(k-1).$$

2. $P(M_k^{i,j}|\mathbf{y}^{k-1})$ – the predicted model probability for the ith lag, which is calculated as in the standard IMM.

Substituting these calculated values in (7.15), the probability $P(M_k^{i,j}|\mathbf{y}^k)$ is calculated and that gives the probability of $P(\mathbf{M}_k^j|\mathbf{y}^k)$ in (7.13).

7.4.3 State estimation of AS-IMM

The model-based state density update $p(\mathbf{X}_k|\mathbf{M}_k^j, \mathbf{y}^k)$ in (7.13) is the same as in the standard IMM algorithm (3.45). So the updated state density $p(\mathbf{X}_k|\mathbf{y}^k)$ follows the same step as in the IMM filter algorithm, once the model probability $P(\mathbf{M}_k^j|\mathbf{y}^k)$ is calculated through (7.15).

7.4.4 AS-IMM algorithm equations

Algorithm 41 AS-IMM filter recursion equations at time k

1: **for** each model j **do**
   - Predicted model probability vector:

$$\mu_j(k|k-1) = P(\mathbf{M}_k^j|\mathbf{y}^{k-1}) = \sum_{i=1}^{d} \pi_{ij}\mu_i(k-1).$$

   - Mixing probability:

$$\mu_{i|j}(k-1|k-1)$$
$$= P(\mathbf{M}_{k-1}^i|\mathbf{M}_k^j, \mathbf{y}^{k-1})$$
$$= \left[\mu_{i|j}^0(k-1|k-1) \quad \mu_{i|j}^1(k-1|k-1) \quad \cdots \quad \mu_{i|j}^N(k-1|k-1)\right]^T,$$

   where

$$\mu_{i|j}^h(k-1|k-1) = \frac{\pi_{ij}\mu_i^h(k-1)}{\mu_j^h(k|k-1)} \quad h = 0, 1, \ldots, N.$$

- Mixing estimate:

$$\hat{\mathbf{X}}^{0j}_{k-1|k-1} = \sum_{i=1}^{d} \text{diag}\{\mu_{i|j}(k-1|k-1)\}\hat{\mathbf{X}}^{i}_{k-1|k-1}.$$

- Mixing covariance:

$$\mathbf{P}^{0j}_{k-1|k-1} = \sum_{i=1}^{d} \left\{ \mathbf{P}^{i}_{k-1|k-1} + [\mathbf{X}_{k-1} - \hat{\mathbf{X}}^{0j}_{k-1|k-1}] \right.$$

$$\left. \times [\mathbf{X}_{k-1} - \hat{\mathbf{X}}^{0j}_{k-1|k-1}]' \right\} \text{diag}\{\mu_{i|j}(k-1|k-1)\},$$

where

$$\text{diag}\{\mu_{i|j}(k-1|k-1)\} = \left[\mu^{0}_{i|j}(k-1|k-1)\mathbf{I}_n, \ldots, \mu^{d}_{i|j}(k-1|k-1)\mathbf{I}_n\right],$$

and $\mathbf{I}_n$ is a unit matrix of dimension identical to single state $\mathbf{x}$.

2: **end for**
3: **for** each model j **do** {model-based filtering}
 - Predicted state: $\hat{\mathbf{X}}^{j}_{k|k-1} = \mathbf{F}^{j}\hat{\mathbf{X}}^{0j}_{k-1|k-1} + \mathbf{U}_k$.
 - Predicted covariance: $\mathbf{P}^{j}_{k|k-1} = \mathbf{F}^{j}\mathbf{P}^{0j}_{k-1|k-1}\mathbf{F}^{j'} + \mathbf{Q}^{j}$.
4: **end for**
 - Updated state:

$$\hat{\mathbf{X}}^{j}_{k|k} = \hat{\mathbf{X}}^{j}_{k|k-1} + \mathbf{P}^{j}_{k|k-1}\mathbf{H}^{j^T}(\mathbf{H}^{j}\mathbf{P}^{j}_{k|k-1}\mathbf{H}^{j^T} + \mathbf{R}_k)^{-1}(\mathbf{y}^{j}_{k} - \mathbf{H}^{j}\hat{\mathbf{X}}^{j}_{k|k-1}).$$

 - Updated covariance:

$$\mathbf{P}^{j}_{k|k} = \mathbf{P}^{j}_{k|k-1} - \mathbf{P}^{j}_{k|k-1}\mathbf{H}^{j^T}(\mathbf{H}^{j}\mathbf{P}^{j}_{k|k-1}\mathbf{H}^{j^T} + \mathbf{R}_k)^{-1}\mathbf{H}^{j}\mathbf{P}^{j}_{k|k-1}.$$

5: Model probability update:
 - The overall likelihood for the $\mathbf{y}_k$ for model $\mathbf{M}^{j}_k$:

$$\Lambda^{i}_{j}(k) = \sum_{h=1}^{d} \cdots \sum_{w=1}^{d} \sum_{s=1}^{d} \mathbf{N}\left(\mathbf{y}_k; \mathbf{H}^{w}\hat{\mathbf{X}}^{w}_{k|k-1}, \mathbf{H}^{w}\mathbf{P}^{w}_{k|k-1}\mathbf{H}^{w^T} + \mathbf{R}_k\right)\pi_{sw}\cdots\pi_{jh}.$$

 - Model probability update for model $\mathbf{M}_j(k)$:

$$\mu_j(k|k) = \begin{bmatrix} \mu^{0}_{j}(k) \\ \mu^{1}_{j}(k) \\ \vdots \\ \mu^{N}_{j}(k) \end{bmatrix}$$

where

$$\mu_j^i(k) = \frac{\mu_j^i(k|k-1)\Lambda_j^i(k)}{\sum_{l=1}^N \mu_l^i(k|k-1)\Lambda_l^i(k)}.$$

6: Output combination:

$$\hat{\mathbf{X}}_{k|k} = \sum_{j=1}^d \text{diag}\left\{\mu_j(k|k)\right\} \hat{\mathbf{X}}_{k|k}^j,$$

$$\mathbf{P}_{k|k} = \sum_{j=1}^d \left\{\mathbf{P}_{k|k}^j + [\hat{\mathbf{X}}_{k|k} - \hat{\mathbf{X}}_{k|k}^j][\hat{\mathbf{X}}_{k|k} - \hat{\mathbf{X}}_{k|k}^j]'\right\} \text{diag}\left\{\mu_j(k|k)\right\},$$

where

$$\text{diag}\left\{\mu_j(k|k)\right\} = \text{diag}\left[\mu_j^0(k)\mathbf{I}_n, \mu_j^1(k)\mathbf{I}_n, \ldots, \mu_j^d(k)\mathbf{I}_n\right].$$

7.5 Smoothing for object tracking in clutter

PDA (Section 4.3) is the most popular and effective algorithm for tracking objects in clutter. One of the first attempts to extend the algorithm for smoothing was Mahalanabis *et al.* (1990). But later, in Challa *et al.* (2002a,b), the augmented modeling of the target state is used to propose a smoothing algorithm for object tracking in clutter. The algorithm results in the same step as in the standard PDA filter (Section 4.3) and is called the augmented state PDA (AS-PDA) algorithm.

7.5.1 Bayesian model of augmented state PDA smoothing

In a cluttered environment, the sensor may collect measurements from sources other than the targets. Moreover, the sensor may also miss the object. These two constraints are modeled as follows:

- The collected measurements are given by

$$\mathbf{y}_k = \{\mathbf{y}_k(1), \mathbf{y}_k(2), \ldots, \mathbf{y}_k(n)\},$$

  where n is the number of validated measurements received at time $t = k$.
- The sensor detects the object with a certain probability of detection $0 < P_D <= 1$.

The object dynamic and sensor model are the same as given in (7.7) and (7.9) respectively.

7.5.2 Gating in AS-PDAS

The measurement validation process starts with the previous state estimate $\mathbf{X}_{k-1|k-1}$ and associated covariance matrix $\mathbf{P}_{k-1|k-1}$. The standard KF prediction steps are then followed to derive the densities of the predicted state and measurement. Based on the predicted measurement $\hat{\mathbf{y}}_k$ and associated covariance $\mathbf{S}_k$, each obtained measurement undergoes a test given by

$$[\mathbf{y}_k(n) - \hat{\mathbf{y}}_k]^T \mathbf{S}_k^{-1}[\mathbf{y}_k(n) - \hat{\mathbf{y}}_k] < \gamma, \tag{7.17}$$

where n denotes the nth measurment received.

The chi-square test threshold γ is chosen to ensure a certain required probability of the object-originated measurement to be within the validation gate. This probability is called the "gating probability" and is given by P_G. The region where the test is satisfied is called the "validation region" or "gate." Under Gaussian assumption, theoretically this ellipsoidal region has a volume $\mathbf{V}_k$ and is given by

$$\mathbf{V}_k = c_{n_z}|\gamma|\mathbf{S}_k||^{1/2}, \tag{7.18}$$

where n_z is the dimension of the measurement and c_{n_z} is the volume of the n_z dimensional unit hypersphere ($c_1 = 1, c_2 = \pi, c_3 = 4\pi/3$, etc.). The measurements that satisfy the test, or in other words that are within the gate, are considered as "valid" measurements and are used for the state update. Therefore, out of m received measurements, m_k valid measurements are chosen.

This subset of m_k measurements is used for the Bayesian recursion that results in the PDA formulation of object tracking in clutter.

The Bayesian formulation for the augmented state approach to object tracking in clutter is given by

$$p(\mathbf{X}_k|\mathbf{y}^k, m^k) = p(\mathbf{X}_k|\mathbf{y}_k, m_k, \mathbf{y}^{k-1}, m^{k-1})$$

$$= p(\mathbf{X}_k|\mathbf{y}_k(1), \mathbf{y}_k(2), \ldots, \mathbf{y}_k(m_k), m_k, \mathbf{y}^{k-1}, m^{k-1}), \tag{7.19}$$

where $\mathbf{y}_k(n)$ is the nth validated measurement at time $t = n$, $\mathbf{y}^k$ is the collection of all validated measurements up to $t = k$, m_k is the number of validated measurements at $t = k$ and m^k is the collection of the number of validated measurements until $t = k$.

The Bayesian definition of smoothing in clutter in (7.19) is the same as the definition of the standard PDAF in (4.2) with the exception that the standard state vectors are replaced by augmented ones. As a result the derivation will also be the same as the standard PDAF. The result will be an algorithm for tracking in clutter with smoothing of the past states along with filtering of the current one.

7.5.3 Augmented state PDA smoothing equations

In an iterative manner, starting from the previous state estimate $\hat{\mathbf{X}}_{k-1|k-1}$ and associated error covariance $\mathbf{P}_{k-1|k-1}$ at time $t = k - 1$, the augmented state approach updates the state by following the steps similar to the standard PDA filter. One iteration of the AS-PDA smoother is described in Algorithm 42.

Algorithm 42 AS-PDA filter recursion equations at time k

1: Prediction:

$$\left[\hat{\mathbf{X}}_{k|k-1}, \mathbf{P}_{k|k-1}\right] = \text{KF}_\text{P}\left[\hat{\mathbf{X}}_{k-1|k-1}, \mathbf{P}_{k-1|k-1}, \mathbf{F}_k, \mathbf{Q}_k\right].$$

2: Measurement selection:

$$[\mathbf{y}_k, V_k] = \text{MS}_1\left[\mathbf{y}_k, \hat{\mathbf{X}}_{k|k-1}, \mathbf{P}_{k|k-1}, \mathbf{H_k}, \mathbf{R_k}\right].$$

3: Likelihoods of all selected measurements i:

$$\left[\{p_k(i)\}_i\right] = \text{ML}_1\left[\{\mathbf{y}_k(i)\}_i, \hat{\mathbf{X}}_{k|k-1}, \mathbf{P}_{k|k-1}, \mathbf{H_k}, \mathbf{R_k}\right].$$

4: **if** non-parametric tracking **then**
5:    V_k calculated by equation:

$$V_k = \frac{\pi^{n/2}}{\Gamma(n/2+1)}\sqrt{|\mathbf{S}(k)|}\gamma^{1/2},$$

   where $|\mathbf{S}(k)|$ is the determinant of $\mathbf{S}(k)$.
6:    Clutter measurement density estimation:

$$\rho = m_k / V_k.$$

7: **end if**
8: Single-object data association (sàns object existence):

$$\left[\{\boldsymbol{\beta}_k(i)\}_{i=0}^{m_k}\right] = \text{STDA}\left[\{p_k(i)\}_{i=1}^{m_k}\right].$$

9: Estimation/merging:

$$\left[\hat{\mathbf{X}}_{k|k}, \mathbf{P}_{k|k}\right] = \text{PDA}_\text{E}\left[\hat{\mathbf{X}}_{k|k-1}, \mathbf{P}_{k|k-1}, \{\mathbf{y}_k(i)\}_{i=1}^{m_k}, \{\boldsymbol{\beta}_k(i)\}_{i=0}^{m_k}, \mathcal{H}_\mathbf{k}, \mathcal{R}_\mathbf{k}\right].$$

10: Output trajectory estimate:
   • track mean value $\hat{\mathbf{X}}_{k|k}$ and covariance $\mathbf{P}_{k|k}$.

7.6 Smoothing with object existence uncertainty

IPDA and its derivatives (Sections 5.7.3 and 5.8.3) are the most effective algorithms for tracking objects with existence uncertainty. Moreover, its theoretical connection with random set formalism (Section 6.4.4 and Appendix) has extended the applicability of such algorithms to multiple-target tracking and other more complex scenarios. Chakravorty and Challa (2004, 2005, 2006) developed a smoothed version of the IPDA algorithm.

The IPDA smoothing follows the augmented state model for object dynamics and sensor observations. For linear systems, the resultant models are as in (7.7) and (7.9). The sensor receives clutter measurements and also detects objects with a certain detection probability. These model parameters also hold the same as in AS-PDA smoothing, described in Section 7.5.1. The IPDA filter models the object existence as a Markov chain with two (Markov chain 1 model) alternative events – "existence" ($E_k = 1$) and "non-existence" ($E_k = 0$). (For clarity, only Markov chain 1 is discussed here. The following discussion can easily be extended to Markov chain 2.) In each time scan, an object can switch between the two events with some certain probability. The switching probability matrix is given by

$$\Gamma = \begin{bmatrix} \Gamma_{00} & \Gamma_{01} \\ \Gamma_{10} & \Gamma_{11} \end{bmatrix},$$

where $\Gamma_{ij} = p(E_k = j | E_{k-1} = i)$, $i, j \in \{1, 0\}$.

But smoothing spans N time intervals and an object may have several alternative histories associated with its existence. Therefore, the core part of augmented state IPDA is to model the object existence augmentations and to recursively update the probabilities of these augmented existences. The state augmentation and recursion is carried out in the same fashion as in the standard IPDA filter.

7.6.1 Recursion of augmented object existences

The object existence event denoted by $E_k = 1$ and non-existence event $E_k = 0$ for lag N are augmented to give rise to the augmented object existences given by

$$\mathbf{E}_k = \begin{bmatrix} E_k = i, \, E_{k-1} = j, \, \ldots, \, E_{k-N} = l \end{bmatrix}^T,$$

where $i, j \in 1, 0$. Therefore, the possible realizations for $\mathbf{E}_k$ can be obtained from all combinations of $i, j, \ldots, l$, taking values of either 1 or 0. However, in IPDA, an existing object may cease to exist and an object would not come back to existence once it becomes "non-existent." This is reflected in the transition probability matrix Γ, with $\Gamma_{01} = 0$ and $\Gamma_{00} = 1$. These conditions have an impact on the augmentation of object existences. For example, let us consider a combination of the augmented existence:

$$\ldots, E_{k-a} = 1, \ldots, E_{k-b} = 0, \ldots.$$

Here, if $k - a > k - b$, the probability of this augmented existence will be zero. As a result, there exist $N + 2$ permissible combinations for the augmentation of object existence. These are as follows:

- *Object existed from time $t = k - N$ to $t = k - m$ but not from $t = k - m + 1$ to $t = k$,*

$$\mathbf{E}_k^m = \begin{bmatrix} E_k = 0, \ldots, E_{k-m+1} = 0, E_{k-m} = 1, \ldots, E_{k-N} = 1 \end{bmatrix}^T, \quad (7.20)$$

where $m = 0, 1, 2, \ldots, N$.
- *Object did not exist anytime within the interval $t = k - N$ to $t = k$,*

$$\mathbf{E}_k^{N+1} = \begin{bmatrix} E_k = 0, \ldots, E_{k-N} = 0 \end{bmatrix}^T.$$

The augmented state IPDA smoother updates the $p(\mathbf{E}_k^m | \mathbf{y}^{k-1})$ in time in order to obtain the updated probability $p(\mathbf{E}_k^m | \mathbf{y}^k)$. This is carried out using the Bayesian recursion formula

$$p(\mathbf{E}_k^m | \mathbf{y}^k) = p(\mathbf{E}_k^m | \mathbf{y}_k, \mathbf{y}^{k-1})$$
$$= \frac{p(\mathbf{y}_k | \mathbf{E}_k^m, \mathbf{y}^{k-1}) \cdot p(\mathbf{E}_k^m | \mathbf{y}^{k-1})}{\Delta}, \quad (7.21)$$

where

$$\Delta = \sum_{m=0}^{N+1} p(\mathbf{E}_k^m | \mathbf{y}^k).$$

The predicted probabilities of augmented existences in (7.21), $p(\mathbf{E}_k^m | \mathbf{y}^{k-1})$, are obtained as below:

$$p(\mathbf{E}_k^0 | \mathbf{y}^{k-1}) = \gamma_{11} p(\mathbf{E}_{k-1}^0 | \mathbf{y}^{k-1}),$$
$$p(\mathbf{E}_k^1 | \mathbf{y}^{k-1}) = \gamma_{10} p(\mathbf{E}_{k-1}^0 | \mathbf{y}^{k-1}),$$
$$p(\mathbf{E}_k^m | \mathbf{y}^{k-1}) = p(\mathbf{E}_{k-1}^{m-1} | \mathbf{y}^{k-1}),$$

where $m = 2, \ldots, N$, and

$$p(\mathbf{E}_k^{N+1} | \mathbf{y}^{k-1}) = p(\mathbf{E}_{k-1}^N | \mathbf{y}^{k-1}) + p(\mathbf{E}_k^{N+1} | \mathbf{y}^{k-1}).$$

These prediction equations for augmented existence probabilities can be summarized by

$$
\begin{bmatrix}
p(E_k^0|y^{k-1}) \\
p(E_k^1|y^{k-1}) \\
p(E_k^2|y^{k-1}) \\
p(E_k^3|y^{k-1}) \\
\vdots \\
p(E_k^N|y^{k-1}) \\
p(E_k^{N+1}|y^{k-1})
\end{bmatrix}
= J
\begin{bmatrix}
p(E_{k-1}^0|y^{k-1}) \\
p(E_{k-1}^1|y^{k-1}) \\
p(E_{k-1}^2|y^{k-1}) \\
p(E_{k-1}^3|y^{k-1}) \\
\vdots \\
p(E_{k-1}^N|y^{k-1}) \\
p(E_{k-1}^{N+1}|y^{k-1})
\end{bmatrix},
\tag{7.22}
$$

where J is the Markov transition probability matrix, defined by

$$
J =
\begin{bmatrix}
\gamma_{11} & 0 & 0 & 0 & 0 & 0 & 0 \\
\gamma_{10} & 0 & 0 & 0 & 0 & 0 & 0 \\
0 & 1 & 0 & 0 & 0 & 0 & 0 \\
0 & 0 & 1 & 0 & 0 & 0 & 0 \\
\vdots & \cdots & \ddots & \cdots & \cdots & \vdots & \vdots \\
0 & 0 & \cdots & 0 & 1 & 0 & 0 \\
0 & 0 & \cdots & 0 & 0 & 1 & 1
\end{bmatrix}.
$$

The likelihood term in (7.21), $p(y_k|E_k^m, y^{k-1})$, is dependent on the existence of the object. For $m = 0$, the object is assumed to be existing and therefore the likelihood becomes

$$
p(y_k|E_k^0, y^{k-1}) = p(y_k|E_k = 1, y^{k-1}) = \delta = \left(\frac{1}{V_k}\right)^{m_k} P_0(m_k)(1 - \delta_k),
$$

where:

-
$$
\delta_k =
\begin{cases}
P_D P_G \left[1 - \dfrac{V_k}{\hat{m}_k} \displaystyle\sum_{i=1}^{m_k} \Lambda_k^i\right], & \text{if } m_k > 0, \\
P_D P_G, & \text{if } m_k = 0;
\end{cases}
\tag{7.23}
$$

- P_D, P_G are the object detection and measurement gate probabilities respectively;
- m_k is the total number of validated measurements;
- V_k is the gate volume;
- $\hat{m}_k$ is the expected number of validated measurements, given by

$$
\hat{m}_k =
\begin{cases}
m_k - P_D P_G p(E_k^0|y_{k-1}), & \text{if } m_k > 0, \\
0, & \text{if } m_k = 0;
\end{cases}
\tag{7.24}
$$

- Λ_k^i is the likelihood of the ith validated measurement;
- $P_0(m_k)$ is the Poisson probability density function of the number of clutter measurements.

For $p(\mathbf{y}_k|\mathbf{E}_k^m, \mathbf{y}^{k-1})$, where $m = 2, \ldots, N + 1$, the object is assumed to be non-existing at k and hence the likelihood will contain only the clutter likelihoods. This is given by

$$p(\mathbf{y}_k|\mathbf{E}_k^m, \mathbf{y}^{k-1}) = \left(\frac{1}{V_k}\right)^{m_k} P_0(m_k). \tag{7.25}$$

Substituting these results in (7.21), the recursion of augmented existences is completed:

$$p(\mathbf{E}_k^m|\mathbf{y}^k) = \begin{cases} \frac{1}{\Delta}\delta \times p(\mathbf{E}_k^0|\mathbf{y}^{k-1}), & \text{if } m = 0, \\ \frac{1}{\Delta}(\frac{1}{V_k})^{m_k} P_0(m_k) \times p(\mathbf{E}_k^m|\mathbf{y}^{k-1}), & \text{if } N + 1 > m > 0, \end{cases} \tag{7.26}$$

where

$$\Delta = \delta \times p(\mathbf{E}_k^0|\mathbf{y}^{k-1}) + \left(\frac{1}{V_k}\right)^{m_k} P_0(m_k) \times p(\mathbf{E}_k^m|\mathbf{y}^{k-1}).$$

This operation is summarized in a functional form below:

$$\left[\left\{p(\mathbf{E}_k^m|\mathbf{y}^k)\right\}_{m=0}^{N+1}\right] = \text{ASTEX}_{\text{E}}\left[\left\{p(\mathbf{E}_{k-1}^m|\mathbf{y}^{k-1})\right\}_{m=0}^{N+1}\right].$$

7.6.2 Recursion of augmented state AS-IPDA smoothing

The state density conditioned on object existence (similar to IPDA formulation) for smoothing estimation can be defined as

$$p(\mathbf{X}_k|\mathbf{E}_k^0, \mathbf{y}^k). \tag{7.27}$$

Using Bayes' theorem the state density can be calculated as

$$p(\mathbf{X}_k|\mathbf{E}_k^0, \mathbf{y}^k) = p(\mathbf{X}_k|\mathbf{E}_k^0, \mathbf{y}_k, \mathbf{y}_{k-1})$$

$$= \frac{p(\mathbf{y}_k|\mathbf{X}_k, \mathbf{E}_k^0, \mathbf{y}_{k-1}) p(\mathbf{X}_k|\mathbf{E}_k^0, \mathbf{y}_{k-1})}{p(\mathbf{y}_k|\mathbf{E}_k^0, \mathbf{y}_{k-1})}.$$

This expansion is exactly same as the standard IPDA filter. Therefore, the state estimation in AS-IPDA is done in the same manner as for the IPDA filter.

7.6.3 AS-IPDA smoothing equations

Algorithm 43 AS-IPDA filter recursion equations at time k

1: Time k inputs:
  - set $\mathbf{Y}_k$ of measurements delivered by the sensor;
  - probability of object existence $\left\{ p(\mathbf{E}_{k-1}^m | \mathbf{y}^{k-1}) \right\}_{m=0}^{N+1}$; and
  - object trajectory estimate mean value $\hat{\mathbf{X}}_{k|k-1}$ and covariance $\mathbf{P}_{k|k-1}$.

2: Track state propagation:

$$p(\mathbf{E}_k^0 | \mathbf{y}^{k-1}) = \gamma_{11} p(\mathbf{E}_{k-1}^0 | \mathbf{y}^{k-1}),$$

$$\left[\hat{\mathbf{X}}_{k|k-1}, \mathbf{P}_{k|k-1} \right] = \text{KF}_{\text{P}} \left[\hat{\mathbf{X}}_{k-1|k-1}, \mathbf{P}_{k-1|k-1}, \mathcal{F}_k, \mathcal{Q}_k \right].$$

3: Measurement selection:

$$[\mathbf{y}_k, V_k] = \text{MS}_1 \left[\mathbf{y}_k, \hat{\mathbf{X}}_{k|k-1}, \mathbf{P}_{k|k-1}, \mathcal{H}_\mathbf{k}, \mathcal{R}_\mathbf{k} \right].$$

4: Likelihoods of all selected measurements i:

$$\left[\{ p_k(i) \}_i \right] = \text{ML}_1 \left[\{ \mathbf{y}_k(i) \}_i, \hat{\mathbf{X}}_{k|k-1}, \mathbf{P}_{k|k-1}, \mathcal{H}_\mathbf{k}, \mathcal{R}_\mathbf{k} \right].$$

5: **if** non-parametric tracking **then**
6:    V_k calculated by (5.77).
7:    Clutter measurement density estimation:

$$\rho = \hat{m}_k / V_k = (m_k - P_D P_G p(\mathbf{E}_k^0 | \mathbf{y}^{k-1})) / V_k.$$

8: **end if**
9: Single-object data association:

$$\left[-, \{ \boldsymbol{\beta}_k(i) \}_{i=0}^{m_k} \right] = \text{STDA} \left[p(\mathbf{E}_k^0 | \mathbf{y}^{k-1})), \{ p_k(i) \}_{i=1}^{m_k} \right].$$

10: Augmented existence probability update:

$$\left[\left\{ p(\mathbf{E}_k^m | \mathbf{y}^k) \right\}_{m=0}^{N+1} \right] = \text{ASTEX}_{\text{E}} \left[\left\{ p(\mathbf{E}_{k-1}^m | \mathbf{y}^{k-1}) \right\}_{m=0}^{N+1} \right].$$

11: Estimation/merging:

$$\left[\hat{\mathbf{X}}_{k|k}, \mathbf{P}_{k|k} \right] = \text{PDA}_{\text{E}} \left[\hat{\mathbf{X}}_{k|k-1}, \mathbf{P}_{k|k-1}, \{ \mathbf{y}_k(i) \}_{i=1}^{m_k}, \{ \boldsymbol{\beta}_k(i) \}_{i=0}^{m_k}, \mathcal{H}_\mathbf{k}, \mathcal{R}_\mathbf{k} \right].$$

12: Output trajectory estimate:
  - track mean value $\hat{\mathbf{X}}_{k|k}$ and covariance $\mathbf{P}_{k|k}$.

7.7 Illustrative example

7.7.1 Simulation scenario

A two-dimensional surveillance region with area 1000 m long and 400 m wide is considered for simulation. Each simulation experiment consists of $N = 1000$ runs, where each run consists of 30 scans. The scan interval $T = 1$ s.

In each run, a single object moves in the area with a constant speed perturbed by a zero mean noise that accounts for small maneuvers. The object dynamics is, in Cartesian coordinates, modeled by

$$\mathbf{x}_k = \mathbf{F}\mathbf{x}_{k-1} + \mathbf{w},$$

where:

- the object state consists of position and velocity in two dimensions, $\mathbf{x}_k = \begin{bmatrix} x_k & \dot{x}_k & y_k & \dot{y} \end{bmatrix}^T$;
- a transition matrix captures the constant velocity model by

$$\mathbf{F} = \begin{bmatrix} 1 & T & 0 & 0 \\ 0 & 1 & 0 & 0 \\ 0 & 0 & 1 & 1 \\ 0 & 0 & 0 & T \end{bmatrix};$$

- $\mathbf{w}$ is the zero mean, white Gaussian noise with covariance

$$\mathbf{Q} = q \begin{bmatrix} T^4/4 & T^3/2 & 0 & 0 \\ T^3/2 & T^2 & 0 & 0 \\ 0 & 0 & T^4/4 & T^3/2 \\ 0 & 0 & T^3/2 & T^2 \end{bmatrix},$$

and q is assumed to be 0.75.

At the start of each simulation run, the object reappears with state

$$\mathbf{x}_1 = \begin{bmatrix} 130\,\text{m} & 35\,\text{m/s} & 200\,\text{m} & 0\,\text{m/s} \end{bmatrix}^T.$$

The sensor, if it detects the object, collects the position of the object in each scan. The sensor observation model is given by

$$\mathbf{y}_k = \mathbf{H}\mathbf{x}_k + \mathbf{v}_k,$$

where:

-
$$\mathbf{H} = \begin{bmatrix} 1 & 0 & 0 & 0 \\ 0 & 0 & 1 & 0 \end{bmatrix};$$

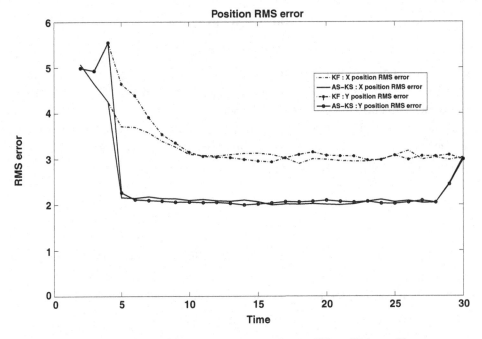

Figure 7.5   RMS position error: augmented state KS vs. Kalman filter.

- $\mathbf{v}_k$ is the zero mean, white Gaussian noise with covariance

$$\mathbf{R} = r \begin{bmatrix} 1 & 0 \\ 0 & 1 \end{bmatrix} m^2,$$

and r is assumed 25.

For each of the smoother algorithms, a fixed lag of 3 has been used. The performances of the algorithms are measured in RMS error. Tracks are initiated using a two-point differencing method. The RMS errors are integrated over 1000 runs.

7.7.2 Augmented state Kalman smoother

The performance of the AS-KS is compared against the standard KF filter. The performance measure is given in Figures 7.5 and 7.6 for each component of the state. Note that, as the simulation was performed usings a fixed lag of 3, the last two time points did not have three measurements to obtain smoothed estimates. Hence they exhibit higher RMS errors than the estimates at earlier time points. (This applies to Figures 7.5–7.11.)

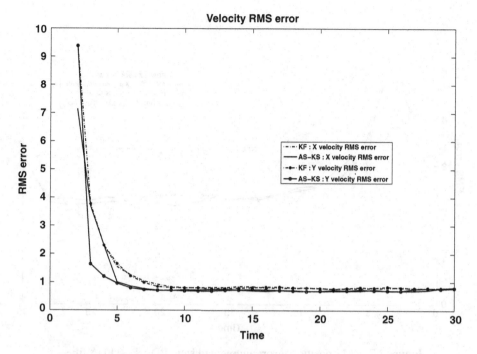

Figure 7.6 RMS velocity error: augmented state KS vs. Kalman filter.

7.7.3 Augmented state PDA smoother

The clutter is generated in each scan using a Poisson distribution with density $\lambda = 1 \times 10^{-4}/\text{m}^2/\text{scan}$. The clutter is scattered uniformly across the entire surveillance region. The sensor detects the object with a detection probability $P_D = 0.9$ (which means the sensor may also miss the object at a certain scan). The probability of the object originated measurement to be within the validation is set to be $P_G = 0.99$ and thereby the validation gate threshold is 9 for a two-dimensional ellipsoid.

The performance of AS-PDAS in comparison with the standard PDA filter is reported in Figures 7.7 and 7.8.

7.7.4 Augmented state IPDA smoother

The clutter and the sensor detection model are the same as in PDA simulation. In view of the discussion in Section 7.6.1, the only model parameter for the object existence switching probability matrix Γ is $\gamma_1 1$. For the purpose of the simulation the parameter is assumed to be $\gamma_{11} = 0.98$, which thereby sets $\gamma_{10} = 1 - \gamma_{11} = 0.02$.

Figures 7.9 and 7.10 demonstrate the performance of AS-IPDA and compare it with the IPDA filter.

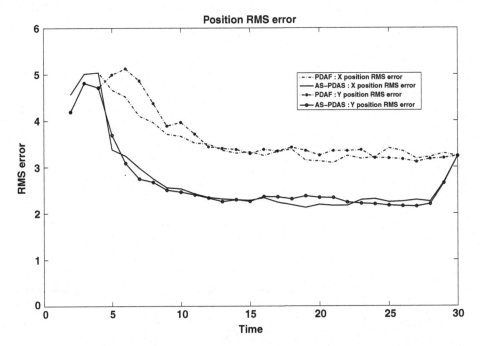

Figure 7.7 RMS position error: augmented state PDAS vs. PDA filter.

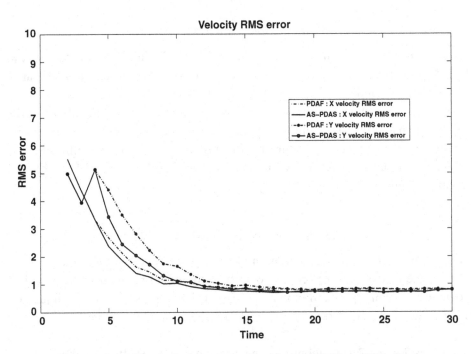

Figure 7.8 RMS velocity error: augmented state PDAS vs. PDA filter.

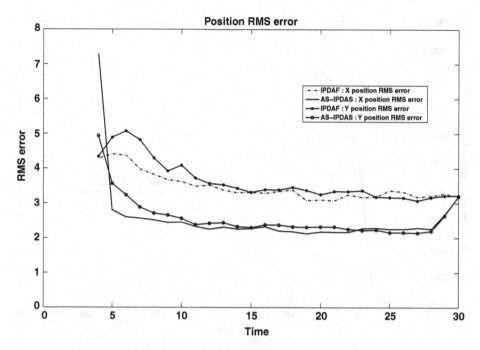

Figure 7.9 RMS position error: augmented state IPDAS vs. IPDA filter.

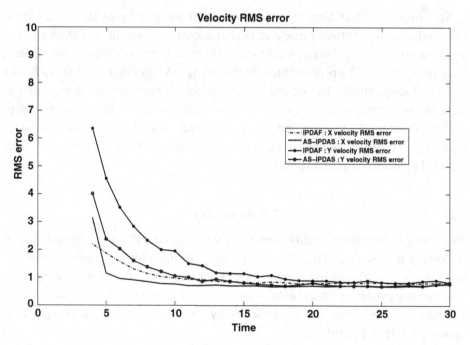

Figure 7.10 RMS velocity error: augmented state IPDAS vs. IPDA filter.

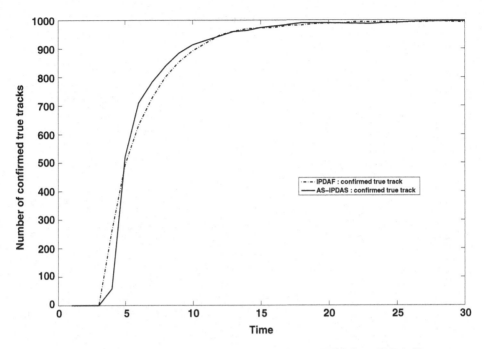

Figure 7.11 Confirmed true tracks: augmented state IPDAS vs. IPDA filter.

Moreover, the IPDA algorithm (both smoother and filter) provides a track measure to distinguish between true and false tracks. In the simulation, 0.95 and 0.1 are the existence probabilities used for the IPDA filter to confirm and terminate a track respectively. The probabilities for the AS-IPDA smoother were set to be 0.3 and 0.001 respectively. The setting of these probabilities is carried out in order to make the false track statistics (469 and 497 per run for filter and smoother respectively) so that the RMS error and true track detection statistics remain consistent for comparison between algorithms. Figure 7.11 shows the confirmed true tracks detected by each of the algorithms.

7.8 Summary

Smoothing is introduced in this chapter. First the optimal Bayesian formulation of smoothing is presented. Then the smoothing algorithms for important scenarios – maneuvering and non-maneuvering object tracking, object tracking in clutter with and without existence uncertainty – have been discussed. Several example sets also illustrate the improvements obtained by smoothing algorithms compared to standard filtering algorithms.

8

Object tracking with time-delayed, out-of-sequence measurements

Object tracking using delayed, out-of-sequence measurements is a problem of growing importance due to an increased reliance on networked sensors interconnected via complex communication network architectures. In such systems, it is often the case that measurements are received out-of-time-order at the processing computer. This problem has appeared in the literature under various names such as the out-of-sequence measurements (OOSM) problem (Blackman and Popoli, 1999; Bar-Shalom, 2000), the problem of tracking with random sampling and delays (Marcus, 1979; Hilton *et al.*, 1993; Thomopoulos and Zhang, 1994), and the problem of incorporating random time delayed measurements (Ravn *et al.*, 1998). In this chapter, we present a Bayesian solution to this problem and provide approximate, implementable algorithms for both cluttered and non-cluttered scenarios involving single and multiple time-delayed measurements. Under linear Gaussian assumptions, the Bayesian solution reduces to an augmented state Kalman filter (AS-KF) for scenarios devoid of clutter and an augmented state probabilistic data association (AS-PDA) for scenarios involving clutter with modified measurement equations and likelihoods. These smoothing algorithms were presented in the previous chapter (Challa *et al.*, 2002c, 2003; Wang and Challa, 2003).

8.1 Optimal Bayesian solution to the OOSM problem

8.1.1 Object dynamics, sensor measurement equations

Usually the object dynamics are modeled by a discrete-time state equation,

$$\mathbf{x}_{k+1} = \mathbf{F}\mathbf{x}_k + \mathbf{w}_{k+1}, \tag{8.1}$$

and sensor measurements are modeled by

$$\mathbf{y}_k = \mathbf{H}\mathbf{x}_k + \mathbf{v}_k, \tag{8.2}$$

where $\mathbf{H}$ is the sensor measurement matrix and $\mathbf{F}$ is the object transition matrix.

Denoting a "standard" measurement sequence, $\mathbf{y}^k = \{\mathbf{y}_1, \mathbf{y}_2, \ldots, \mathbf{y}_k\}$, the object "standard" tracking problem reduces to the problem of computing the conditional mean estimate of the object state,

$$\hat{\mathbf{x}}_{k|k} \triangleq E(\mathbf{x}_k | \mathbf{y}^k), \tag{8.3}$$

and its associated error covariance,

$$\mathbf{P}_{k|k} \triangleq E[(\mathbf{x}_k - \hat{\mathbf{x}}_{k|k})(\mathbf{x}_k - \hat{\mathbf{x}}_{k|k})' | \mathbf{y}^k]. \tag{8.4}$$

A key problem arising when dealing with multiple interconnected sensors with communication links is the time delay between the sensor and tracking computer. This problem can be defined as follows: when a measurement corresponding to time τ, expressed as

$$\mathbf{y}_\tau = \mathbf{H}_\tau \mathbf{x}(\tau) + \mathbf{v}(\tau), \qquad \tau < k, \tag{8.5}$$

arrives at time t_k after (8.3) and (8.4) have been computed, one faces the problem of updating the state estimate and its covariance with the delayed measurement (8.5), i.e., to compute

$$\hat{\mathbf{x}}_{\tau|k} \triangleq E(\mathbf{x}_k | \mathbf{y}^k, \mathbf{y}_\tau), \tag{8.6}$$

and

$$\mathbf{P}_{\tau|k} \triangleq E[(\mathbf{x}_k - \hat{\mathbf{x}}_{k|k})(\mathbf{x}_k - \hat{\mathbf{x}}_{k|k})' | \mathbf{y}^k, \mathbf{y}(\tau)]. \tag{8.7}$$

8.1.2 Optimal Bayesian filter

Let $\mathbf{x}_k$ be the object state at time t_k, $\mathbf{y}_\tau$ be the set of delayed sensor measurements, corresponding to time τ, and $\mathbf{y}^k$ be the set of sensor measurement sequence received up to time t_k.

Having processed all the measurements $\mathbf{y}^k$, the complete information about the object state $\mathbf{x}_k$ is described in the probability density function $p(\mathbf{x}_k | \mathbf{y}^k)$. The OOSM problem arises as a consequence of receiving a measurement set $\mathbf{y}_\tau$ at time t_k that corresponds to time $\tau < t_k$. The solution to this OOSM problem seeks to update $p(\mathbf{x}_k | \mathbf{y}^k)$ with $\mathbf{y}_\tau$ to obtain $p(\mathbf{x}_k | \mathbf{y}^k, \mathbf{y}_\tau)$.

Invoking Bayes' rule,

$$p(\mathbf{x}_k | \mathbf{y}^k, \mathbf{y}_\tau) = \frac{p(\mathbf{y}_\tau | \mathbf{x}_k, \mathbf{y}^k) p(\mathbf{x}_k | \mathbf{y}^k)}{p(\mathbf{y}_\tau | \mathbf{y}^k)}. \tag{8.8}$$

Consider the numerator of (8.8) and introducing the object state at time τ, x_τ,

$$p(\mathbf{y}_\tau | x_k, \mathbf{y}^k) p(\mathbf{x}_k | \mathbf{y}^k) = \int p(\mathbf{y}_\tau, \mathbf{x}_\tau | x_k, \mathbf{y}^k) p(\mathbf{x}_k | \mathbf{y}^k) d\mathbf{x}_\tau$$

$$= \int p(\mathbf{y}_\tau | \mathbf{x}_\tau, \mathbf{x}_k, \mathbf{y}^k) p(\mathbf{x}_\tau | \mathbf{x}_k, \mathbf{y}^k) p(\mathbf{x}_k | \mathbf{y}^k) d\mathbf{x}_\tau.$$

Since $p(\mathbf{x}_\tau, \mathbf{x}_k | \mathbf{y}^k) = p(\mathbf{x}_\tau | \mathbf{x}_k, \mathbf{y}^k) p(\mathbf{x}_k | \mathbf{y}^k)$, we have

$$p(\mathbf{y}_\tau | \mathbf{x}_k, \mathbf{y}^k) p(\mathbf{x}_k | \mathbf{y}^k) = \int p(\mathbf{y}_\tau | \mathbf{x}_\tau, \mathbf{x}_k, \mathbf{y}^k) p(\mathbf{x}_\tau, \mathbf{x}_k | b y^k) d\mathbf{x}_\tau. \tag{8.9}$$

Substituting (8.9) back into (8.8) yields

$$p(\mathbf{x}_k | \mathbf{y}^k, \mathbf{y}_\tau) = \frac{\int p(\mathbf{y}_\tau | \mathbf{x}_\tau, \mathbf{x}_k, \mathbf{y}^k) p(\mathbf{x}_\tau, \mathbf{x}_k | \mathbf{y}^k) d\mathbf{x}_\tau}{p(\mathbf{y}_\tau | \mathbf{y}^k)}$$

$$= \int \frac{p(\mathbf{Y}(\tau) | \mathbf{x}(\tau), \mathbf{x}_k, \mathbf{Y}^k) p(\mathbf{x}(\tau), \mathbf{x}_k | \mathbf{Y}^k)}{p(\mathbf{Y}(\tau) | \mathbf{Y}^k)} d\mathbf{x}(\tau).$$

Using the inverse form of Bayes' rule,

$$p(\mathbf{x}_k | \mathbf{Y}^k, \mathbf{Y}(\tau)) = \int p(\mathbf{x}(\tau), \mathbf{x}_k | \mathbf{Y}^k, \mathbf{Y}(\tau)) dx(\tau). \tag{8.10}$$

It is thus clear that solving the OOSM problem involves *consideration of the joint density of the current object state and the object state corresponding to the delayed measurement.*

Generalizing this, the OOSM problem involving multiple delays can be stated as follows: let the delayed measurements received at time t_k be denoted by $Y(\tau) = \{Y(\tau_1), Y(\tau_2), \ldots, Y(\tau_d)\}$, where $\tau_i < t_k, \forall i \in \{1, \ldots, d\}$, and τ_d is the time corresponding to the maximum time delay. Then the solution to the OOSM problem is to determine the density,

$$p(\mathbf{x}_k | \mathbf{Y}^k, \mathbf{Y}(\tau)) = \int_{\mathbf{x}(\tau_1)} \int_{\mathbf{x}(\tau_2)} \cdots \int_{\mathbf{x}(\tau_d)} p(\mathbf{x}_k, \mathbf{x}(\tau_1), \mathbf{x}(\tau_2), \ldots, \mathbf{x}(\tau_d) | \mathbf{Y}^k, \mathbf{Y}(\tau))$$

$$\times d\mathbf{x}(\tau_1), d\mathbf{x}(\tau_2), \ldots, d\mathbf{x}(\tau_d), \tag{8.11}$$

which indicates that, in general, the solution involves a Bayes recursion for the joint probability density of an augmented state vector $\mathbf{X}_k = [\mathbf{x}(t_k), \mathbf{x}(\tau_1), \ldots, \mathbf{x}(\tau_d)]^T$, i.e.,

$$p(\mathbf{x}_k, \mathbf{x}(\tau_1), \mathbf{x}(\tau_2), \ldots, \mathbf{x}(\tau_d) | \mathbf{Y}^k, \mathbf{Y}(\tau)) = p(\mathbf{X}_k | \mathbf{Y}^k, \mathbf{Y}(\tau)). \tag{8.12}$$

Consider a discrete time system where

$$\tau_1 = t_{k-1}, \tau_2 = t_{k-2}, \ldots, \tau_d = t_{k-d}.$$

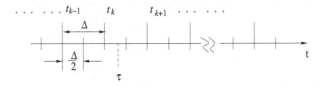

Figure 8.1 Engineering approximation to delayed time τ.

Note that, in general, the delayed measurement time τ need not correspond exactly to the times $t_{k-1}, t_{k-2}, \ldots, t_{k-d}$, i.e., τ might fall within the intervals around t_{k-i}, $i \in [1, d]$. If τ falls inside the interval $[t_k - \frac{\Delta}{2}, t_k + \frac{\Delta}{2})$ centered at t_k, as shown in Figure 8.1, we approximate $\tau = t_k$. $\Delta = t_k - t_{k-1} \; \forall \, k$ gives the time lag between successive states of the augmented state. The error caused by this approximation is called the discretization error. For small Δ, discretization errors are small, which leads to the consideration of more lags in the augmented state, and hence to an increased computational load to cover the maximum time delay. Therefore, striking a balance between the choice of Δ and the algorithm computational load becomes an important issue when the augmented state approach is used to solve the OOSM problem.

From Bayes' rule, (8.12) becomes

$$p(\mathbf{x}_k, \mathbf{x}(t_{k-1}), \ldots, \mathbf{x}(t_{k-d})|\mathbf{Y}^k, \mathbf{Y}(\tau))$$

$$= \frac{1}{\delta} p(\mathbf{Y}(\tau)|\mathbf{x}_k, \mathbf{x}(t_{k-1}), \ldots, \mathbf{x}(t_{k-d}), \mathbf{Y}^k) p(\mathbf{x}_k, \mathbf{x}(t_{k-1}), \ldots, \mathbf{x}(t_{k-d})|Y^k),$$

$$(8.13)$$

for a normalizing constant δ.

Denoting $[\mathbf{x}_k, \ldots, \mathbf{x}(t_{k-d})]^T$ as $\mathbf{X}_k^d$, the Bayes recursion for (8.13) becomes

$$p(\mathbf{X}_k^d|\mathbf{Y}^k, \mathbf{Y}(\tau)) = \frac{1}{\delta} p(\mathbf{Y}(\tau)|\mathbf{X}_k^d, \mathbf{Y}^k) p(\mathbf{X}_k^d|\mathbf{Y}^k). \qquad (8.14)$$

If the two densities on the right-hand side of (8.14) are Gaussian, then the posterior density on the left-hand side of (8.14) is also Gaussian and the solution reduces to a standard KF with an augmented state (Ho and Lee, 1964).

In order to simplify the derivation/implementation of a number of algorithms proposed in this chapter, it is desirable to include the current measurement along with the delayed measurements even if it is not delayed, and thus the measurement vector becomes

$$\mathbf{Y}_k = [\mathbf{y}(k), \mathbf{y}(k-1), \ldots, \mathbf{y}(k-d)]^T. \qquad (8.15)$$

Hence, by changing notation, (8.14) can be rewritten as

$$p(\mathbf{X}_k^d|\mathbf{Y}^{k-1}, \mathbf{Y}_k) = \frac{1}{\delta} p(\mathbf{Y}_k|\mathbf{X}_k^d, \mathbf{Y}^{k-1}) p(\mathbf{X}_k^d|\mathbf{Y}^{k-1}). \qquad (8.16)$$

Equation (8.16) is a fundamental relationship which leads to the development of the AS-KF and AS-PDA algorithms proposed in this chapter.

8.2 Single- and multi-lag OOSM algorithms

8.2.1 Y-algorithm

This approach assumes that the measurement delay is less than one sampling period, i.e., $t_{k-1} \leq \tau < t_k$. Define a joint Gaussian random variable z_k,

$$\mathbf{z}_k = \begin{bmatrix} \mathbf{x}(t_k) \\ \mathbf{y}(\tau) \end{bmatrix} \quad \text{with covariance} \quad \mathbf{P}_z = \begin{bmatrix} \mathbf{P}_{xx} & \mathbf{P}_{xy} \\ \mathbf{P}_{yx} & \mathbf{P}_{yy} \end{bmatrix}, \quad (8.17)$$

where

$$\mathbf{P}_{xx} = E[(\mathbf{x}_k - \hat{\mathbf{x}}_{k|k})(\mathbf{x}_k - \hat{\mathbf{x}}_{k|k})^T | \mathbf{Y}^k] = \mathbf{P}_{k|k}, \quad (8.18)$$

$$\mathbf{P}_{yy} = E[(\mathbf{y}(\tau) - \hat{\mathbf{y}}(\tau))(\mathbf{y}(\tau) - \hat{\mathbf{y}}(\tau))^T | \mathbf{Y}^k] = \mathbf{S}_{\tau|k}, \quad (8.19)$$

$$\mathbf{P}_{xy} = E[(\mathbf{x}_k - \hat{\mathbf{x}}_{k|k})(\mathbf{y}(\tau) - \hat{\mathbf{y}}(\tau))^T | \mathbf{Y}^k] = \mathbf{P}_{yx}^T. \quad (8.20)$$

The solution to the OOSM problem requires the conditional density $p(\mathbf{x}_k | \mathbf{y}(\tau), \mathbf{Y}^k)$. Using the results in Bar-Shalom and Fortmann (1988), this density is known to be Gaussian with mean

$$\hat{\mathbf{x}}(t_{k|\tau,k}) = \hat{\mathbf{x}}_{k|k} + \mathbf{P}_{xy}\mathbf{P}_{yy}^{-1}(\mathbf{y}(\tau) - \hat{\mathbf{y}}(\tau)), \quad (8.21)$$

and covariance

$$\mathbf{P}(t_{k|\tau,k}) = \mathbf{P}_{xx} - \mathbf{P}_{xy}\mathbf{P}_{yy}^{-1}\mathbf{P}_{yx}, \quad (8.22)$$

where the backward predicted measurement is expressed as

$$\hat{\mathbf{y}}(\tau) = \mathbf{H}_\tau \mathbf{F}_{\tau|k}[\hat{\mathbf{x}}_{k|k} - \mathbf{Q}_k(\tau)\mathbf{H}_\tau^T \mathbf{S}_{\tau|k}^{-1}(y_k - \hat{\mathbf{y}}(t_{k|k-1}))]. \quad (8.23)$$

In this expression, $\mathbf{H}_\tau$ is the observation matrix at time τ, $\mathbf{F}_{\tau|k}$ is the system backward transition matrix[1] from t_k to τ, the last term, which is ignored in Hilton *et al.* (1993), Blackman and Popoli (1999) and Mallick *et al.* (2001a), accounts for the effect of process noise (with covariance $Q_k(\tau)$) on the estimate $\hat{x}_{k|k}$.

The cross covariance P_{xy} in (8.20) is given by

$$\mathbf{P}_{xy} = [\mathbf{P}_{k|k} - \mathbf{P}_{x\tilde{y}}]\mathbf{F}_{\tau|k}^T \mathbf{H}_\tau^T, \quad (8.24)$$

where

$$\mathbf{P}_{x\tilde{y}} \triangleq \text{Cov}\{\mathbf{x}_k, \mathbf{w}_k(\tau) | \mathbf{y}^k\} = \mathbf{Q}_k(\tau) - \mathbf{P}(t_{k|k-1})\mathbf{H}_\tau^T \mathbf{S}_k^{-1}\mathbf{H}_\tau \mathbf{P}(t_{k|k-1}). \quad (8.25)$$

The Y-algorithm, as pointed out in Bar-Shalom (2000), requires storage of the last innovation and can be interpreted as a type of non-standard smoothing.

---

[1] Although τ represents time, it also is used to indicate corresponding time index whenever no confusion arises.

8.2.2 M-algorithm

The M-algorithm, proposed in Mallick *et al.* (2001a, 2001b), extends the Y-algorithm, in an approximate way, to account for multiple delays. The key idea of this approach is to determine the cross covariance of (8.20) for each delayed measurement and at each time interval. However, it fails to account for the non-zero conditional mean of the process noise covariance and hence its extension of the Y-algorithm is approximate. By expressing the delayed measurement $\mathbf{y}(\tau)$ as a function of the current state x_k, the multiple-lag OOSM problem can be solved by computing the cross covariance P_{xy} in a recursive manner for each time-delayed measurement. For example, when the delay time τ is more than n sampling intervals, we have

$$\mathbf{P}_{xy|n} = -M_{k-n+1}\mathbf{Q}(k-n+1,k;k-n+1,k)$$

$$-\sum_{i=1}^{n} M_{k-i+1}\mathbf{Q}(k-i+1,k-i;k-i+1,k), \qquad (8.26)$$

where

$$M_{k-i+1} = \begin{cases} B_k, & i = 1, \\ C_k C_{k-1} \cdots C_{k-i+2} B_{k-i+1}, & i = 2, \ldots, n, \end{cases} \qquad (8.27)$$

$$B_i = I - K_i H_i, \qquad (8.28)$$

$$C_i = B_i F_{i-1|i}, \qquad (8.29)$$

and the covariance of the process noise is

$$\mathbf{Q}(k-i+1,k-i;k-i+1,k)$$

$$\triangleq E\{\mathbf{w}(k-i+1,k-i;k-i+1)\mathbf{w}^T(k-i+1,k-i;k-i+1)\}. \qquad (8.30)$$

Clearly, in the calculation of the covariance in (8.26), one needs to evaluate the process noise from the time when measurement delay occurs to the current time and all time steps in between. One also has to evaluate all corresponding filter gains. The AS-KF solution presented below does not need to evaluate the process noise explicitly. All that is needed is an augmented state and the standard KF computing steps.

8.3 Augmented state Kalman filter for multiple-lag OOSM

For multiple delays, the measurement vector has the form (8.15), and the Bayes' recursion of (8.16) reduces to the AS-KF with an augmented state $\mathbf{X}_k^d$. The system dynamics model using the augmented state can be constructed from (8.1) and (8.2) based on methods in Anderson and Moore (1979),

$$\mathbf{X}_k^d = \mathbf{F}_k \mathbf{X}_k^d + \mathbf{W}_k,$$
$$\mathbf{Y}_k = \mathbf{H}_k \mathbf{X}_k^d + \mathbf{V}_k, \tag{8.31}$$

where

$$\mathbf{F}_k = \begin{bmatrix} F_{t_k|\tau} & 0 & \cdots & 0 & 0 \\ I & 0 & \cdots & 0 & 0 \\ 0 & \ddots & 0 & \vdots & \vdots \\ \vdots & 0 & \ddots & 0 & 0 \\ 0 & \cdots & 0 & I & 0 \end{bmatrix}. \tag{8.32}$$

F is the system transition matrix in discrete form, the observation matrix is

$$\mathbf{H}_k = \begin{bmatrix} H_k & 0 & \cdots & 0 \\ 0 & H_{\tau_1} & 0 & \vdots \\ \vdots & \cdots & \ddots & 0 \\ 0 & \cdots & 0 & H_{\tau_d} \end{bmatrix}, \tag{8.33}$$

and the noise covariance matrix is

$$\mathbf{R}_k = \begin{bmatrix} R_k & 0 & \cdots & 0 \\ 0 & R_{\tau_1} & 0 & \vdots \\ \vdots & \cdots & \ddots & 0 \\ 0 & \cdots & 0 & R_{\tau_d} \end{bmatrix}. \tag{8.34}$$

The predicted density and the likelihood are given by

$$p(\mathbf{X}_k^d|\mathbf{Y}^{k-1}) = \mathcal{N}(\mathbf{X}_k^d; \widehat{\mathbf{X}}_{k|k-1}^d, \mathbf{P}_{k|k-1}^d), \tag{8.35}$$

$$p(\mathbf{Y}_k|\mathbf{X}_k^d, \mathbf{Y}^{k-1}) = \mathcal{N}(\mathbf{Y}_k; \mathbf{H}_k\widehat{\mathbf{X}}_{k|k-1}^d, \mathbf{S}_k^d), \tag{8.36}$$

and the updated density (Ho and Lee, 1964) is given by

$$p(\mathbf{X}_k^d|\mathbf{Y}^k) = \mathcal{N}(\mathbf{X}_k^d; \widehat{\mathbf{X}}_{k|k}^d, \mathbf{P}_{k|k}^d), \tag{8.37}$$

with mean and covariance

$$\widehat{\mathbf{X}}_{k|k}^d = \widehat{\mathbf{X}}_{k|k-1}^d + \mathbf{K}_k\widetilde{\mathbf{Y}}_k, \tag{8.38}$$

$$\mathbf{P}_{k|k}^d = (\mathbf{I} - \mathbf{K}_k\mathbf{H}_k)\mathbf{P}_{k|k-1}^d, \tag{8.39}$$

where the innovation is

$$\widetilde{\mathbf{Y}}_k = \mathbf{Y}_k - \mathbf{H}_k \widehat{\mathbf{X}}^d_{k|k-1}, \tag{8.40}$$

with covariance

$$\mathbf{S}_k = \mathbf{H}_k \mathbf{P}_{k|k-1} \mathbf{H}_k^T + \mathbf{R}_k, \tag{8.41}$$

and the Kalman gain matrix is

$$\mathbf{K}_k = \mathbf{P}^d_{k|k-1} \mathbf{H}_k^T \mathbf{S}_k^{-1}. \tag{8.42}$$

Clearly, (8.35)–(8.42) are the standard Kalman filter equations for an augmented state space model (Section 2.2). This approach solves the problem of incorporating delayed measurements and also provides smoothed outputs (Anderson and Moore, 1979).

8.3.1 Iterative AS-KF

Apart from direct computation via the augmented vector and associated matrix form, the AS-KF can also be implemented in a nested form, i.e., iteratively computing (8.38)–(8.42) using measurements corresponding to different delays independently. This is because the gain matrix is column independent with respect to the time indices (see Appendix A for a proof), i.e.,

$$\mathbf{K}_k = \begin{bmatrix} K_k & K_{k-1} & \cdots & K_{k-d} \end{bmatrix}. \tag{8.43}$$

Then, given the measurement set $\mathbf{Y_k} = [y_k, y(t_{k-1}), \ldots, y(t_{k-d})]^T$ received at time t_k, the update equations of the AS-KF state estimate and its covariance are given by (8.38) and (8.39). These can be simplified into an iterative form given by

$$\widehat{\mathbf{X}}^d_{k|k} = \widehat{\mathbf{X}}^d_{k|k-1} + \sum_{i=k}^{k-d} \mathbf{K}^i \widetilde{\mathbf{Y}}^i, \tag{8.44}$$

$$\mathbf{P}^d_{k|k} = \mathbf{P}^d_{k|k-1} - \sum_{i=k}^{k-d} \mathbf{K}^i \mathbf{H}^i \mathbf{P}^d_{k|k-1}, \tag{8.45}$$

where $\mathbf{H}^i$ is given by (8.33) with zero everywhere except for the ith partition component. $\widetilde{\mathbf{Y}}^i$ is the same dimension as $\mathbf{Y}_k$ but produced by assuming only the ith component of $\mathbf{Y}_k$ is received and $\mathbf{K}^i$ is the Kalman gain produced using $\mathbf{H}^i$, i.e.,

$$\mathbf{K}^i = \mathbf{P}^d_{k|k-1} \mathbf{H}^{i^T} \mathbf{S}_k^{-1}. \tag{8.46}$$

The above equations provide an equivalent but efficient implementation of (8.38) and (8.39). The equivalence is shown in Appendix A.

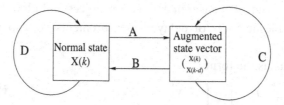

Figure 8.2 The VDAS-KF process.

8.3.2 *Variable dimension augmented state Kalman filter*

The AS-KF presented in Section 7.4.2 uses an augmented state involving past states up to the maximum delay. However, augmenting all the past states up to the maximum delay leads to increased computational complexity. A more efficient solution can be obtained that adaptively augments only essential past states and removes the states that need not be a part of the augmented state. This variable dimension AS-KF (VDAS-KF) is considered in this section.

The idea of the VDAS-KF is that the augmented state only carries the current state and the past state for which there was a missing measurement. The filter will reduce to a normal Kalman filter if there is no OOSM. The VDAS-KF processes measurements are shown in Figure 8.2 for the following four cases:

- *Case A*: object state is augmented if current measurement is delayed.
- *Case B*: the augmented state reduces to the standard state vector if the delayed measurement $y(\tau)$ is received and has been processed.
- *Case C*: the augmented state will keep its dimension unchanged when either:
  - the delayed data $y(\tau)$ has not arrived yet; or
  - the delayed data has been received but the current data is delayed.
- *Case D*: no augmented state (standard KF) if there is no measurement delay.

For all the above cases, changes to the state dimension will lead to changes in the associated covariance matrix.

The VDAS-KF algorithm is computationally more efficient than the standard AS-KF. However, it is more complicated to implement and has a tracking performance and computational load that are comparable with the Y-algorithm.

8.4 Augmented state PDA filter for multiple-lag OOSM in clutter

8.4.1 *Clutter modeling for the augmented state PDA filter with OOSM*

The OOSM problem in clutter, involving probabilistic data association (PDA), presents some interesting challenges. In such situations, the measurement at time

t_k may include[2] a set of measurements (from object and clutter) corresponding to the current time and sets of delayed measurements corresponding to earlier times.

Thus, the data has the form

$$[Y_k, Y(\tau)]^T! = \begin{bmatrix} Y_k \\ \vdots \\ Y_{k-d} \end{bmatrix} = \left\{ \begin{matrix} \mathbf{y}_k^1, & \mathbf{y}_k^2, & \cdots, & \mathbf{y}_k^m \\ \vdots & \vdots & \cdots & \vdots \\ \mathbf{y}^1(t_{k-d}), & \mathbf{y}^2(t_{k-d}), & \cdots, & \mathbf{y}^n(t_{k-d}) \end{matrix} \right\}. \quad (8.47)$$

All possible combinations of the current and delayed measurements form the "measurements" in the augmented measurement space. For example, at time t_k, let us assume that the following four measurements are received – two for each time index:

$$\left[Y_k, Y(t_{k-d}) \right] = \begin{bmatrix} \mathbf{y}_k^1, & \mathbf{y}^1(t_{k-d}) \\ \mathbf{y}_k^2, & \mathbf{y}^2(t_{k-d}) \end{bmatrix}.$$

Then the total measurement set formed by exploring all possible combinations is

$$\mathbf{Y}_k = \{\mathbf{Y}^1, \mathbf{Y}^2, \mathbf{Y}^3, \mathbf{Y}^4\}$$

$$= \left\{ \begin{bmatrix} \mathbf{y}_k^1 \\ \mathbf{y}^1(t_{k-d}) \end{bmatrix}, \begin{bmatrix} \mathbf{y}_k^2 \\ \mathbf{y}^2(t_{k-d}) \end{bmatrix}, \begin{bmatrix} \mathbf{y}_k^1 \\ \mathbf{y}^2(t_{k-d}) \end{bmatrix}, \begin{bmatrix} \mathbf{y}_k^2 \\ \mathbf{y}^1(t_{k-d}) \end{bmatrix} \right\},$$

which should be used for computing the combined innovations and the data association probabilities for the augmented state vector. Here we use $\mathbf{Y}^i$ to denote the ith combination of current and past measurements, i.e., it is "the measurement" in the augmented space. The set of all such combinations is given by $\mathbf{Y}_k$ and is formed from the measurements received at time t_k, i.e., $\mathbf{Y}_k \subset \{Y_k, Y(\tau_1), \ldots, Y(\tau_d)\}$. The augmented state approach can handle data association via the augmented state probabilistic data association described below.

8.4.2 Augmented state PDA filter

Once the measurements in the augmented space are obtained, the standard PDA technique shown in Section 4.3 can be used to obtain the state estimates for the augmented state. Similar to the standard PDA (Section 4.3), the association probability parameter $\beta(k)$ is defined based on the following mutually exclusive and exhaustive events:

---

[2] In this chapter, we use t_k to denote time with time index k. Sometimes we will use the time index on its own, e.g., t_k is equivalent to k.

θ_0: None of the validated measurements is object originated and they are all from clutter.

θ_I: The Ith combination of the measurement set in augmented space is object originated, and the others are from clutter.

$$I = 0, 1, \ldots, M_k. \tag{8.48}$$

Thus, the $\boldsymbol{\beta}(k)$ can be expressed as

$$\beta_I(k) \triangleq P(\boldsymbol{\theta}_I(k)|\mathbf{Y}^k). \tag{8.49}$$

M_k is the total number of measurements in the augmented space and $\beta_I(k)$ is the data association probability for the Ith measurement vector in the augmented space, which are calculated using the standard PDA formula as in Bar-Shalom and Fortmann (1988) by replacing the standard measurement with the measurement vector in augmented state space.

The AS-PDA state update equation is then given by

$$\widehat{\mathbf{X}}^d_{k|k} = \widehat{\mathbf{X}}^d_{k|k} + \mathbf{K}_k \widetilde{\mathbf{Y}}_k, \tag{8.50}$$

where

$$\widetilde{\mathbf{Y}}_k = \sum_{I=1}^{M_k} \beta_I(k)(\mathbf{Y}^I_k - \widehat{\mathbf{Y}}_k).$$

The covariance update is given by

$$\mathbf{P}^d_{k|k} = \beta_0(k)\mathbf{P}^d_{k|k-1} + (\mathbf{I} - \beta_0(k))\mathbf{P}^c_{k|k} + \widetilde{\mathbf{P}}_k, \tag{8.51}$$

where

$$\mathbf{P}^c_{k|k} = (\mathbf{I} - \mathbf{K}_k \mathbf{H}_k)\mathbf{P}^d_{k|k-1}, \tag{8.52}$$

and

$$\widetilde{\mathbf{P}}_k = \mathbf{K}_k \left[\sum_{I=1}^{M_k} \beta_I(k)\widetilde{\mathbf{Y}}^I_k \widetilde{\mathbf{Y}}^{I^T}_k - \widetilde{\mathbf{Y}}_k \widetilde{\mathbf{Y}}^T_k \right] \mathbf{K}^T_k. \tag{8.53}$$

Clearly, this implementation of AS-PDA requires combinations of all different time-indexed measurements. The computational complexity grows exponentially as the number of delayed measurements grows. An approximate, yet computationally efficient method of implementing AS-PDA is considered later in this chapter.

8.4.3 Iterative AS-PDA algorithm

Note that the hypothesis $\boldsymbol{\theta}_I(k)$ (bold face) in (8.48) corresponds to a set of hypotheses which occur at their corresponding times denoted by their superscripts, i.e.,

$$\boldsymbol{\theta}_I(k) \equiv \{\theta_k^i, \theta_{k-1}^j, \ldots, \theta_{k-d}^l\}. \tag{8.54}$$

Thus the association probability $\beta_I(k)$ is a probability of the joint event (8.54), which may be written as

$$\beta_I(k) \triangleq P(\boldsymbol{\theta}_I|\mathbf{Y}^k) = P(\theta_k^i, \theta_{k-1}^j, \ldots, \theta_{k-d}^l|\mathbf{Y}^k) \quad i, j, l \in \{0, 1, 2, \ldots\}. \tag{8.55}$$

The subscript I on β denotes a particular combination of the measurement vector given by (8.47). Similar to the notation θ^i, we use a superscript i on β to represent the data association probability associated with the (sub) event θ^i, i.e., the probability that the ith validated measurement in the measurement subset Y_{k-n} ($n = 0, 1, \ldots, d$) is from the object. Under the assumption that the measurement noises are white, and that the association events $\{\theta_k^i, \theta_{k-1}^j, \ldots, \theta_{k-d}^l\}$ are independent, we have

$$\beta_I(k) = \frac{1}{\delta_c} p(\mathbf{Y}_k|\boldsymbol{\theta}_I(k), \mathbf{Y}^{k-1}) P(\boldsymbol{\theta}_I(k)|\mathbf{Y}^{k-1})$$

$$= \frac{1}{\delta_c} p(y_k^i, y_{k-1}^j, \ldots, y_{k-d}^l|\theta_k^i, \theta_{k-1}^j, \ldots, \theta_{k-d}^l, \mathbf{y}^{k-1})$$

$$\times P(\theta_k^i, \theta_{k-1}^j, \ldots, \theta_{k-d}^l|\mathbf{Y}^{k-1})$$

$$= \frac{1}{\delta_c} p(y_k^i|\theta_k^i, \mathbf{y}^{k-1}) P(\theta_k^i|\mathbf{Y}^{k-1}) p(y_{k-1}^j|\theta_{k-1}^j, \mathbf{y}^{k-1}) P(\theta_{k-1}^j|\mathbf{Y}^{k-1})$$

$$\cdots p(y_{k-d}^l|\theta_{k-d}^l, \mathbf{y}^{k-1}) P(\theta_{k-d}^l|\mathbf{Y}^{k-1})$$

$$= \beta_k^i \beta_{k-1}^j \cdots \beta_{k-d}^l, \tag{8.56}$$

where δ_c is a normalization constant based on the fact that $\sum_{I \in M_k} \beta_I(k) = 1$.

Based on the relation (8.56) and given measurement set (8.47) at time k, an equivalent evaluation of (8.50) and (8.51) may be written in the following iterative form:

$$\hat{\mathbf{X}}_{k|k}^d = \hat{\mathbf{X}}_{k|k-1}^d + \sum_{i=k}^{k-d} \mathbf{K}^i \tilde{Y}_i, \tag{8.57}$$

$$\mathbf{P}_{k|k}^d = \mathbf{P}_{k|k-1}^d - \sum_{i=k}^{k-d} [(\mathbf{K}^i \mathbf{H}^i - \beta_i^0 \mathbf{K}^i \mathbf{H}^i) \mathbf{P}_{k|k-1}^d - \tilde{\mathbf{P}}^i], \tag{8.58}$$

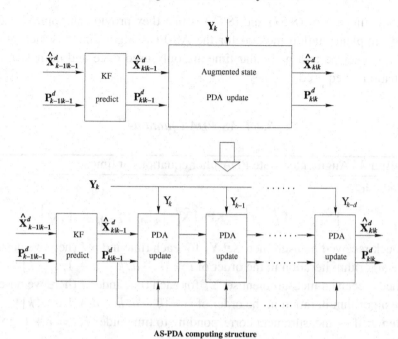

AS-PDA computing structure

Figure 8.3 Augmented state PDA and its computation structure.

where (superscript and subscript) i serves as the time index and β_i^0 is the probability that *none of the validated measurements from the subset Y_i of (8.47) is object-originated at time i*:

$$\tilde{Y}_i = \sum_{j=1}^{h} \beta_i^j (y_i^j - \hat{y}_i), \tag{8.59}$$

$$\sum_{i=k}^{k-d} \tilde{\mathbf{P}}^i = \begin{bmatrix} \tilde{P}_{\mathbf{k}} & 0 & \cdots & 0 \\ 0 & \tilde{P}_{\mathbf{k-1}} & 0 & 0 \\ \vdots & 0 & \ddots & \vdots \\ 0 & \cdots & 0 & \tilde{P}_{\mathbf{k-d}} \end{bmatrix}, \tag{8.60}$$

where

$$\tilde{P}_{\mathbf{i}} = K_{\mathbf{i}} \left[\sum_{j=1}^{h} \beta_{\mathbf{i}}^j \tilde{y}_{\mathbf{i}}^j \tilde{y}_{\mathbf{i}}^{j^T} - \tilde{Y}_{\mathbf{i}} \tilde{Y}_{\mathbf{i}}^T \right] K_{\mathbf{i}}^T. \tag{8.61}$$

The result in (8.60) is established in Appendix B. The difference between the ways that a standard AS-PDA and an iterative AS-PDA treat the measurements is shown in Figure 8.3.

The significance of (8.57) and (8.58) is that they provide an approximate, yet efficient, implementation method for the AS-PDA algorithm in which all measurements received at a particular time are only used once and no measurement combination is required.

8.4.4 AS-PDA equations

Algorithm 44 Augmented state PDA filter equations at time k

1: Prediction:

$$\left[\widehat{\mathbf{X}}_{k|k-1}^{d}, \mathbf{P}_{k|k-1}^{d}\right] = \text{KF}_{\text{P}}\left[\widehat{\mathbf{X}}_{k-1|k-1}^{d}, \mathbf{P}_{k-1|k-1}^{d}, \mathbf{F}, \mathbf{Q}\right].$$

2: Check received measurement set $\mathbf{Y}_k$ for each time index i (here we describe the algorithm iteration in the order of $i = \{k-d, k-d+1, \ldots, k\}$.

3: Check received measurement set $\mathbf{Y}_k$ for each time index i (here we describe the algorithm iteration in the order of $i = \{k-d, k-d+1, \ldots, k\}$.

4: **Step 3:** If no measurement corresponding to time index i, $i = i + 1$, go to Step 2.

5: Generate $\mathbf{H}^i$, $\mathbf{R}^i$ according to (8.33) and (8.34), respectively, with zeros everywhere except for the ith partition components. For example,

$$\mathbf{R}^{k-d} = \text{Diag}\left[0 \quad \cdots \quad 0 \quad R_{k-d}\right],$$

where R_{k-d} is the usual observation covariance matrix for normal state KF.

6: Measurements validation (gating):

$$\left[\{\boldsymbol{\beta}_k(i)\}_{i=0}^{m_k}\right] = \text{STDA}\left[\{p_k(i)\}_{i=1}^{m_k}\right].$$

7: PDA update:

$$\left[\widehat{X}_{k|k}^{d}, \mathbf{P}_{k|k}^{d}\right] = \text{PDA}_{\text{E}}\left[\widehat{X}_{k|k-1}^{d}, \mathbf{P}_{k|k-1}^{d}, \{\mathbf{y}_k(i)\}_{i=1}^{m_k}, \{\boldsymbol{\beta}_k(i)\}_{i=0}^{m_k}, \mathbf{H}, \mathbf{R}\right].$$

8.5 Simulation results

In this section three numerical examples are presented. Algorithm performance for OOSM tracking is compared. The tracking performance is characterized by the root mean square (RMS) error over 500 Monte Carlo runs for each specific scenario. In Example 8.1, it is assumed that the OOSM has a delay of only one lag. The performance of the Y-algorithm, the VDAS-KF and a two-lag AS-KF are compared. In Example 8.2, a more general scenario is considered where OOSM tracking with multiple delays is allowed and the performance is compared between the M-algorithm and the standard AS-KF. The OOSM problem in clutter

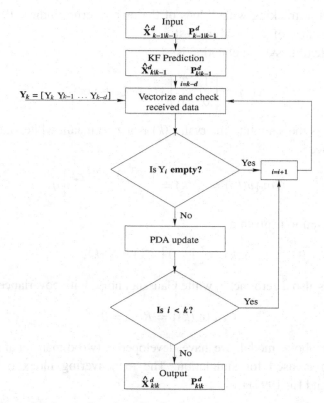

Figure 8.4 Logic flowchart for the single-cycle iterative AS-PDA.

is presented in the final example, where the performance of the proposed AS-PDA is investigated.

Note that the output of an augmented state filter can be taken either from the first component of the augmented state, or from the last component of the augmented state. The former involves "filtering" only[3] and the corresponding algorithm is denoted using a suffix "F," while the latter involves smoothing and is denoted using a suffix "S." For example, AS-KF2-F denotes an AS-KF with two lags and its output is taken from the first component of the augmented state.

8.5.1 Example 8.1

This numerical example has been extensively used in many object tracking algorithm comparisons, such as Bar-Shalom and Fortmann (1988), Bar-Shalom (2000) and Mallick *et al.* (2001a, 2001b), because it involves the most commonly used

---

[3] Despite the original meaning, we use "filtering" here to signify the filter output corresponding to the current time.

motion model in tracking with values of the maneuvering indices that cover the entire motion range of practical interest.

The discrete time system equation is

$$x(k) = \begin{bmatrix} 1 & T \\ 0 & 1 \end{bmatrix} x(k-1) + v(k), \tag{8.62}$$

where $T = 1$ is the sampling interval, $v(k)$ is a zero mean, white Gaussian noise with covariance

$$Cov\{v(k)\} = Q(k) = \begin{bmatrix} T^3/3 & T^2/2 \\ T^2/2 & T \end{bmatrix} q, \tag{8.63}$$

and the observation is given by

$$y(k) = \begin{bmatrix} 1 & 0 \end{bmatrix} x(k) + w(k), \tag{8.64}$$

where $w(k)$ is also a zero mean, white Gaussian noise with covariance

$$Cov\{w(k)\} = R(k) = 1. \tag{8.65}$$

Based on the above model, we have developed a two-dimensional object state model which is used for simulation. The maneuvering index is defined (in Bar-Shalom and Li, 1993) by

$$\lambda = \sqrt{\frac{qT^3}{R}}. \tag{8.66}$$

In the following, two cases (process noise, $q = 0.1$ and 1) corresponding to $\lambda = 0.3$ and 1 are examined, i.e., the underlying object performs straight line motion, or is highly maneuvering. Data are generated randomly for each run starting with an initial state,

$$x(0) = \begin{bmatrix} 200 \text{ km}, & 0.5 \text{ km/sec}, & 100 \text{ km}, & -0.08 \text{ km/sec} \end{bmatrix}. \tag{8.67}$$

A two-data point method is used to initialize the filters with

$$P(0|0) = \begin{bmatrix} P_0 & 0 \\ 0 & P_0 \end{bmatrix} \quad \text{where} \quad P_0 = \begin{bmatrix} R & R/T \\ R/T & 2R/T^2 \end{bmatrix}, \tag{8.68}$$

for the a priori error covariance or to form the initial error covariance for the augmented state.

In this example, we assume that the OOSM can only have a maximum delay of one lag, and the data delay is uniformly distributed within the whole simulation period with a probability P_r that the current measurement is delayed.

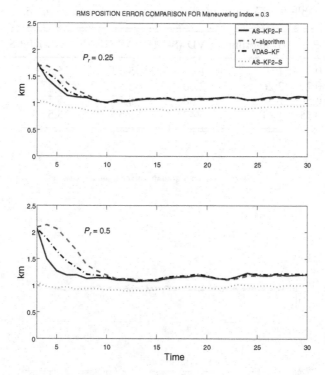

Figure 8.5 RMS performance comparison in the case of a straight line motion object ($\lambda = 0.3$) with single-delay OOSM ($P_r = 0.25$ and 0.5) for the Y-algorithm, VDAS-KF and AS-KF2.

Figures 8.5 and 8.6 show the simulation results for Example 8.1, where the performance of the Y-algorithm, VDAS-KF and AS-KF2 are compared over 500 runs. A computational load comparison for these algorithms is listed in Table 8.1 in terms of the number of floating point operations normalized to that of a standard Kalman filter.

The following observations can be drawn:

1. The Y-algorithm and VDAS-KF have similar RMS error performance within the whole range of the maneuvering indices.
2. The AS-KF2-F always outperforms both the Y-algorithm and VDAS-KF. When the probability of measurement delay (P_r) increases, this performance difference is observed to be greater.
3. The AS-KF2-S (smoothed AS-KF output) is superior to all other methods tested and has the least RMS error.
4. As shown in Table 8.1, the computational load of the VDAS-KF is comparable to the Y-algorithm and the AS-KF algorithm needs twice the computation.

Table 8.1 *Computational comparison for Example 8.1.*

P_r	Y-algorithm	VDAS-KF	AS-KF2-F	AS-KF2-S
0	1	1	5.57	5.57
0.25	2.26	2.60	5.57	5.57
0.5	2.30	2.68	5.57	5.57
0.75	4.41	5.47	5.57	5.57

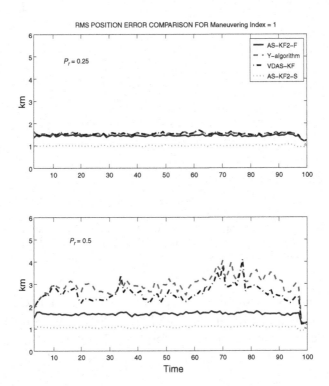

Figure 8.6 RMS performance comparison in the case of a highly maneuverable object ($\lambda = 1$) with single-delay OOSM ($P_r = 0.25$ and 0.5) for the Y-algorithm, VDAS-KF and AS-KF2.

8.5.2 Example 8.2

This example is the same as Example 8.1, except that multiple delays are allowed for an OOSM. The delayed measurement sequence is generated randomly by assuming they can be delayed by up to a maximum of three sampling periods, i.e., $t_k - \tau_{max} = 3T$. The distribution of the delayed measurements is assumed to be uniform in 1, 2, 3 lags with a probability P_r that the measurements at time k will be delayed.

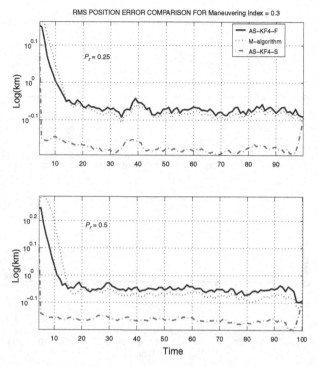

Figure 8.7 RMS performance comparison in the case of a straight line motion object ($\lambda = 0.3$) with multiple delay OOSM ($P_r = 0.25$ and 0.5) for AS-KF4-F, AS-KF4-S and M-algorithm.

Simulation results are shown in Figures 8.7 and 8.8. It is observed that:

1. As shown in Figure 8.7, for non-maneuvering object tracking, both AS-KF4-F and the M-algorithm have similar RMS performance regardless of OOSM. In other words, the OOSM problem is not critical. This can also be seen from Figure 8.5 in Example 8.1.

2. AS-KF-F overall outperforms the M-algorithm, while the performance of AS-KF-S is better than both because AS-KF-S corrects all components of its augmented state vector using each delayed measurement rather than M-algorithm, which can only make a correction to the current state.

3. For maneuvering object tracking ($\lambda = 1$), the average RMS error of the M-algorithm is larger than AS-KF4. Such a performance difference is large when the data delay probability (P_r) increases as shown in Figure 8.8.

4. The computational load of the standard four-lag AS-KF is about 11 times that of the M-algorithm in the case of $P_r = 0.25$. While the AS-KF remembers past states, the M-algorithm needs to compute past gain sequences and non-standard process noises in order to make a correction to the current state.

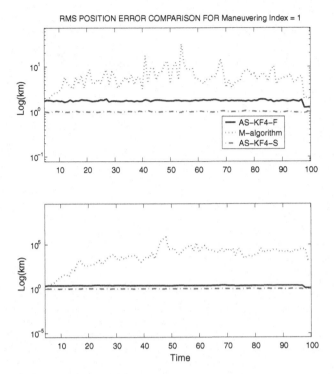

Figure 8.8 RMS performance comparison in the case of a highly maneuverable object ($\lambda = 1$) with multiple delay OOSM ($P_r = 0.25$ and 0.5) for AS-KF4-F, AS-KF4-S and the M-algorithm.

The performance of the M-algorithm as presented in Mallick *et al.* (2001a) is expected to improve if the conditional mean of the process noise is appropriately calculated.

8.5.3 *Example 8.3*

Example 8.3 focuses on the following aspects:

- Compare the performance of the proposed AS-PDA algorithm with a (fixed lag) smoother of the same dimension. When there is no OOSM problem, AS-PDA will reduce to the PDA smoother, from which better performance can be expected because of smoothing.
- Compare the performance differences between the AS-PDA filtering (no delay) output and its smoothing (delayed) output.
- Computational comparison between PDAF and AS-PDAF.

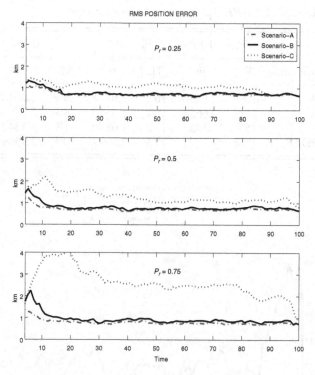

Figure 8.9 AS-PDA algorithm RMS performance.

The simulation scenario is the same as that in Example 8.2 with a maneuvering index of 0.7. In addition, clutter is added. Clutter is assumed to be uniformly distributed over the surveillance region with a density $C_D = 0.037/\text{km}^2$.

An AS-PDA filter with four lags is implemented and used in the simulation. The simulation is based on 500 Monte Carlo runs for each of the following scenarios:

1. An AS-PDA using "full" measurement sequence (without OOSM), denoted as *Scenario-A*.
2. An AS-PDA using OOSM with the probability of measurement delay $Pr = 0.25,\ 0.5$ and 0.75 respectively, denoted as *Scenario-B*.
3. An AS-PDA which treats the delayed measurements as missing measurements, denoted as *Scenario-C*.

Figure 8.9 compares the root mean squared (RMS) error for the three scenarios. It is clear that:

- as the value of P_r increases, the track performance becomes poor for all filters;
- the performances for *Scenario-A* and *Scenario-B* are nearly identical.

Table 8.2 *AS-PDA performance comparison.*

Scenario	$P_r = 0.25$		$P_r = 0.5$		$P_r = 0.75$	
	TL	$\bar{m}$	TL	$\bar{m}$	TL	$\bar{m}$
A	16%	2.6256	16%	2.6411	8%	2.8613
B	15%	2.6658	16%	2.6610	8%	2.8860
C	15%	2.1453	17%	1.6879	11%	1.3342
PDAF	15%	2.7821	12%	2.8942	10%	2.9509

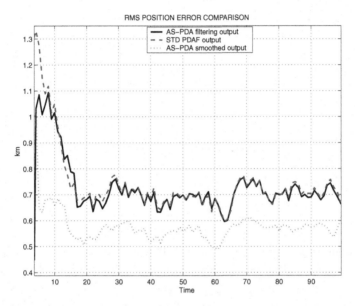

Figure 8.10 RMS performances of the filtered and smoothed estimation error of an AS-PDA.

The statistics of tracking in clutter are listed in Table 8.2, where it can be seen that the AS-PDA using OOSM is very close to a standard PDA in terms of the average expected number of validated measurements $\bar{m}$ and track loss. A track is deemed to be lost if there was no measurement update for eight consecutive sampling intervals. In fact, a lost track in our simulations results in fairly large RMS error and is unlikely to follow the true object trajectory. Therefore, we only count the remaining tracks in the calculation of the RMS errors shown in Figure 8.9.

It is interesting to compare both the performance of the filtered output and the smoothed output from an augmented state filter. The former acts like the output from a normal state estimation algorithm (with certain memory), while the latter is the output from the last lag of an augmented state estimation algorithm, which, obviously, has a delay of $d_{\max}$ (here $d_{\max} = 4T$) sample steps. Figure 8.10 provides the RMS error performance comparison for such a case. Results show that the improvement of the tracking performance after smoothing is significant.

In order to highlight the performance of the proposed AS-PDA algorithm, the RMS performance of a standard PDAF without OOSM is also shown on the figure. It is worth noting that the performance of AS-PDA with OOSM is very close to that of standard PDAF without OOSM.

8.6 Summary

This chapter discusses a solution to the very practical problem of real-time object tracking. In a scenario where a large number of sensors are connected through a network, measurements are received in a delayed time for a number of reasons. Therefore, assumptions of measurement availability at a particular time are not ensured and there is a need to incorporate that measurement when it is received at some later time. This problem is addressed in this chapter through the proposal of a Bayesian formulation of the solution. Several major algorithms – AS-KF, AS-PDA, etc. – are presented along with some illustrative results.

9

Practical object tracking

Chapters 1 to 8 introduced optimal (and suboptimal) Bayes tracking recursions and associated approximations.

This chapter covers some points that are important when considering the practical implementation of object tracking. It can be viewed as a collection of separate sections, each section dealing with a specific practical issue. Although object existence is often mentioned in this chapter, with due diligence and prudence the material presented can also apply to, or provide infrastructure for, other algorithms.

Section 9.2 introduces the linear multi-target method for suboptimal multi-object tracking in clutter. As the name implies, the additional numerical complexity of the linear multi-target method is linear in the number of targets and the number of measurements. This is followed by some practical methods for the clutter measurement density estimation in Section 9.3. Bayes recursion needs to be initialized; in the absence of prior target information, tracks are initialized using available measurements. Some track initialization methods and trade-offs are discussed in Section 9.4. For various reasons, multiple tracks may end following the same sequence of measurements; in Section 9.5 the track merging procedure detects and solves this situation. Finally, Section 9.6 presents some (simulated) surveillance situations and automatic target tracking solutions.

9.1 Introduction

In complex situations, involving a large number of objects and/or heavy clutter, algorithms based on the optimal multi-object approach (Section 5.5.4) may not be feasible due to its excessive computational requirements. The linear multi-target procedure to efficiently convert single-object trackers into multi-object trackers is detailed in Section 9.2. It is a sub-optimal multi-object tracking approach (Mušicki and La Scala, 2008) that has been tested in situations with a large (Mušicki and La Scala, 2005) number of objects and substantial clutter. As implied by the name, linear multi-target procedure overheads are linear in the number of tracks and the number of selected measurements. Therefore, it does not suffer from the combinatorial number of possible joint (global) feasible measurements of track allocations. A fringe benefit of linear multi-target tracking is that its algorithms retain the simple and elegant structure of the single-object tracking algorithms. This means simplified (software) implementation, reliable programs and easy maintenance, which all contribute to an improved bottom line.

The clutter measurement density was assumed to be known throughout Chapter 5. In a majority of applications it is not correct, however, and the clutter measurement density has to be estimated using measurements. Section 9.3 describes some approaches to estimating the clutter measurement density for single- and multi-object tracking algorithms.

The object tracking algorithms are derived using the Bayes' equation. The Bayes' recursion assumes the existence of an initial track state. In the (usual) absence of prior knowledge on object existence and position, new tracks have to be initialized using measurements. Some useful approaches are presented in Section 9.4.

Tracks are initialized using measurements. As a consequence, every object may initialize a number of tracks, one per measurement scan, following it. Another possibility is that a track (false or true) starts following an object already being followed by another track. Whatever the reason, we may and often do end up in a situation where a number of tracks follow the same object. The track merging procedure in Section 9.5 describes the procedures used to recognize and correct this situation.

Section 9.6 contains some well-chosen examples that show the effectiveness of the techniques and algorithms presented in Chapters 5 and 9.

9.2 Linear multi-target tracking

Multi-object issues arise when tracks are close together, so that one or more measurements are selected by more than one track. The optimal multi-object approach described in Section 5.5.4 suffers from the combinatorial explosion in the number of joint measurement to track assignments, as exemplified in (5.46). Thus the optimal multi-object approach may easily become computationally non-feasible when the number of tracks (objects) and the number of selected measurements increase. A number of sub-optimal approaches which reduce the multi-object tracking computational requirements have been proposed. Linear multi-target tracking (Mušicki and La Scala, 2008) is one such approach.

Linear multi-target (LM) tracking is a procedure to convert single-object trackers into multi-object trackers with a number of operations that are linear in the number of tracks and the number of measurements. Furthermore, the overheads of the conversion are negligible in most situations, amounting to just a couple of percent of the processor's time requirements. So far, only object-existence-based trackers from IPDA to IMM-ITS have been used in conjunction with the LM conversion. However, given proper consideration, other single-object tracking algorithms may also be used.

The prefix LM is used to indicate that LM has been applied. Thus, LMIPDA is the multi-object tracker obtained by applying the LM procedure to IPDA.

Linear multi-target tracking measurement assumes that all measurements are produced by the sensor(s) having infinite resolution. In other words, each measurement can have only one source. When a measurement is used to update a track, it can be one of the following:

1. measurement (detection) of the object being tracked;
2. clutter measurement; or
3. measurement (detection) of an object being followed by some other track.

A measurement produced by an object not being followed by the track τ being updated is a spurious measurement and as such it really is part of the clutter "observed" by track τ. This additional clutter is not Poisson distributed. This is the essence of the linear multi-target approach to tracking: "when updating one track, clutter measurement density is modulated by the possible contributions from objects being followed by other tracks." The word "possible" is used as the other objects may or may not exist, and even if they do their measurements exist only with a certain probability of detection.

As in previous sections, denote by $p_k^\tau(i)$ the likelihood of measurement $\mathbf{y}_k(i)$ with respect to track τ, and denote by P_D^τ and P_G^τ the detection and selection probability of track τ respectively.

Denote by $P^\tau(i)$ the a priori probability that measurement $\mathbf{y}_k(i)$ is the detection of object τ at time k. This probability is approximated by

$$P^\tau(i) \triangleq p(\theta_k^\tau(i)|\mathbf{Y}^{k-1}) \approx p(\chi_k^\tau|\mathbf{Y}^{k-1}) P_D^\tau P_G^\tau \frac{p_k^\tau(i)/\rho_k(i)}{\displaystyle\sum_{j=1}^{m_k} p_k^\tau(j)/\rho_k(j)}. \tag{9.1}$$

The a priori probability $p(\overline{\theta_k^\tau(0)}|\mathbf{Y}^{k-1}) = p(\chi_k^\tau|\mathbf{Y}^{k-1}) P_D^\tau P_G^\tau$ that object τ will cause a selected detection at time k equals the probability that the object exists, that it is detected given that it exists and that it is selected given that it exists and is detected. This probability is split in (9.1) between the selected measurements proportionally to their likelihoods with respect to track τ. Please note that (9.1) implies that a priori detection events are mutually exclusive given a single track (which is correct), and mutually independent with respect to different tracks (which is the essence of the linear multi-target approximation).

From a strictly mathematical point of view, (9.1) seems to carry a contradiction. How can a priori probabilities be calculated using values of measurements in the current scan? Part of the explanation can be that it is only an approximation. The other part lies in the way this expression is used. The values of $P^\tau(i)$ are used to update other tracks, but not track τ. This prevents the multiple use of the same data in an independent manner, or data incest as it is often termed.

Probabilities $P^\tau(i)$ are calculated for all tracks τ and all measurements i. If track τ does not select measurement i, then

$$P^\tau(i) = p_k^\tau(i) = 0.$$

Single-object tracking algorithms, such as IPDA and ITS, expressions for the a posteriori probability of object existence and a posteriori data association probabilities depend on the ratio

$$\frac{P_D^\tau P_G^\tau p_k^\tau(i)}{\rho_k(i)},$$

in other words on the ratio of object measurement density and clutter measurement density at the measurement coordinates.

Consider here the update of track τ, and concentrate on the measurement $\mathbf{y}_k(i)$. Define by Φ the set of all tracks excluding track τ. The event that measurement $\mathbf{y}_k(i)$ is the detection of object τ now includes the event that measurement $\mathbf{y}_k(i)$ is not a detection of any object from Φ:

$$P_D^\tau P_G^\tau p_k^\tau(i) \xrightarrow{LM} P_D^\tau P_G^\tau P^{0,\Phi}(i) p_k^\tau(i),$$

where

$$P^{0,\Phi}(i) = \prod_{\phi \in \Phi} \left(1 - P^\phi(i)\right)$$

denotes the probability that measurement $\mathbf{y}_k(i)$ is not a detection of any object being followed by any track from Φ. When updating track τ, the total clutter measurement density at $\mathbf{y}_k(i)$ is defined as the measurement density of all sources excluding object τ, and is given by

$$\rho_k(i) \xrightarrow{LM} P^{0,\Phi}(i)\rho_k(i) + \sum_{\eta \in \Phi} P^{\eta,\Phi}(i) p_k^\eta(i) = P^{0,\Phi}(i)\Omega^\tau(i),$$

where:

- $P^{0,\Phi}(i)\rho_k(i)$ is the clutter measurement density multiplied by the probability that measurement $\mathbf{y}_k(i)$ is not a detection of any object from set Φ;

- $P^{\eta,\Phi}(i)\,p_k^{\eta}(i)$ is the contribution to the total clutter (as observed by track τ) of track η at point $\mathbf{y}_k$ (i);
- $P^{\eta,\Phi}(i)$ is the a priori probability that measurement $\mathbf{y}_k$ (i) is the detection of object η from track set Φ:

$$
P^{\eta,\Phi}(i) = P^{\eta}(i) \prod_{\substack{\phi \in \Phi \\ \phi \neq \eta}} \left(1 - P^{\phi}(i)\right) = P^{0,\Phi}(i)\frac{P^{\eta}(i)}{1 - P^{\eta}(i)},
$$

and

$$
\Omega^{\tau}(i) \overset{\triangle}{=} \rho_k(i) + \sum_{\eta \neq \tau} p_k^{\eta}(i)\frac{P^{\eta}(i)}{1 - P^{\eta}(i)}. \tag{9.2}
$$

Therefore, applying LM is obtained by the following transition:

$$
\frac{P_D^{\tau} P_G^{\tau} p_k^{\tau}(i)}{\rho_k(i)} \overset{LM}{\longrightarrow} \frac{P_D^{\tau} P_G^{\tau} P^{0,\Phi}(i) p_k^{\tau}(i)}{P^{0,\Phi}(i)\Omega^{\tau}(i)} = \frac{P_D^{\tau} P_G^{\tau} p_k^{\tau}(i)}{\Omega^{\tau}(i)},
$$

which is equivalent to simply replacing the clutter measurement density $\rho_k(i)$ with the modulated (with respect to track τ) clutter measurement density $\Omega^{\tau}(i)$, when updating the state of track τ. In this context, $\Omega^{\tau}(i)$ is termed the equivalent LM clutter observed by track τ at the coordinates of measurement $\mathbf{y}_k$ (i).

To recap, applying LM to a single-object tracker consists of two LM steps, performed before the single-object data association step:

- For each track τ and its selected measurements, calculating prior probabilities $P^{\tau}(i)$, using (9.1). Please note that for measurement j not selected by track τ, $P^{\tau}(j) = p_k^{\tau}(j) = 0$.
- When updating track τ, replace the clutter measurement densities of selected measurements $\rho_k(i)$ by the LM modulated clutter measurement densities $\Omega^{\tau}(i)$, calculated using (9.2).

Another way of viewing LM is as a predictor/corrector method. The initial probabilities of measurement allocations are calculated under the single-object assumptions for each track (9.1), which are then corrected by applying the modulated clutter measurement density equation (9.2).

To illustrate how LM works in practice, assume that two tracks, τ and η, share one measurement $\mathbf{y}_k$ (i). Further assume that track η is a "strong" track which selects only measurement $\mathbf{y}_k$ (i), thus $P_i^{\eta} \to 1$. When updating track τ, the modulated clutter measurement density $\Omega^{\tau}(i) \to \infty$. Thus, track τ will effectively ignore measurement $\mathbf{y}_k$ (i), which is a desirable outcome.

LM approach has been applied to both single-object single-scan trackers, such as IPDA and IMM-IPDA (to obtain LMIPDA and IMM-LMIPDA) (Mušicki and

Suvorova, 2008), and to single-object multi-scan trackers, such as ITS and IMM-ITS (to obtain LMITS and IMM-LMITS) (Mušicki *et al.*, 2005a; Mušicki and Evans, 2008). The results obtained so far indicate a negligible, and in some cases undetectable, loss of performance between LM-based trackers, and trackers based on the optimal ("joint") multi-object approach. The difference is almost undetectable for the single-scan-based trackers (LMIPDA, IMM-LMIPDA versus JIPDA, IMM-JIPDA), and is (barely) noticeable for multi-scan-based trackers (LMITS and IMM-LMITS versus JITS and IMM-JITS). Of course, due to the combinatorial explosion in the computational requirements of the "joint" multi-object trackers, in many situations "joint" multi-object tracking is not feasible, and LM (or some other sub-optimal approach) remains the only published possibility.

Both additional LM steps are linear in the number of tracks and the number of measurements (hence the name linear multi-target) and with a minimal amount of careful implementation add only a small percentage to computational requirements. LM has been successfully tested in difficult situations with large numbers of (maneuvering) objects and heavy clutter (Mušicki and La Scala, 2005).

Given a fixed number of objects and fixed clutter statistics, the peak to average ratio of required computational requirements for LM is very low compared to the optimal, joint multi-object approach. Given that hardware computational capabilities have to conform to peak demands, this is a very significant benefit of LM. Additionally, LM implementation logic is much simpler compared to the optimal, joint multi-object approach; e.g., one does not have to implement joint hypothesis enumeration and evaluation. This simplicity results in faster development and a more robust application which is less difficult to maintain.

9.3 Clutter measurement density estimation

Estimation of the clutter measurement density is an often neglected aspect of tracking in clutter. Yet, as the examples in Section 9.6 show, better knowledge of the clutter measurement density may result in turning near-useless results into a success. It is almost impossible to over-emphasize the importance of better clutter measurement density knowledge.

As is detailed below, given no prior knowledge of clutter measurement density (non-parametric object tracking), clutter measurement density is estimated on-line using selected measurements. In this case, the volume V_k of the selection gate is important. The bigger the selection gate, the more precise our estimate of the clutter measurement density will be, as, statistically speaking, more selected clutter measurements will give us a better clutter measurement density estimate. On the other hand, if the clutter is non-homogeneous, the larger gate volumes will give us more biased, and therefore detrimental, estimates of the clutter measurement density. These effects on the correctness of the clutter measurement density estimates in many environments far outweigh other considerations on the selection gate size, including the probability that object detection may not be selected.

One may suggest having two selection gates. One selection gate, usually smaller, selects measurements used to actually update tracks. The other selection gate, usually larger, selects measurements used to estimate the clutter measurement density. The parameters of both can be independently adjusted to optimize both the clutter measurement density estimation, and the number of measurements used to update the track state. These considerations, however, will not be further pursued in this book.

The selection gate volumes of different object tracking algorithms will be different, even given the same probability of section P_G. Gate volume depends also on the shape of the a priori measurement pdf, as well as on the various approximations used. As an example, Mušicki (1994) compares the selection gate volumes of IPDA and IMM-PDA. A fair comparison between these (non-parametric) algorithms is only achieved once the gate sizes are corrected to be (statistically) similar.

Generally speaking, the clutter measurement density is more often than not a priori unknown. For example, one may be able to calculate the a priori probability of a false alarm and therefore the clutter measurement density in the rare case when the only source of false signals is the thermal noise (the radar receiver antenna points to empty space). These and similar situations do not happen often.

If the clutter is "almost stationary," meaning that the parameters of clutter signal change slowly in time, the clutter measurement density in the surveillance space may be estimated (averaged) using a clutter map (Mušicki et al., 2005b). In many cases this assumption is not possible. Some examples include rapidly moving sensors, and/or a fast-changing environment, fast warm-up time, etc. In these cases clutter mapping is not possible and one must assume no a priori clutter measurement density information. Tracking in this situation is described here as non-parametric object tracking, and the clutter measurement density estimation in this situation is the subject of this section. Although the assumption is that tracking is performed by one of the object-existence-based trackers, which are described in Chapter 5, with the possible LM extension described in Chapter 9, these results with appropriate modifications can also be used with other algorithms.

Non-parametric object tracking carries an additional assumption: "the clutter measurement density within the surveillance area of interest is uniform." The "surveillance area of interest" depends on the object tracking algorithm. For single-object tracking, the area of interest is the selection gate area of the track under consideration. For multi-object tracking, the area of interest is the current cluster area.

In non-parametric tracking, the clutter measurement density estimation is performed after the measurement selection operation, and the surveillance area of interest is established. The estimated clutter measurement density is used by the data association operation (Sections 5.5.3 and 5.5.4).

Inputs to the clutter measurement density estimation at time k are, for each object trajectory model σ of each component c_{k-1} of each track τ:

- the number $m_k^\tau(c_{k-1}, \sigma)$ of selected measurements; and
- the volume $V_k^\tau(c_{k-1}, \sigma)$ of the selection gate; and
- the a priori probability of object existence, $p(\chi_k^\tau | \mathbf{Y}^{k-1})$.

The output of the clutter measurement density estimation operation at time k for each surveillance area of interest is:

- the clutter measurement density at the location of each measurement $\mathbf{y}_k(i)$, which is in this case identical for each measurement in the surveillance area of interest, $\rho \overset{\triangle}{=} \rho_k(i)$.

The clutter measurement density is estimated as

$$\rho = \frac{\hat{m}_k}{V_k}, \tag{9.3}$$

where $\hat{m}_k$ is the estimate (statistical mean) of the number of clutter measurements in the surveillance area of interest, and V_k is the volume of the surveillance area of interest.

The track selection gate is a union of the selection gates of the individual components, each of which is the union of the selection gates of the individual models of the component. For single-object tracking, the track selection gate is the surveillance area of interest. For multi-object tracking, the cluster area is the surveillance area of interest, and is the union of the selection gates of constituent tracks. Thus, in both cases, the surveillance area of interest boils down to a union of individual selection gates of trajectory models of individual track components of individual tracks; which we term the "elementary selection gate" in this section.

As discussed in Section 5.5.2 on measurement selection, each elementary selection gate is a hyper-ellipsoid in the observation space. Thus the volume V_k is the volume of a union of intersecting hyper-ellipsoids. To the best of the authors' knowledge there is no closed expression for the volume of the surveillance area of interest. Instead, we use an approximation, published in Mušicki and Evans (2002) and Mušicki and Morelande (2004). Assume that the volume V_k is the union of H intersecting hyper-ellipsoids, indexed here by h, and denote by $V_k(h)$ and $m_k(h)$ the volume and the number of selected measurements of the elementary hyper-ellipsoid h. Then, with a slight abuse of notation, the total number of selected

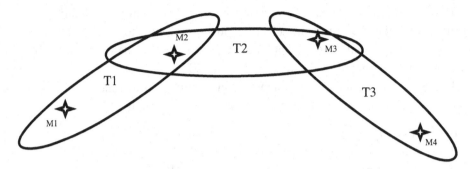

Figure 9.1 Multi-object situation.

measurements is

$$m_k = \bigcup_h m_k(h) \le \sum_h m_k(h),$$

and the V_k approximation is

$$V_k \approx \max \left(\sum_h V_k(h) \frac{m_k}{\sum\limits_h m_k(h)}, \max_h V_k(h) \right), \qquad (9.4)$$

where the max operations ensure that the approximated volume is at least as big as the volume of the largest elementary selection gate, and

$$o = \frac{m_k}{\sum\limits_h m_k(h)} \qquad (9.5)$$

approximates the overlapping ratio defined as the ratio of volume of union of overlapped hyper-ellipsoids divided by the sum of volumes of individual hyper-ellipsoids. The larger the number of selected measurements (the higher the true clutter measurement density is), the more precise approximation (9.4) becomes. Nevertheless, Mušicki and Evans (2002) and Mušicki and Morelande (2004) show that the approximation works well enough even with a small number of selected measurements.

To illustrate, consider the multi-object situation with one cluster, depicted in Figure 5.4, repeated here for convenience as Figure 9.1. In this situation:

- there are four selected measurements in the cluster, $m_k = 4$;
- each track selects two measurements, $m_k(h) = 2, h = 1, \ldots, 3$;
- the overlapping factor (9.5) is $o = 4/6 = 2/3$; and

- given that the volumes of individual selection gates are equal, $V_k(1) = V_k(2) = V_k(3)$, cluster volume is approximated by $V_k \approx 2\ V_k(1)$, which is a reasonable approximation.

Clutter measurement density is of interest only in the case where the track selects at least one measurement. Thus in this section the case of $m_k = 0$ is ignored, and the denominator in (9.4) is never zero.

For single-object tracking, (9.4) becomes

$$V_k = V_k^\tau = \bigcup_{\xi_{k-1}^\tau} \bigcup_\sigma V_k^\tau(c_{k-1}, \sigma)$$

$$\approx \max\left(\sum_{\xi_{k-1}^\tau, \sigma} V_k^\tau(c_{k-1}, \sigma) \frac{m_k^\tau}{\displaystyle\sum_{\xi_{k-1}^\tau, \sigma} m_k^\tau(c_{k-1}, \sigma)}, \ \max_{c_{k-1}, \sigma} V_k^\tau(c_{k-1}, \sigma) \right), \quad (9.6)$$

and for multi-object tracking, (9.4) becomes

$$V_k = \bigcup_\tau V_k^\tau = \bigcup_\tau \bigcup_{\xi_{k-1}^\tau} \bigcup_\sigma V_k^\tau(c_{k-1}, \sigma)$$

$$\approx \max\left(\sum_\tau \sum_{\xi_{k-1}^\tau, \sigma} V_k^\tau(c_{k-1}, \sigma) \frac{m_k}{\displaystyle\sum_\tau \sum_{\xi_k^\tau, \sigma} m_k^\tau(c_{k-1}, \sigma)}, \ \max_{\tau, c_{k-1}, \sigma} V_k^\tau(c_{k-1}, \sigma) \right), \quad (9.7)$$

where τ indexes all tracks in the cluster under observation.

The estimated number of selected clutter measurements, needed to complete (9.3), is calculated in a different manner for single- and multi-object tracking.

Single-object tracking ignores the possibility of other objects. The a priori probability that one of the selected measurements is object detection is

$$P_D P_G p(\chi_k | \mathbf{Y}^{k-1}),$$

which is also the statistical mean of the number of selected object measurements. Therefore, for single-object tracking (IPDA (Mušicki, 1994; Mušicki *et al.*, 1994), IMM-IPDA (Mušicki *et al.*, 2004a; Mušicki and Suvorova, 2008), ITS (Mušicki *et al.*, 2003, 2007) and IMM-ITS (Mušicki *et al.*, 2004b, 2007))

$$\hat{m}_k = m_k - P_D P_G p(\chi_k | \mathbf{Y}^{k-1}). \quad (9.8)$$

Multi-object tracking (JIPDA (Mušicki and Evans, 2002, 2004b), IMM-JIPDA (Mušicki and Suvorova, 2008), JITS (Mušicki *et al.*, 2003; Mušicki and Evans,

2008) and IMM-JITS (Mušicki and Evans, 2008)) assumes that the events of measurement $\mathbf{y}_k\,(i)$ not being the object detection are independent across tracks which select the measurement $\mathbf{y}_k\,(i)$. The probability that measurement $\mathbf{y}_k\,(i)$ is a detection of track τ is approximated by (see Section 9.2; (9.1))

$$P_{k,i}^{\tau} \approx P_D\,P_G\,p(\chi_k^{\tau}|\mathbf{Y}^{k-1})\frac{p_k^{\tau}(i)}{\sum\limits_{j} p_k^{\tau}(j)}, \tag{9.9}$$

where the prior probability $P_D\,P_G\,p(\chi_k^{\tau}|\mathbf{Y}^{k-1})$ that one measurement is the detection of track τ is divided among the selected measurements, proportionally to the measurement likelihood. The probability that measurement $\mathbf{y}_k\,(i)$ is not a detection of object τ becomes $1 - P_{k,i}^{\tau}$, the probability that $\mathbf{y}_k\,(i)$ is a detection of no object becomes (due to the assumed/approximated event independence)

$$P_{k,i}^{0} \approx \prod_{\tau}\left(1 - P_{k,i}^{\tau}\right),$$

and the total number of expected clutter measurements equals the sum of the clutter probabilities of individual tracks in the surveillance area of interest:

$$\hat{m}_k = \sum_{j=1}^{m_k} P_{k,j}^{0} \approx \sum_{j=1}^{m_k}\prod_{\tau}\left(1 - P_{k,j}^{\tau}\right). \tag{9.10}$$

Equations (9.9)–(9.10) are denoted by the pseudo-function

$$\hat{m}_k = \mathrm{MTT}_{\mathrm{MK}}(\{p(\chi_k^{\tau}|\mathbf{Y}^{k-1})\}_{\tau},\,\{p_k^{\tau}(i)\}_{\tau,i}). \tag{9.11}$$

For a single-track cluster, the multi-object tracking formulae for V_k and $\hat{m}_k$ – (9.7) and (9.10) – revert to the single-object tracking formulae, (9.6) and (9.8) respectively, as they should.

As originally published, some of the algorithms detailed in Chapter 5 use somewhat different formulae for the clutter measurement density estimation. The difference was highlighted, and the authors are of the opinion that the formulae presented here yield better results in the majority of implementations.

9.4 Track initialization

To start the Bayes' recursion, tracks have to be initialized. Barring some human intervention, the only way to initialize tracks is using measurements. The new track initialization sometimes bears the highly scientific name "track birth process."

Track initialization is even less of an exact science than the track update. In this section some, by no means exhaustive, possibilities are explored. The practitioner needs to carefully evaluate the various approaches, preferably first by detailed simulations. Choosing the wrong way to initialize tracks may result in either the saturation of computational resources (too many initialized false tracks), or poor tracking performance (true tracks not initialized or not timely initialized). Either result spells object tracking problems.

The track initialization procedure should take place at each measurement time k after the track state update operations. At that point, the measurement usage by existing tracks is known and may be utilized to tune the track initialization.

Inputs to the new track initialization process at time k are (not all of them are used by all procedures suggested):

- The sets of measurements $\mathbf{Y}_{k-1}$ and $\mathbf{Y}_k$ returned by the sensor at previous time $k-1$ and at current time k respectively. In this case $\mathbf{Y}_{k-1}$ and $\mathbf{Y}_k$ refer to all measurements returned by the sensors.
- For each measurement $\mathbf{Y}_{k-1}(j)$ and $\mathbf{Y}_k(i)$ from $\mathbf{Y}_{k-1}$ and $\mathbf{Y}_k$ respectively, the a posteriori data association probabilities $P_{k-1|k-1,j}^0$ and $P_{k|k,i}^0$ that the measurements are not detections of any object followed by existing tracks.

Outputs of the new track initialization process are new tracks (indexed by superscript τ), defined by their probability of object existence, $p(\chi_k^\tau|\mathbf{Y}^k)$, and the object trajectory state pdf, $p(\mathbf{x}_k^\tau|\chi_k^\tau, \mathbf{Y}^k)$.

This section describes two track initialization techniques that are often implemented. One is termed "one-point track initialization," and the other "two-point differencing." The one-point track initialization is conceptually simpler, and requires only measurements $\mathbf{Y}_k$ from the current scan k. However, the new tracks will have no a priori speed information, which results in a (relative to two-point differencing) large selection gate area at the subsequent scan $k+1$. This may be detrimental in a heavy clutter situation as potentially more clutter measurements will be selected. These additional clutter measurements will not only increase the computational requirements, they will increase the estimation errors and may cause the true track to "lose" its object or may cause the confirmation of false tracks. The two-point differencing is more complex; however, the initial speed information results in a much smaller selection gate area in the subsequent scan $k+1$ for newly initialized tracks, which helps in a heavy clutter environment. There is no universal clear winner between these techniques; depending on the situation sometimes one and sometimes the other will be advantageous.

The trajectory state estimate $\mathbf{x}_k$ usually consists of position $\mathbf{r}_k$, speed $\mathbf{v}_k$ and sometimes acceleration $\mathbf{a}_k$, each of which has n components. For two-dimensional tracking $n=2$, and for three-dimensional tracking $n=3$, and (omitting the track superscript)

$$\mathbf{x}_k^T = \begin{bmatrix} \mathbf{r}_k^T & \mathbf{v}_k^T & \mathbf{a}_k^T \end{bmatrix}.$$

Therefore, the a posteriori object trajectory estimate at time k, given track component c_k and trajectory model σ is defined by

$$\hat{\mathbf{x}}_{k|k}(c_k, \sigma) = \begin{bmatrix} \hat{\mathbf{r}}_{k|k}(c_k, \sigma) \\ \hat{\mathbf{v}}_{k|k}(c_k, \sigma) \\ \hat{\mathbf{a}}_{k|k}(c_k, \sigma) \end{bmatrix},$$

and

$$\mathbf{P}_{k|k}(c_k, \sigma) = \begin{bmatrix} \mathbf{P}_{rrk|k}(c_k, \sigma) & \mathbf{P}_{rvk|k}(c_k, \sigma) & \mathbf{P}_{rak|k}(c_k, \sigma) \\ \mathbf{P}_{vrk|k}(c_k, \sigma) & \mathbf{P}_{vvk|k}(c_k, \sigma) & \mathbf{P}_{vak|k}(c_k, \sigma) \\ \mathbf{P}_{ark|k}(c_k, \sigma) & \mathbf{P}_{avk|k}(c_k, \sigma) & \mathbf{P}_{aak|k}(c_k, \sigma) \end{bmatrix}.$$

If the acceleration is not used, the corresponding elements from $\mathbf{x}_k$, $\hat{\mathbf{x}}_{k|k}(c_k, \sigma)$ and $\mathbf{P}_{k|k}(c_k, \sigma)$ should be simply removed.

9.4.1 One-point track initialization

As the name implies, each measurement used for track initialization at time k independently results in a new track. We thus concentrate here on one measurement $\mathbf{Y}_k(i)$, with corresponding measurement covariance matrix $\mathbf{R}$. The procedure is repeated for each measurement in $\mathbf{Y}_k(i)$.

Based on one position measurement, one has no information on object speed and acceleration. If the maximum object speed and acceleration are v_{max} and a_{max}, the uniform distribution of object speed and acceleration with appropriate bounds may reflect our ignorance. This uniform distribution is replaced by a Gaussian pdf with mean zero and covariance $v_{max}^2 \mathbf{I}_n/3$ and $a_{max}^2 \mathbf{I}_n/3$ for speed and acceleration respectively:

$$\hat{\mathbf{x}}_{k|k}(c_k, \sigma) = \begin{bmatrix} \mathbf{Y}_k(i) \\ \mathbf{0}_{n,1} \\ \mathbf{0}_{n,1} \end{bmatrix}$$

$$\mathbf{P}_{k|k}(c_k, \sigma) = \begin{bmatrix} \mathbf{R} & \mathbf{0}_{n,n} & \mathbf{0}_{n,n} \\ \mathbf{0}_{n,n} & v_{max}^2 \mathbf{I}_n/3 & \mathbf{0}_{n,n} \\ \mathbf{0}_{n,n} & \mathbf{0}_{n,n} & a_{max}^2 \mathbf{I}_n/3 \end{bmatrix}$$

(9.12)

with $\mathbf{0}_{n,m}$ denoting an $n \cdot m$ zero matrix, and $\mathbf{I}_n$ denoting the identity matrix of order n. For $n = 3$, (9.12) may be modified in an obvious manner for the case where the maximum speed and acceleration in the vertical direction are different from the maximum speed and acceleration in the horizontal plane. One should note that (initially) position, speed and acceleration errors are mutually uncorrelated.

The new track has only one component, with relative probability $p(c_k|\chi_k, \mathbf{Y}_k(i)) = 1$. All trajectory models σ have an identical state estimate pdf with mean and covariance defined by (9.12). The a posteriori trajectory model probabilities $\mu_{k|k}(c_k, \sigma)$ may be proportional to their expected life; another often reasonable initiation choice is $\mu_{k|k}(c_k, \sigma) = 1/M, \sigma = 1, \ldots, M$.

The initial probability of object existence, $p(\chi_k|\mathbf{Y}_k(i))$, for a new track, can be calculated by

$$p(\chi_k|\mathbf{Y}_k(i)) = P^0_{k|k,i} \frac{\rho_t(\mathbf{Y}_k(i))}{\rho(\mathbf{Y}_k(i)) + \rho_t(\mathbf{Y}_k(i))}, \tag{9.13}$$

where $P^0_{k|k,i}$ is the probability that measurement $\mathbf{Y}_k(i)$ is not a detection of any object being followed by already existing tracks (Section 9.4.3), and $\rho_t(\mathbf{y})$ denotes the new object spatial density at $\mathbf{y}$. However, in most cases the new object density $\rho_t(\mathbf{y})$ is not known, and the prior clutter measurement density estimate $\rho(\mathbf{y})$ is also often either unknown, or calculated with large relative errors. Thus, it is reasonable to ignore the second factor in (9.13). The initial probability of object existence then becomes

$$p(\chi_k|\mathbf{Y}_k(i)) = P^0_{k|k,i} P_0, \tag{9.14}$$

with P_0 denoting the initial probability of object existence for measurements which are not detections of objects being followed by existing tracks. In practice, P_0 becomes a tuning parameter for the false track discrimination procedure.

9.4.2 Two-point differencing

Two-point differencing (Bar-Shalom *et al.*, 1990) uses measurements from consecutive scans to initialize tracks. The object speed estimate may be initialized using measurements at two different times, but not the acceleration.

This procedure is repeated by all measurements from time $k - 1$, we thus concentrate on tracks initialized based on measurement $\mathbf{Y}_{k-1}(j)$, received at time $k - 1$. In the following scan, at time k, a rectangular selection gate centered at $\mathbf{Y}_{k-1}(j)$, is created with a side at dimension $d = 1, \ldots, n$ equal to

$$2\left(\Delta T_k \cdot \mathbf{v}_{\max} + 2\sqrt{\mathbf{R}(d, d)}\right),$$

where ΔT_k is the time interval between scan k and $k - 1$, and $\mathbf{R}(d, d)$ is the d-th diagonal element of $\mathbf{R}$. In this case the gating probability is $P_G \approx 1$. Denote by V_k the volume of the rectangular selection gate, and by $\mathbf{y}_k$ the set of measurements selected by the rectangular selection gate at time k. In this case

$$\{\mathbf{Y}^{k-1}\} = \{\mathbf{Y}_{k-1}(j)\}, \quad \{\mathbf{Y}^k\} = \{\mathbf{Y}_{k-1}(j), \mathbf{y}_k\}.$$

A new track is initialized by measurement $\mathbf{Y}_{k-1}(j)$ and each selected measurement $\mathbf{y}_k(i)$; $i = 1, \ldots, m_k$. The a posteriori object trajectory state estimate of the new track, given measurements $\mathbf{Y}_{k-1}(j)$ and $\mathbf{y}_k(i)$ has a Gaussian pdf, with mean

$$\hat{\mathbf{x}}_{k|k}(c_k, \sigma) = \begin{bmatrix} \mathbf{y}_k(i) \\ (\mathbf{y}_k(i) - \mathbf{Y}_{k-1}(j))/\Delta T_k \\ \mathbf{0}_{n,1} \end{bmatrix}, \tag{9.15}$$

and covariance

$$\mathbf{P}_{k|k}(c_k, \sigma) = \begin{bmatrix} \mathbf{R} & \mathbf{R}/\Delta T_k & \mathbf{0}_{n,n} \\ \mathbf{R}/\Delta T_k & 2\mathbf{R}/\Delta T_k^2 & \mathbf{0}_{n,n} \\ \mathbf{0}_{n,n} & \mathbf{0}_{n,n} & a_{\max}^2 \mathbf{I}_n/3 \end{bmatrix}. \tag{9.16}$$

Our no a priori knowledge of object velocity is modeled by assuming the uniform a priori measurement pdf at time k,

$$p(\mathbf{y}|\chi_k, \mathbf{Y}^{k-1}) = 1/V_k = p_k(i).$$

The probability of object existence for the first measurement is given by (9.14):

$$p(\chi_{k-1}|\mathbf{Y}^{k-1}) = P_{k-1|k-1,j}^0 P_0,$$

which will then propagate as (Section 5.3.1)

$$p(\chi_k|\mathbf{Y}^{k-1}) = \text{TEX}_{\text{P}}\big[p(\chi_{k-1}|\mathbf{Y}^{k-1}), \gamma \big].$$

The a posteriori probability of object existence $p(\chi_k|\mathbf{Y}^k)$ and a posteriori data association probabilities conditioned on object existence $\beta_k(i)$, $i = 0, \ldots, m_k$, are calculated by the data association operations (Mušicki and Evans, 2004a). Whether we use single- or multi-object tracking for the existing track update, for the computational requirements reason we recommend using the single-object data association here (5.82):

$$\Big[p(\chi_k|\mathbf{Y}^k), \{\beta_k(i)\}_{i \geq 0} \Big] = \text{STDA}\Big[p(\chi_k|\mathbf{Y}^{k-1}), \Big\{ p_k(i) P_{k|k,i}^0 \Big\}_{i>0} \Big].$$

Multi-scan trackers

Multi-scan trackers retain a number of a posteriori track components. They initialize one track for each measurement $\mathbf{Y}_{k-1}(j)$, and one track component for each selected measurement in the subsequent scan $\mathbf{y}_k(i)$. Number of components at time k equals $m_k + 1$, one for each selected measurement $\mathbf{y}_k(i)$, and one for the "null" measurement, $i = 0$.

For each component, all object trajectory models σ are initialized with identical state estimate pdfs. Thus, the initialized component state estimate pdf is identical to the state estimate pdfs of its object trajectory models.

Component $i = 0$ still has a (mostly) uniform position pdf as it was created by the single measurement $\mathbf{y}_{k-1}(j)$. For various practical reasons it is advantageous to now convert its pdf into a Gaussian pdf with mean

$$\hat{\mathbf{x}}_{k|k}(c_k, \sigma) = \begin{bmatrix} \mathbf{Y}_{k-1}(j) \\ \mathbf{0}_{n,1} \\ \mathbf{0}_{n,1} \end{bmatrix}, \tag{9.17}$$

and covariance

$$\mathbf{P}_{k|k}(c_k, \sigma) = \begin{bmatrix} \mathbf{R} + (\Delta T_k \mathbf{v}_{max})^2 \mathbf{I}_n/3 & \mathbf{0}_{n,n} & \mathbf{0}_{n,n} \\ \mathbf{0}_{n,n} & \mathbf{v}_{max}^2 \mathbf{I}_n/3 & \mathbf{0}_{n,n} \\ \mathbf{0}_{n,n} & \mathbf{0}_{n,n} & \mathbf{a}_{max}^2 \mathbf{I}_n/3 \end{bmatrix}. \tag{9.18}$$

The mean and covariance of the object trajectory state estimate pdf of component $i > 0$ are specified by (9.15) and (9.16) respectively.

The relative probabilities of the track components are calculated as in Section 5.5.5; using (5.90), the relative probability of the a posteriori component i equals $\beta_k(i)$, $i = 0, \ldots, m_k$.

Using (5.91), the a posteriori probability of the object trajectory model σ is equal to the single-scan initialization σ model probability; a reasonable initiation choice often is $\mu_{k|k}(c_k, \sigma) = 1/M$, $\sigma = 1, \ldots, M$. Another often reasonable choice is to initialize the trajectory model probabilities $\mu_{k|k}(c_k, \sigma)$ to be proportional to their expected life.

Single-scan trackers

Single-scan (PDA-based) trackers in principle merge all the a posteriori components into one, which would completely negate any advantages of the two-point differencing. Instead of track component merging we implement "track component splitting," where each a posteriori track component incorporating $\mathbf{y}_k$ $(i > 0)$ is split to form an independent track. This procedure is described in Mušicki and Evans (2004a).

After performing the multi-scan tracker initialization, each component $i \geq 0$ becomes a new single-component track with probability of existence equal to $\beta_k(i) p(\chi_k | \mathbf{Y}^k)$. In most cases the new track corresponding to $i = 0$ is terminated.

9.4.3 Calculation of $P_{k|k,i}^0$

$P_{k|k,i}^0$ is the probability that measurement $\mathbf{Y}_k(i)$ is not a detection at time k of any object being followed by already existing tracks. If no existing track selects measurement $\mathbf{Y}_k(i)$, then $P_{k|k,i}^0 = 1$.

Single-object tracking

Single-object tracking processes each track separately. The probability $P_{k|k,i}^{\tau,0}$ that measurement $\mathbf{Y}_k\,(i)$ is not a detection of the object followed by track τ is the complement of the a posteriori probability that the object exists and that its detection is measurement $\mathbf{Y}_k\,(i)$,

$$P_{k|k,i}^{\tau,0} = 1 - p\big(\chi_k^\tau|\mathbf{Y}^k\big)\boldsymbol{\beta}_k^\tau(i).$$

Single-object tracking assumes the independence of object non-detection and

$$P_{k|k,i}^0 = \prod_\tau P_{k|k,i}^{\tau,0} = \prod_\tau \big(1 - p\big(\chi_k^\tau|\mathbf{Y}^k\big)\boldsymbol{\beta}_k^\tau(i)\big).$$

Multi-object tracking

Multi-object tracking processes all tracks in a cluster jointly. Allocations of measurement $\mathbf{Y}_k\,(i)$ to various tracks are mutually exclusive (rather than independent as in the single-object tracking case above). The logic of Section 5.5.4 is also followed here and extends the joint multi-object data association operation.

Set $\Xi(\tau, i)$ is defined as the set of feasible joint events which allocate measurement i to track τ; the set $\Xi(0, i)$ is therefore the set of feasible joint events which does not allocate measurement $\mathbf{y}_k\,(i)$ to any existing track. The event that no track in a cluster selects measurement $\mathbf{y}_k\,(i)$ is the union of all (mutually exclusive) feasible joint events in $\Xi(0, i)$, and thus

$$P_{k|k,i}^0 = \sum_{\varepsilon \in \Xi(0,i)} p(\varepsilon|\mathbf{Y}^k)$$

$$= 1 - \sum_\tau \sum_{\varepsilon \in \Xi(\tau,i)} p(\varepsilon|\mathbf{Y}^k) = 1 - \sum_\tau p(\chi_k^\tau, \theta_k^\tau(i)|\mathbf{Y}^k)$$

$$= 1 - \sum_\tau p(\chi_k^\tau|\mathbf{Y}^k)\boldsymbol{\beta}_k^\tau(i),$$

as in the multi-object tracking case, elementary joint events are mutually exclusive.

9.4.4 New track initialization measurement choice

Which measurements should be used to initialize new tracks? So far, the obvious answer to this question is to use all measurements returned by the sensor(s). If the measurement is also used by a (strong) existing track, the newly initialized track will start with a low initial probability of object existence and will (most likely) be quickly terminated.

Things are usually not this simple, although we would sure like them to be!

First consider the case of single-object tracking, where each track is updated separately. Further assume that there is one object in the surveillance area. In each scan when that object is detected, a new track is initialized following the same object. As the new tracks are true tracks, and each track is updated separately, their probability of object existence is going to quickly grow until they are confirmed. We end up with a number of confirmed tracks following the same object. Additional logic must be employed to clear up this undesirable situation.

Next consider the number of initialized false tracks. In the heavy clutter, we would like to employ two-point differencing. If we are using single-scan (PDA-based) algorithms, the number of initialized false tracks per scan is proportional to the square of the clutter measurement density, and is also proportional to v_{max}^n. Pretty soon we may find ourselves in danger of saturating available computational resources.

These problems can be alleviated, and in some cases solved, by a proper choice of measurements to use for initialization of new tracks.

One may imagine a number of approaches briefly listed below. In practical design, they should be tested by simulating the most difficult conditions in which the tracker is expected to perform (the highest clutter measurement density, the highest object maximum speed v_{max}, maximum number of objects, ...).

Some of the possible approaches to the choice of measurements used for new track initialization are:

- brute force, use all available measurements in all scans;
- use all available measurements, apart from the "nearest neighbor measurements" used by confirmed tracks, and apart from the "nearest neighbor measurements" used by tentative tracks (H. A. P. Blom and E. Bloem, personal correspondence with D. Mušicki, 2007). The nearest neighbor measurement of track τ is the measurement $\mathbf{y}_k(i)$ with highest $\beta_k^\tau(i)$;
- use all available measurements, apart from the "nearest neighbor measurements" used by confirmed tracks;
- use all available measurements, apart from the measurements selected by confirmed tracks;
- use all available measurements, apart from the measurements selected by existing tracks;
- for two-point differencing procedure used in single-scan (PDA-based) tracking use all available pairs of measurements $\mathbf{y}_{k-1}(j)$ and $\mathbf{y}_k(i)$, provided that they are not both selected by the same existing track.

9.5 Track merging

Multiple tracks may end up following one object.

As described in Section 9.4.4, in single-object tracking one existing object may initialize a new track in every scan. All these tracks are true tracks (they will follow the object) and are therefore likely to get confirmed eventually.

In multi-object tracking a similar situation will also arise. However, due to joint data association, the strongest track (usually the first initialized track following the object will have the highest probability of object existence) will generally be allocated the object measurements with highest probability. Due to the mutual exclusivity of the measurement allocation to tracks, the weaker tracks will be allocated the object measurements with reduced probability. This will in effect force these weaker tracks to ignore object measurements. As a consequence, the probability of object existence for these weaker tracks will decrease in time, forcing the false track discrimination procedure to eventually terminate them.

Nevertheless, even in the multi-object tracking case, the additional tracks use up valuable computational resources. This is especially problematic if the optimal ("joint") multi-object tracking approach is used on all tracks. Computational requirements for the joint multi-object trackers grow combinatorially with the number of tracks. Additional tracks can cause saturation of available computational resources, even when only a small number of objects exist in the surveillance volume.

There are many other reasons why we may end up with multiple tracks following the same object. A false track may start to follow an object already being followed by an existing track, or a true track may switch objects and start to follow an object being followed by another existing track. This may happen when tracks cross trajectories, or due to unfavorable object detection/clutter measurement positions combination. If the additional track is strong, even in multi-object tracking it may take some time for the tracks to sort out which is to continue, and which is to be terminated. In the meantime, valuable computational resources are being wasted, and (even worse) the operators get a wrong picture of the objects in the surveillance area.

This is obviously undesirable, and something should be done about it – enter the track merging procedure. The track merging procedure involves the detection of track pairs that are likely to follow the same object, and merging the tracks. It is potentially a surprisingly computationally intensive operation, especially if implemented in a theoretically optimal manner. We limit ourselves here to present this procedure from a more practical viewpoint.

The track merging procedure has two conceptual parts: merge test and (only if the pair of tracks has passed the merge test) the actual track merging. The track merge test is applied to every pair of tracks. If the number of tracks is denoted by T, the number of pairs of tracks is $T(T-1)/2$, potentially a large number. In many practical situations it is therefore imperative to make the track merge test as efficient as possible. Here we illustrate the procedure as applied to two tracks, denoted by superscripts τ and η. Of course, this procedure must be repeated for every pair of tracks.

9.5.1 Track merging test

Optimal track merging test should find out how close the a posteriori trajectory estimate pdfs of track τ and track η are. As both pdfs are Gaussian mixtures, a closed form for, say, Kullback–Liebler distance (relative entropy) (Cover and Thomas, 2006) may be applied. In real-life applications, with numerous tracks, this would inflict prohibitive computational penalty. We recommend using just mean $\hat{\mathbf{x}}_{k|k}$ and covariance $\mathbf{P}_{k|k}$ of tracks τ and η. These values are calculated for track output (5.93).

A proper, although computationally intensive, track merging test is

$$\left(\hat{\mathbf{x}}_{k|k}^{\tau} - \hat{\mathbf{x}}_{k|k}^{\eta}\right)^{T} \left(\mathbf{P}_{k|k}^{\tau} + \mathbf{P}_{k|k}^{\eta}\right)^{-1} \left(\hat{\mathbf{x}}_{k|k}^{\tau} - \hat{\mathbf{x}}_{k|k}^{\eta}\right) < \tau_{m},$$

where τ_{m} is a suitably chosen track merging threshold. A conceptual problem with this test is that, especially due to the presence of clutter measurements, a track may end up having a large covariance matrix, which would let it pass the merging test prematurely. Additional concern is due to the fact that inversion of the state covariance matrix is a computationally intensive operation, which needs to be performed a large number of times (once for each pair of existing tracks).

For reasons of efficiency, the track merge test may be split in two: the crude track merge test and the fine track merge test. The fine track merge test is applied only if the pair of tracks passes the crude track merge test.

The fast and crude track merge test may be testing the distance between estimated means of tracks τ and η,

$$\left(\mathbf{H}_{x}\hat{\mathbf{x}}_{k|k}^{\tau} - \mathbf{H}_{x}\hat{\mathbf{x}}_{k|k}^{\eta}\right)^{T} \left(\mathbf{H}_{x}\hat{\mathbf{x}}_{k|k}^{\tau} - \mathbf{H}_{x}\hat{\mathbf{x}}_{k|k}^{\eta}\right) < R_{m}^{2},$$

where R_{m}^{2} is the minimal distance test, and $\mathbf{H}_{x}$ selects just the position coordinates from $\hat{\mathbf{x}}_{k|k}$. This test may also be repeated for the velocity components of $\hat{\mathbf{x}}_{k|k}$-s.

The small number of tracks which pass the crude track merging test are subjected to the fine track merging test,

$$\left(\mathbf{H}_{m}\hat{\mathbf{x}}_{k|k}^{\tau} - \mathbf{H}_{m}\hat{\mathbf{x}}_{k|k}^{\eta}\right)^{T} \left(\mathbf{H}_{m}\mathbf{P}_{m}\mathbf{H}_{m}^{T}\right)^{-1} \left(\mathbf{H}_{m}\hat{\mathbf{x}}_{k|k}^{\tau} - \mathbf{H}_{m}\hat{\mathbf{x}}_{k|k}^{\eta}\right) < \tau_{M},$$

where $\mathbf{H}_{m}$ selects the states which are used in the test. Position and velocity are often used (and are often part of the track state), whereas in most applications the acceleration estimate is deemed not accurate enough to be used for this purpose. Using a value of $\mathbf{P}_{m}$ independent of actual $\mathbf{P}_{k|k}^{\tau}$ and $\mathbf{P}_{k|k}^{\eta}$ prevents a premature pass of the test by tracks with large covariance matrices, as discussed above. The value of $\mathbf{P}_m$ may be chosen a priori and depend on the average distance (and angles) between tracks τ and η and the sensor. Using precomputed values of $(\mathbf{H}_m \mathbf{P}_m \mathbf{H}_m^T)^{-1}$ further increases the computational efficiency of the procedure.

The merging threshold τ_M may be chosen as a trade-off between early detection of the track merging condition, and premature merging of legitimate tracks which follow different objects with close trajectories. The threshold is therefore application dependent, although, all other things being equal, multi-object tracking algorithms require smaller values of τ_M than their single-object tracking counterparts.

This two-tier structure, with suitable modifications, may also be applied to increase the computational efficiency of the measurement selection procedure.

9.5.2 Track merging execution

Optimal track merging also becomes a computationally intensive operation, particularly for multi-scan object tracking algorithms, whose a posteriori trajectory state estimate pdf is a Gaussian mixture. Tracks τ and η have passed the track merging test, and are assumed to follow the same object. Over the last number of scans, tracks τ and η have been updated, one would presume, with an almost identical set of measurements, and thus their estimation errors are highly correlated.

If this is true, then optimal merging will not reduce the estimation errors much, at least not enough to justify extensive use of computational resources. The following approach may be used instead:

- Only one track is kept, the other is simply terminated.
- If one or both of the tracks are confirmed, than the remaining track also remains or acquires the confirmed status.
- The higher probability of object existence is used for the remaining track.

One possibility is to retain the track with the higher probability of object existence. The authors have found, however, that retaining the track with a smaller (determinant of) estimation error covariance matrix often yields better results.

Thus the track merging procedure as described here is actually a misnomer, as it is actually more of an elimination of surplus tracks. Track merging is the operation which *could* be done in a proper theoretic sense. In a practical world the optimal approach is often not justified.

9.6 Illustrative examples

Two simulation studies are presented here. Both are designed to highlight the false track discrimination properties of the object-existence-based algorithms presented in Chapter 5, and the second one also demonstrates the capabilities of the linear multi-target procedure, described in Section 9.2.

Each simulation experiment consists of a number of simulation runs, and each simulation run consists of a number of scans. In each simulation run, the objects repeat their trajectories; however, the detection sequence, measurement noise sample and clutter measurements are random and independent for each simulation run, and for each simulated scan within simulation runs. Random number generators are initialized, so that each simulated algorithm in a given simulation study receives an identical set of measurements.

In both instances, automatic object tracking situations are simulated. The object tracking algorithms have no prior information on the existence, number and position of possible objects within the surveillance area. Tracks are initialized using measurements, as proposed in Section 9.4. The two-point differencing (Section 9.4.2) procedure is used to initialize tracks. The first scan probability of object existence for new tracks is calculated by (9.14). After initialization, each track state is updated according to the algorithm simulated.

A simple but efficient false track discrimination procedure is applied, as described in Section 5.1 and depicted in Figure 5.1. After each scan, false track discrimination is applied and track status updated. Terminated tracks are then removed from computer memory. At the end of each simulation run, true tracks are removed from memory, and false tracks are retained. The purpose of this is to simulate a continuous surveillance operation and establish a stationary field of false tracks.

The false track discrimination procedure is defined by three parameters:

- P_0 denoting the initial probability of object existence for measurements which are not detections of objects being followed by existing tracks (9.14);
- the track confirmation threshold t_c (Section 5.1); and
- the track termination threshold t_t (Section 5.1).

For a given algorithm and given surveillance situation, the false track discrimination performance becomes a function of these three parameters. For each simulation experiment, and for each simulated object tracking algorithm, false track discrimination is optimized. The goal of this optimization was to find the set of false track discrimination parameters which maximizes the confirmed true track statistic, subject to a predefined and acceptable false track statistic common to all algorithms in an experiment. One exception to this strategy is noted in Section 9.6.2.

The optimization constraint of the false track statistic is the number of scans in which a confirmed false track exists. For the system to be actually used, particularly by the human operators, this number has to be small. Each occurrence of a confirmed false track amounts to a false alarm, and too many false alarms will easily lead to the operators distrusting the results. In extreme cases, including some defence applications, each confirmed false track may trigger a costly and unnecessary response. Here we aim for a value of less than one such scan in a thousand scans. This is achievable in the first simulation study described in Section 9.6.1. However, in the multi-object simulation (Section 9.6.2), this objective is much more difficult to achieve. In the multi-object study, confirmed false tracks are mostly initialized and maintained by cross-pollinating different objects. These confirmed false tracks appear together with a large number of existing confirmed true tracks. Thus, whilst still undesirable, these confirmed false tracks will not cause a general false alarm situation by themselves, and the confirmed false track constraint can be somewhat relaxed. The lesson is clear: one must deduce the object tracking requirements from the application and avoid any unnecessary grief caused by over-designing the tracker.

Please also note that each confirmed false track will exist for a number of scans before it gets automatically terminated by the false track discrimination procedure. Thus, typically, the total number of confirmed false tracks is about an order of magnitude smaller than this constraint.

The true confirmed track statistic that is to be optimized in the case of the single-object tracking comparison (Section 9.6.1) is the total number of scans when the confirmed true track is present. For the multi-object tracking comparison (Section 9.6.2), this statistic equals the sum across objects of the total number of scans when the corresponding confirmed true track is present.

As noted in Mušicki and Suvorova (2004), it is not easy to optimize false track discrimination. To start with, the false track discrimination, as a function of the three parameters, has numerous local peaks, where the optimization process may prematurely end. Also, the only way to obtain the false track discrimination performance is via lengthy simulations.

In all simulation experiments, track merging has also been applied to remove duplicate tracks. The pragmatic track merging procedure, described in Section 9.5, has been implemented.

Both simulation studies assume a two-dimensional surveillance situation. One linear sensor is assumed in both cases. The object measurements contain object Cartesian coordinates, each corrupted by mutually independent, zero mean, white Gaussian measurement noise, with covariance matrix $\mathbf{R} = 25\mathbf{I}_2 m^2$. Objects are detected with a fixed and known probability of detection. Clutter measurements follow a Poisson distribution.

9.6.1 Single-object tracking simulation study

This single-object tracking in clutter simulation study is designed to demonstrate:

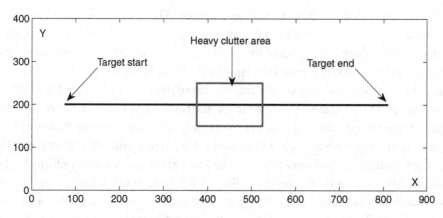

Figure 9.2 Simulation 1 environment.

- advantages that may be accrued using multi-scan object tracking;
- the importance of improved knowledge of clutter measurement density; and
- the false track discrimination trade-off between true track confirmation success rate and false track confirmation rate.

In a single-object tracking environment, we compare IPDA and ITS, the principal single-object trackers described in Chapter 5, across a range of detection probabilities, $P_D = 0.7, 0.8, 0.9$. The number of ITS components is limited to 20 per track. Both non-parametric and parametric versions of the algorithms are included in this simulation study.

In the experiments described in this section, track initiation is carried out using all measurement pairs from consecutive scans, subject to the maximum speed constraint, and given that both measurements were not selected by the same already existing track. The latter constraint assumes that measurements not satisfying the constraint are likely to be detections of an object already being followed by an existing track. Thus this criterion is designed to reduce the number of duplicate tracks. Whilst the logic seems to be working in a sparse object environment, as noted in Section 9.6.2, it seems to be inadequate in a dense object environment where measurements from objects in close mutual proximity may systematically initialize and confirm false tracks. The maximum object speed constraint is $v_{max} = 25$ m/s. Object acceleration is not part of the object state, thus no maximum object acceleration value was required.

The simulated environment is depicted in Figure 9.2. A single object is moving with uniform velocity during 50 scans of each simulation run, each scan time being 1 s. Initial object position in Cartesian coordinates is $[75 \quad 200]^T$ m, and initial velocity is $[15 \quad 0]^T$ m/s.

Clutter measurement density is non-uniform. The base clutter measurement density equals 10^{-4}, with a heavy clutter area, shown on Figure 9.2, having a clutter measurement density equal to 7×10^{-4}. This is a high clutter measurement density, which would, comparatively speaking, favor ITS over IPDA.

During each simulation experiment, the simulation run is repeated 500 times, for a total of 25 000 scans. The false track discrimination parameters are tuned to deliver a confirmed false track in approximately 25 scans per simulation experiment. The tuning procedure was terminated when an insignificant difference in the true track statistics was found between the last experiments with confirmed false track statistics crossing the desired value of the false track statistic. Due to the small number of the false track statistic, the simulation results should be taken with a grain of salt. The only way to increase the confidence in simulation results was to significantly increase the number of simulation runs per simulation experiment, although this would result in a somewhat exorbitant use of resources. However, whilst the simulation results themselves could and should be regarded critically, the reported performance differences between algorithms are of such magnitude that it is safe to believe in the relative algorithm ranking delivered by this simulation study, at least in this situation.

The false track statistic amounts to having one confirmed false track visible every 1000 scans, which gives an approximate confirmed false tracks rate of one in 4 000 scans. This should also be compared to the number of initialized false tracks of approximately 450 000, or approximately 18 per scan. A vast majority of false tracks get terminated without being confirmed in the meantime.

You will note that, even in the case of $P_D = 0.9$, the success rate of confirmed true tracks for the non-parametric version of IPDA leaves something to be desired. Additional experiments have been performed for non-parametric IPDA with the confirmed false tracks statistic relaxed to approximately 60 scans during which a confirmed false track exists.

The true track confirmation success rates are shown in Figures 9.3, 9.4 and 9.5, for the case of $P_D = 0.9$, $P_D = 0.8$ and $P_D = 0.7$ respectively.

- The curves "ITS exact" and "IPDA exact" correspond to the parametric ITS and IPDA algorithms respectively, using perfect prior knowledge of clutter measurement density.
- The curves "ITS np" and "IPDA np" correspond to non-parametric ITS and IPDA respectively, where the clutter measurement density is estimated based on the selected measurements in the current scan.
- The curve "IPDA np+" corresponds to non-parametric IPDA with the increased number of confirmed false tracks constraint.

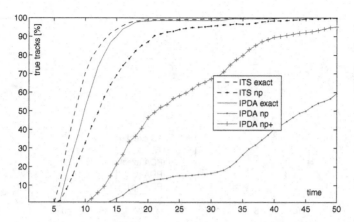

Figure 9.3 Confirmed true tracks success, $P_D = 0.9$.

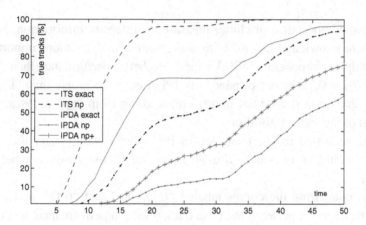

Figure 9.4 Confirmed true tracks success, $P_D = 0.8$.

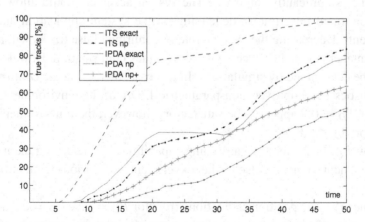

Figure 9.5 Confirmed true tracks success, $P_D = 0.7$.

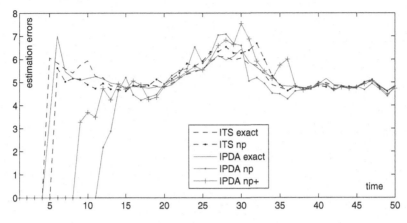

Figure 9.6 Root mean square errors, $P_D = 0.9$.

First note the importance of clutter measurement density information. For $P_D = 0.9$ and 0.8, knowledge of clutter measurement density is more important than using multiple components; "IPDA exact" has better performance than "ITS np." Only for $P_D = 0.7$, the performance of "IPDA exact" is worse than "ITS np." In all cases, there is a substantial improvement going from "np" to "exact" implementation of the same algorithm.

Second, note that the performance of IPDA deteriorates faster than the ITS. Indeed, it would seem that *in this environment*, IPDA should not be used for $P_D < 0.8$.

Finally, comparing the curves labeled "IPDA np" and "IPDA np+," one can observe the trade-off between the false track rejection performance, and true track confirmation. By allowing more confirmed false tracks, the true track confirmation performance significantly improves. The system designer should allow a certain rate of confirmed false tracks, depending on the system application and overall requirements. Decreasing the average rate of confirmed false tracks will also have a detrimental effect on the success rate of true track confirmation. How detrimental needs to be determined by simulations. In this case, increasing the confirmed false track statistics has improved non-parametric IPDA in this environment, and the results for $P_D = 0.9$ appear to be satisfactory; however, the improvement was not enough for $P_D \leq 0.8$.

For low P_D, ITS offers further tuning capabilities by increasing the number of track components. One can also further experiment with various track initialization methods.

Root mean square position estimation errors for $P_D = 0.9$, $P_D = 0.8$ and $P_D = 0.7$ are shown in Figures 9.6–9.8. For a given P_D, the estimation errors do not

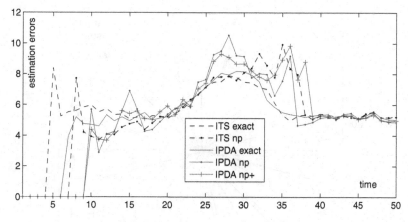

Figure 9.7 Root mean square errors, $P_D = 0.8$.

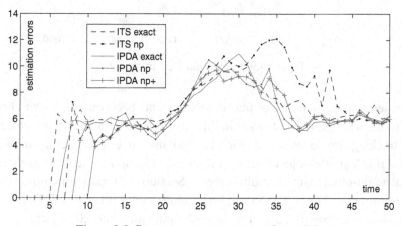

Figure 9.8 Root mean square errors, $P_D = 0.7$.

differ between algorithms in any significant manner. The estimation errors are, as expected, larger for smaller values of P_D.

9.6.2 Multi-object tracking simulation study

This multi-object tracking in clutter simulation study is designed to demonstrate false track discrimination and track retention capabilities in a multi-object situation. The environment is depicted in Figure 9.9.

A multi-object situation with a significant number (20) of objects with intersecting trajectories is simulated. This number of objects simultaneously being in proximity of each other precludes tracking of objects using any of the "joint" multi-object trackers presented in Chapter 5. The problem lies in the computational

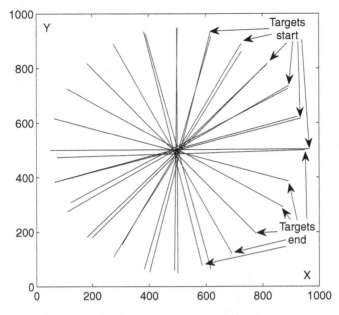

Figure 9.9 Simulation 2 environment.

requirements, as the number of feasible joint events becomes huge. Our choice is either to use a single-object tracker in this situation, or to use a sub-optimal multi-object tracking approach. As the objects are not maneuvering, we compare here the ITS and IPDA single-object trackers presented in Chapter 5, as well as their sub-optimal multi-object (linear multi-target – Section 9.2) transformations, namely linear multi-target ITS (LMITS) and linear multi-target IPDA (LMIPDA).

The scan time was fixed to 1 s. In each simulation run 20 objects start their trajectory from the edge of a circle with radius of 450 m. The objects start with an angular separation of 15 degrees. Each object follows a uniform (constant velocity vector) motion toward the center of the circle, scheduled to reach in 20 scans, and then continue trajectories for a total of 40 scans in each simulation run. Random components are added to the initial speed vectors of each object, to ensure a more difficult situation where all objects trajectories do not intersect at the same point, but are close enough to form one big cluster.

Each object is detected with the probability of detection $P_D = 0.9$. A significant clutter in the surveillance area is present with a uniform clutter measurement density of 10^{-4}. All object tracking algorithms have perfect knowledge of the value of the clutter measurement density.

The maximum object speed constraint on the track initialization procedure is defined to be 30 m/s. The object acceleration is not part of the object state, thus no maximum object acceleration value was required.

Table 9.1 *Multi-object algorithm summary.*

Filter	CFtStats	nCTtracks	nOk [%]	nSwitch [%]
LMITS	73	4518	97.6	2
LMIPDA	143	4292	97.1	2.7
ITS	294	4450	57	9
IPDA	285	3812	73	5

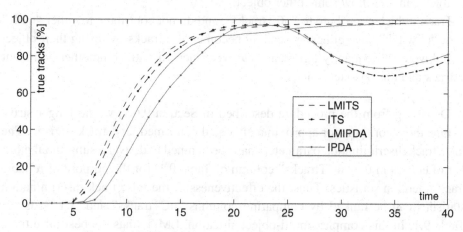

Figure 9.10 Confirmed true tracks success, simulation 2.

In this environment, a large number of false tracks may be initialized by measurements (detections) from close objects when they converge near the center of the surveillance area. A large number of these tracks can get confirmed by moving "sideways" from object to object (particularly in the case of the single-object trackers IPDA and ITS). Initializing new tracks using only measurements which are not selected by any existing track *in this situation* greatly reduces the number of initialized and confirmed false tracks. All algorithms simulated in this study use this logic for initializing tracks using the two-point differencing procedure described in Section 9.4.2. Despite this precaution, a total of approximately 280 000 false tracks are initialized, or approximately 28 per scan.

Trackers are compared with respect to the false track discrimination and track retention in this environment. Results are integrated over the simulation experiments consisting of 250 simulation runs for each tracker. False track discrimination and track retention results are presented in Figure 9.10 and Table 9.1. The columns of Table 9.1 show:

- "CFtStats": confirmed false track statistic, consist of the number of scans in the simulation experiment with existing confirmed false track (out of the total of $250 * 40 = 10\,000$ scans).

- "nCTtracks" denotes the total number of confirmed true tracks at scan 16. Theoretical maximum number equals $20 * 250 = 5000$. Each track contributing to this number is noted together with its object, and at scan 39 it is examined whether this track is still following the original object.
- "nOK" denotes the percentage of tracks contributing to "nCTtracks" which after scan 39 still follow the original object.
- "nSwitch" denotes the percentage of tracks contributing to "nCTtracks" which after scan 39 follow some other object.
- Tracks which contribute to "nCTtracks", and do not contribute to either "nOK" or "nSwitch" have either by scan 39 become false tracks by losing their object and were subsequently terminated, or were merged with some other dominant track and thus also terminated.

Departing from the procedure described in Section 9.6.1, we no longer strive to tune the algorithms for approximately equal confirmed false track statistic. The false track discrimination parameters have been tuned to deliver a substantial number of entries in the "nCTtracks" column of Table 9.1, for the purpose of reliable track retention statistics. Thus, the effectiveness of the false track discrimination statistics can be ranked by comparing columns "nCTtracks" and "CFtStats" of Table 9.1. In this complex multi-object situation, LMITS has the best false track discrimination statistics, followed by LMIPDA. Then comes ITS and, finally, IPDA. The confirmed false tracks statistics should be taken into context by the total number of initialized false tracks (approximately 280 000) and the total number of scans (10 000) in each simulation experiment.

Consulting Figure 9.10, we also conclude that LMITS and LMIPDA do not have significant problems due to mutual proximity of objects. ITS and IPDA, on the other hand, lose a significant percentage of confirmed true tracks after scan 20. This is also confirmed by Table 9.1, where LMITS and LMIPDA have almost perfect track retention statistics, and ITS and IPDA do not. This vindicates the need for multi object tracking, and also vindicates the use of the linear multi-target procedure.

IPDA, although significantly worse than ITS with respect to the false track discrimination procedure, has significantly better track retention properties in this situation. The reason is that, during the objects' cross-over, different ITS track components latch onto different objects, and not always the original object wins this tug of war.

The bottom line is that one should not use single-object tracking algorithms in multi-object situations, particularly as the linear multi-target procedure provides excellent multi-object capabilities with very little cost to either computational or algorithmic (structural) complexity.

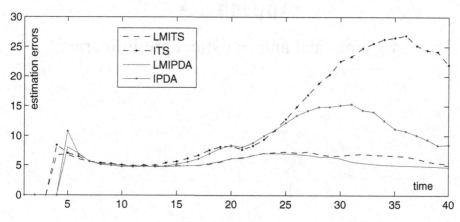

Figure 9.11 Root mean square errors, simulation 2.

Finally, the root mean square position estimation errors over time are presented in Figure 9.11. There is statistically insignificant difference between LMIPDA and LMITS in this respect, whilst single-object trackers have significantly larger estimation errors. These increased errors are mostly due to the large estimation error of tracks which are starting to lose their respective objects. Before these tracks finally lose their true track status, they contribute significantly to the estimation root mean square errors.

9.7 Summary

This chapter deals with a number of technically important subjects of implementation of any object tracking algorithms – such as clutter measurement density estimation, track initialization and merging. Important linear multi-target tracking is also presented in this chapter as a computationally efficient solution to multiple-object tracking. Simulated examples are also presented pointing out the major features of different trackers.

Appendix A

Mathematical and statistical preliminaries

A.1 Probability laws and distributions

A.1.1 Sample space and events

The sample space

The *sample space* is the set of all possible values, or *outcomes*, of a realization that is not known, be it in the past, present or future. In the Bayesian probabilistic framework, every unknown quantity is treated as a *random* quantity.

Examples:
1. When tracking an object in 3D space, the position of the target at some point in time in the future is not known. The 3D space is the sample space.
2. The exact position of that object in the past may not be known. In many tracking situations, the exact position is never observed, only estimated. In that case, although in the past, the exact position of the target is a random quantity and the 3D space is the sample space.
3. Measurements in object tracking are the results of observations by sensing devices. They are subject to random fluctuations. Measurement errors are attached to the measurements, making them random quantities. The focus is on the errors and they are treated as random values. Their sample space is problem dependent, but often is the value space of the measurements.

The sample space is the mathematical set of all values that can be taken by an unknown quantity of interest. One of the simplest examples would be the tossing of a coin. The sample space is $\{H, T\}$, where H is the outcome of a head in the tossing and T is tail.

A random event

An *event* is a subset of the sample space. Any set of values contained in the sample space is the mathematical representation of a corresponding event.

Examples:
1. A target moves along a straight line. The sample space is $\Omega = \mathbb{R} = (-\infty, +\infty)$. Any interval on the real line or union of intervals on the real line represents an event. For example, the target being in the vicinity of 1 can be represented by the event $[0, 2]$. The event of the target being even closer to 1 could be $[0.8, 1.2]$.

2. In the coin-tossing example, $E = H$ is the event that the coin turns up a head.
3. If we consider the case of tossing two coins, then the sample space is made up of the basic outcomes $\{H, H\}, \{H, T\}, \{T, H\}, \{T, T\}$. The event that the first coin turns up a head is $E = (\{H, H\}, \{H, T\})$.

If E and F are two events in the sample space, then the *union* of E and F is $E \bigcup F$ and consists of all the outcomes that are in E **or** in F.

If E and F are two events in the sample space, then the *intersection* of E and F is $E \bigcap F$ and consists of all the outcomes that are both in E **and** in F.

Examples:

1. Rolling a dice can result in 1, 2, 3, 4, 5 or 6. These are the outcomes that make up the sample space. If we let E be the event that the outcome is even, and F the event that the outcome is less than 3, then $E \bigcup F = \{2, 4, 6\} \bigcup \{1, 2\} = \{1, 2, 4, 6\}$.
2. $E \bigcap F = \{2\}$.

E and F are said to be *disjoint events* if $E \bigcap F$ is the empty set $\emptyset$. We write $E \bigcap F = \emptyset$. The empty set $\emptyset$ is the representative of the *null event* $\varnothing$. The null event is the event that cannot occur, the impossible event.

Example:

1. A target moving along a straight line cannot be both at the same time in $[1, 3]$ and $[7, 12]$ since $[1, 3] \bigcap [7, 12] = \emptyset$.

A set of events $E_1, E_2, \ldots, E_n$ are said to be *mutually exclusive* if all the sets are pairwise disjoint, that is $E_i \bigcap F_j = \varnothing$ for all (i, j).

A set of events $E_1, E_2, \ldots, E_n$ are said to be *mutually exclusive and exhaustive* if the sets are mutually exclusive and $E_1 \bigcup E_2 \cdots \bigcup E_n = \Omega$.

A.1.2 Probabilities, conditional probabilities and independence

Probabilities

To measure our uncertainty about a random event, we use *probabilities*. This is the fundamental tenant of the probabilistic framework.

If E is a random event, then given all our knowledge and history background $\mathcal{H}$, we assign a number $p(E|\mathcal{H})$ to all events E, such that:

- $0 \leq p(E|\mathcal{H}) \leq 1$;
- $p(\Omega|\mathcal{H}) = 1$;
- for any sequence of events $E_1, E_2, \ldots$, which are mutually exclusive,

$$p \left(\bigcup_{i=1}^{\infty} E_i | \mathcal{H} \right) = \sum_{i=1}^{\infty} p(E_i | \mathcal{H}).$$

It is through this quantification, the use of numbers called probabilities, that the probabilistic framework allows for the resolution of problems with random events. A large body of scientific work, throughout centuries, was needed to arrive at the above three axioms. They need to be satisfied by probabilities. The discussion of this topic is beyond the scope of this book. It can be found in any literature material treating the definition and meaning of probability. As readers of this book, it is enough to know that $p(E|\mathcal{H})$ defines a probability attached to the event E, given all acquired knowledge $\mathcal{H}$ up to the time of the assessment of $p(E|\mathcal{H})$.

This probabilistic way of representing uncertainty makes it possible for two people with different knowledge to have two different assessments for $p(E|\mathcal{H})$ as $\mathcal{H}$ is different for both of them.

Example:
1. In the case of two stations tracking the same target, the assessment of the probability of the target's position may differ for both stations. This may be due to the difference in information received by both stations, and to the knowledge in tracking of the engineers in the two tracking stations.
2. If I am cheating and tossing a coin, I know that the probability of head, say, is higher. Whereas the person I am cheating may still assume that the coin is fair and that the probability of a head is $1/2$.

In the reminder of the text, $p(E|\mathcal{H})$ will be written $p(E)$. We drop the $(.|\mathcal{H})$ for simplification purposes. $\mathcal{H}$ is always assumed to be present in the assessment of a probability.

Conditional probabilities

$p(E|F)$ is the conditional probability of E *given* F. It is the probability of event E occurring were we to assume that F has occurred:

$$p(E|F) = \frac{p(E \bigcap F)}{p(F)}.$$

Note that F does not need to have actually occurred. We are simply assessing $p(E)$ in the hypothetical case where F has occurred.

Examples:
1. Rolling a dice, if we let E be the event that the outcome is even, and F the event that the outcome is less than 3. Then $E \bigcap F = \{2\}$ leading to $p(E|F) = \frac{p(E \bigcap F)}{p(F)} = \frac{1/6}{1/3} = 1/2$.
2. A more intuitive example is the following: consider drawing a card out of a deck of 52 (ace, king, queen, jack, 10, ..., 2) in the four usual suits; hearts, clubs, diamonds, spades. Consider the event $E = \{K, Q, J\}$, that is drawing that card as either a king, queen or jack. Let F be the event that the card is a diamond. Then,

$$p(E|F) = \frac{p(E \bigcap F)}{p(F)} = \frac{3/52}{1/4} = 3/13.$$

It is clear that the result is intuitive. There are 13 diamond cards and $\{K,Q,J\}$ make up 3 out of 13 possibilities. Note also that $p(E) = 12/52 = 3/13$.

Independent events

Two events E and F are said to be *independent* if

$$p(E \bigcap F) = p(E)p(F).$$

Similarly,

$$p(E|F) = p(E),$$

which has a nice interpretation, since knowing that F has occurred, or in the case it will occur, does not change our assessment of $p(E)$.

Example:

1. Drawing a card out of a deck of 52 (ace, king, queen, jack, 10, ..., 2) in the four usual suits, hearts, clubs, diamonds, spades, we let the event $E = \{K, Q, J\}$ be that of drawing that card as either a king, queen or jack. We let F be the event that the card is diamond. However, in this example, we throw away the ace of clubs before drawing the card. Then,

$$p(E|F) = \frac{p(E \cap F)}{p(F)} = \frac{3/51}{13/51} = 3/13.$$

It is still 3/13. However, $p(E) = 12/51 > 12/52 = p(E|F)$. This means that E and F are not independent. It is clear that the suit of the card affects the probability of getting a king, queen or jack.

Note: at times, and in most of the material in the book, we note $p(E \cap F)$ as $p(E, F)$. This is a standard notation in most textbooks.

A.1.3 The multiplication law

Given two events E and F, the conditional probability of event E given the observation of event F is

$$p(E|F) = \frac{p(E, F)}{p(F)}.$$

It is equal to the joint probability of events E and F, $p(E, F)$, normalized by the unconditional probability of event F, $p(F)$. Using the *multiplication law* in the above equation twice, we arrive at *Bayes' theorem*, which can be written as

$$p(E|F) = \frac{p(F|E)p(E)}{p(F)}. \tag{A.1}$$

Many target tracking quantities of interest such as the number of targets and/or their states can be modeled as event(s) E and many types of sensor outputs, e.g., radar returns or infrared images as event F. Bayes' theorem in (A.1) is applied to obtain the conditional probabilities $p(E/F)$.

A.1.4 Law of total probability

The *Chapman–Kolmogorov equation* or *law of total probability* is essential in target tracking. Let $F_1, F_2, \ldots, F_n$ be a set of mutually exclusive and exhaustive events. Then

$$p(E) = \sum_{i=1}^{n} p(E|F_i)p(F_i).$$

This is also called the *law of extension of conversation* as consideration is extended to the set $\{F_1, F_2, \ldots, F_n\}$ when assessing $p(E)$.

Using the Chapman–Kolmogorov equation, we can rewrite Bayes' theorem,

$$p(E|F) = \frac{p(F|E)p(E)}{\sum_{i=1}^{n} p(F|E_i)p(E_i))},$$

where this time $E_1, E_2, \ldots, E_n$ is a set of mutually exclusive and exhaustive events.

A.1.5 Random variables

A *random variable* is a mathematical function that maps events to real numbers. For example, a random variable can be used to allocate to each face of a rolled dice a number in {1, 2, 3, 4, 5, 6}. By mapping events from the real world into real numbers, we are able to apply the theory of probability to real-world problems.

A random variable $\mathbf{X}$ takes the sample space Ω into the real line

$$\mathbf{X} : \Omega \to \mathbb{R}.$$

Depending on whether the function $\mathbf{X}$ is discrete or continuous, or a mixture of the two, that is $\mathbf{X}(\Omega)$ is discrete/continuous, the random variable $\mathbf{X}$ is said to be a *discrete random variable* or a *continuous random variable*.

Note that most often, the term "random variable" applies to the image of $\mathbf{X}$ rather than $\mathbf{X}$ itself. For example, in the case of the toss of a coin, if we let $\mathbf{X}(\text{head}) = 0$ and $\mathbf{X}(\text{tail}) = 1$, then often we call the set {0,1} the random variable. It is a short cut in notation used by all.

A.1.6 Discrete random variables

In target tracking, we may be interested in identifying a target. For example, the target could be a commercial flight, a surveillance helicopter, a friendly fighter jet or an enemy fighter plane. We model the target state S_k at time k to be 1, 2, 3 or 4 respectively. In this case, S_k is a discrete random variable that takes values in the set {1, 2, 3, 4}. We refer to $p(S_k = i), i = 1, 2, 3, 4$ as the probability that the target is a commercial flight, a surveillance helicopter, a friendly fighter jet or an enemy fighter plane, respectively. These four values constitute the *probability distribution* of the discrete random variable S_k. Often we write this distribution as $p(S_k)$ when there is no ambiguity.

A.1.7 Continuous random variables

Similarly, we may be interested in determining the position of a target. S_k may refer to the coordinates of the target. In the one dimensional case, S_k takes values in the real line. In higher dimensions, S_k is a vector that takes values in $\mathbb{R} \times \mathbb{R}$ or $\mathbb{R}^3$.

In the case where the target state includes the velocity as well as the position, then S_k is a vector that takes values in $\mathbb{R}^n \times \mathbb{R}^n$ where n is the dimension of the space in which the target evolves. We can also augment the vector S_k with the acceleration vector and so on.

Because S_k is continuous, we write $p(S_k = x)$ for the *probability density function* (pdf) of S_k at the point x. Let f be a non-negative function such that

$$\int_{\mathcal{A}} f(x)dx = p(S_k \in \mathcal{A}),$$

where $\mathcal{A}$ is a subset of the sample space, and $\int$ is meant to be a multiple integral if need be. In this case, f is the pdf of S_k. In the book, we often refer to f as $p(S_k)$.

Most introductory mathematical statistics books offer a good explanation of the concepts of a pdf. For example, see Hogg and Craig (1995) for an excellent treatment of the material.

A.1.8 Expected values

Expected value existed before probabilities. Christian Huygens (1629–1695) wrote a book in 1657 that contained the first published study of mathematical expectations. Because of the inherent appeal of games of chance, the first problems in probability theory were treated through the notion of expected value.

In probability theory the expected value (or mathematical expectation) $E(\mathbf{X})$ of a random variable $\mathbf{X}$ is the sum of the probability of each possible outcome of the experiment multiplied by the value of the random variable for that outcome

$$E(\mathbf{X}) = \sum_{\text{all x}} x p(\mathbf{X} = x).$$

For a continuous random variable, the sum is the integral

$$E(\mathbf{X}) = \int_{\mathbf{x}} x f(x) dx.$$

The expected value can also be defined for any function of $\mathbf{X}$, say $g(x)$ as

$$E(g(\mathbf{X})) = \int_{\mathbf{x}} g(x) f(x) dx.$$

See Hogg and Craig (1995) for a comprehensive treatment of the subject.

In the mathematical derivations of the book, we use $E(\cdot)$ to mean the expected value of $(\cdot)$ whenever there is no ambiguity.

A.1.9 Joint, marginal and conditional distributions

The *joint probability distribution* of the discrete random variables $\mathbf{X}_1, \mathbf{X}_2, \ldots, \mathbf{X}_n$ is denoted by $p(\mathbf{X}_1, \mathbf{X}_2, \ldots, \mathbf{X}_n)$ to represent the multivariate function $p(\mathbf{X}_1 = x_1, \mathbf{X}_2 = x_2, \ldots, \mathbf{X}_n = x_n)$, for all $(x_1, x_2, \ldots, x_n)$ possible values.

$p(\mathbf{X}_1, \mathbf{X}_2, \ldots, \mathbf{X}_n)$ is also used to denote the *joint probability density function* in the case of continuous random variables. It is again used in the case when some random variables are discrete and some are continuous. It is a notation that allows for simpler derivation of mathematics where it would be tedious to write everything explicitly. It is a common practice in probabilistic and statistical work. For a more elaborate explanation of the Joint distributions and pdf's, see Hogg and Craig (1995).

The *marginal probability distribution* of $\mathbf{X}_i$ in the case of the discrete random variables $\mathbf{X}_1, \mathbf{X}_2, \ldots, \mathbf{X}_n$ is

$$p(\mathbf{X}_i) = \sum_{\text{all}\mathbf{X}_1} \cdots \sum_{\text{all}\mathbf{X}_{i-1}} \sum_{\text{all}\mathbf{X}_{i+1}} \cdots \sum_{\text{all}\mathbf{X}_n} p(\mathbf{X}_1, \mathbf{X}_2, \ldots, \mathbf{X}_n).$$

All the random variables except $\mathbf{X}_i$ are averaged out of the joint probability distribution to obtain the marginal probability distribution of $\mathbf{X}_i$. A similar averaging is done through integration in the case of continuous variables.

For two discrete random variables, $\mathbf{X}_i$ and $\mathbf{X}_j$, we write $p(\mathbf{X}_i|\mathbf{X}_j)$ to mean the *conditional probability distribution* of $\mathbf{X}_i$ given $\mathbf{X}_j$. It is the probability distribution for the events $(\mathbf{X}_i = x_i|\mathbf{X}_j = x_j)$ for all (x_i, x_j). Recall that

$$p(\mathbf{X}_i = x_i|\mathbf{X}_j = x_j) = \frac{p(\mathbf{X}_i = x_i, \mathbf{X}_j = x_j)}{p(\mathbf{X}_j = x_j)}.$$

For two continuous random variables, $\mathbf{X}_i$ and $\mathbf{X}_j$, we write also $p(\mathbf{X}_i|\mathbf{X}_j)$ to mean the *conditional probability density function* of $\mathbf{X}_i$ given $\mathbf{X}_j$. It is defined as

$$p(\mathbf{X}_i|\mathbf{X}_j) = \frac{p(\mathbf{X}_i, \mathbf{X}_j)}{p(\mathbf{X}_j)}.$$

A.1.10 Bayes' theorem and the Chapman–Kolmogorov equation

Bayes' theorem

Bayes' theorem in the case of two continuous random variables, $\mathbf{X}_i$ and $\mathbf{X}_j$, is

$$p(\mathbf{X}_i|\mathbf{X}_j) = \frac{p(\mathbf{X}_j|\mathbf{X}_i)p(\mathbf{X}_i)}{p(\mathbf{X}_j)}.$$

It is the twice application of the definition of conditional densities. This notation, by choice, applies in all cases, be the random variables both discrete, both continuous or one discrete and the other continuous.

In target tracking we are interested in estimating the target state S_k after knowing all the measurement vectors $y^k = (y_1, \ldots, y_{k-1}, y_k)$. We seek the probability distribution or pdf

$$p(S_k|y^k).$$

It is a conditional distribution and is often derived using Bayes' theorem,

$$p(S^k|y^k) = \frac{p(y^k|S^k)p(S^k)}{p(y^k)}.$$

Using the same theorem on a set of random variables, we can have, for example,

$$p(S_k|r_k, y^k) = \frac{p(y^k|S_k, r_k)}{p(y^k|r_k)}p(S_k|r_k),$$

where r_k stays in the background, that is, r_k is always given.

Sometimes, it is necessary to write explicitly the value of a random variable such as that of r_k in Section 3.2.2:

$$p(x_k|r_k = i, y^k) = p(x_k|r_k = i, y_k, y^{k-1})$$

$$= \frac{p(y_k|x_k, r_k = i, y^{k-1})}{p(y_k|r_k = i, y^{k-1})}p(x_k|r_k = i, y^{k-1}),$$

where this time, it is given y^{k-1}, meaning its value was when it was observed, and given $r_k = i$.

A reader familiar with probabilistic or statistical work will have no difficulty grasping this notation. Other readers can review some fundamentals in Bayesian probabilistic modeling and will find it easy to follow this notation. Starting with Dennis V. Lindley's book on Bayesian statistics (1972), a series of excellent treatises of the subject can be found in the work of Box and Tiao (1973), Bernardo and Smith (1994), Berger (1985), Berry (1996), Gelman *et al.* (1995), and many others.

The Chapman–Kolmogorov equation

The law of total probability can be written as

$$p(\mathbf{X}) = \int_{\mathbf{Y}} p(\mathbf{X}, \mathbf{Y}) d\mathbf{Y},$$

where the integration is taken over the whole range of $\mathbf{Y}$ and $\int_{\mathbf{Y}}$ can be a multiple integration, summation or mixture, depending on the nature of $\mathbf{Y}$. The above can be written as

$$p(\mathbf{X}) = \int_{\mathbf{Y}} p(\mathbf{X}|\mathbf{Y}) p(\mathbf{Y}) d\mathbf{Y}, \tag{A.2}$$

and is often known as the Chapman–Kolmogorov equation.

In target tracking, the conditional density that most algorithms compute or seek to approximate is (1.5)

$$p(S_k|y^k) = \frac{1}{p(y_k|y^{k-1})} p(y_k|S_k) \int_{S_{k-1}} p(S_k|S_{k-1}) p(S_{k-1}|y^{k-1}) dS_{k-1}.$$

The integral

$$\int_{S_{k-1}} p(S_k|S_{k-1}) p(S_{k-1}|y^{k-1}) dS_{k-1},$$

is the *Chapman–Kolmogorov* equation for the problem, where S_k is $\mathbf{X}$ in (A.2), (S_{k-1}, y^{k-1}) is $\mathbf{Y}$. Note that a simplification was made, where

$$p(S_k|S_{k-1}, y^{k-1}) = p(S_k|S_{k-1}).$$

A.2 Markov chains

Markov chains are a class of probability models in a discrete-time with the Markov property. In such a model, the knowledge of the current state is enough for predicting a future state.

Markov chains have a wide variety of applications in the real world. A Markov chain describes a system in a discrete time. The system can be in a number of states. At the next time, the system changes from the present state to a different state, including the present one. The changes of state are called transitions. The Markov property means that the system probabilistic description at the next transition time depends only on its present state, and not its past states.

Let $\{r_k\}_{i=0}^{\infty}$ be a sequence of random variables indexed over time $i, i = 0, 1, 2, \ldots$. $\{r_k\}_{i=0}^{\infty}$ is a stochastic process where $r_i, i = 0, 1, 2, \ldots$ represents the value of a random quantity at time i. For example, in maneuvering target tracking (see Chapter 3), r_k represents the mode of the target dynamics model, that is, the dynamics model in effect at time k. Let us say that after rearrangement and labeling of the values that can be taken by r_i, r_i takes values in $\{1, 2, \ldots, d\}$, and these are called the states of r_i.

$\{r_k\}_{i=0}^{\infty}$ is said to be a Markov chain if

$$p(r_k = i | r_{k-1} = j, r_{k-2} = j_{k-2}, \ldots, r_1 = j_1) = p(r_k = i | r_{k-1} = j) = \Gamma_{ij},$$

for all $k \geq 1$. The conditional probabilities Γ_{ij} are called the transition probabilities. In Chapter 3, they are introduced in the target dynamics model in Section 3.2.1.

The transition matrix $\Gamma = [\Gamma_{ij}]$ characterizes the Markov chain. Along with initial conditions $p(r_0)$, the stochastic process is completely defined:

$$p(r_k = i) = \sum_{j=1}^{d} p(r_k = i | r_{k-1} = j) p(r_{k-1} = j)$$

$$= \sum_{j=1}^{d} \Gamma_{ij} \, p(r_{k-1} = j).$$

For an introduction to the subject, see Ross (2003).

A.3 The delta function

The *delta function* $\delta(x)$ is sometimes called Dirac's delta function or the impulse symbol. The delta function can be viewed as the derivative of a step function $H(x)$,

$$\delta(x) = \frac{d H(x)}{dx}.$$

The main property of the delta function used in the book is

$$\int_x f(x)\delta(x - x^0)dx = f(x^0).$$

For an intuitive understanding of the delta function, one can think of a function which is zero everywhere except at the origin, where it is infinite and such that

$$\int_{-\infty}^{\infty} \delta(x)dx = 1.$$

No function has the above properties, but the delta function can be thought of as the limit of a sequence of functions so that the above two conditions are met.

In Chapter 2, the delta function is used in

$$p(x_k | y^k) \approx \sum_{i=1}^{n} w_k^i \delta(x_k - x_k^i).$$

A.4 Gaussian distribution theorems

The *Gaussian distribution* is the *normal distribution*. For a single variable, it is defined in terms of its mean μ and variance σ^2 as

$$N(x; \mu, \sigma^2) = \frac{1}{\sqrt{2\pi}\sigma} \exp\frac{-(x-\mu)^2}{2\sigma^2}.$$

A Gaussian distribution in an n-dimensional vector x is denoted through its mean vector μ and covariance matrix P in the following way:

$$N(x; \mu, P) = \frac{1}{|P|^{1/2}(2\pi)^{n/2}} \exp\frac{-1}{2}(x-\mu)^T P^{-1}(x-\mu).$$

The following theorems are special cases of the one-dimensional results that the product of Gaussians is another Gaussian, and the integral of a Gaussian is also another Gaussian.

Theorem 1

$$\frac{N(x_2; Hx_1, P_2)N(x_1; \mu_1, P_1)}{N(x_2; H\mu_1, P_3)} = N(x_1; \mu, P),$$

where

$$\mu = \mu_1 + K(x_2 - H\mu_1),$$
$$P = P_1 - KHP_1,$$
$$K = P_1 H^T (HP_1 H^T + P_2)^{-1}.$$

The method of proving the above theorem is relatively well known, being first shown in Ho and Lee (1964) and later appearing in a number of texts. The proof of the next theorem, which deals with the Chapman–Kolmogorov theorem, is given in Challa and Koks (2005).

Theorem 2

$$\int N(x_2; Fx_1, P_2)N(x_1; \mu_1, P_1)dx_1 = N(x_2; \mu, P),$$

where

$$\mu = F\mu_1,$$
$$P = FP_1 F^T + P_2.$$

Appendix B

Finite set statistics (FISST)

B.1 Introduction

The statistics of finitely varying random sets depends on the mathematics of finite sets. In this appendix, the finite set mathematics will be presented. This appendix is reconstructed principally from Mahler (2004b).

B.2 Random set model for target dynamics and sensors

Random set notation for target motion model is give by

$$\Gamma_{k+1} = \Phi_k(X_k, V_k) \cup B_k(X_k), \tag{B.1}$$

where $\Phi(.)$ represents the change of target dynamics from time $t = k$ to $t = k + 1$. $B(.)$ caters for the target birth process in the multiple-target case.

Similarly, the sensor model in random set notation is given by

$$\Sigma = T(X) \cup C(X), \tag{B.2}$$

where $T(.)$ defines the measurements originated from true targets while $C(.)$ accounts for clutter measurements.

The models (B.1) and (B.2) give rise to multi-target Markov transition densities and global likelihood for randomly varying set.

B.3 Belief-mass function of sensor model

The probability mass $p(S|x) = Pr(Z \in S)$ captures the statistical behavior of observation set Z. In random set domain, the statistics of Σ is characterized by its belief-mass function $\beta(S|X)$. The belief mass measure is given by

$$\beta(S|X) = \beta_{\Sigma|\Gamma}(S|X) = Pr(\Sigma \subseteq S). \tag{B.3}$$

This belief mass measure is the *total probability that all observations in a sensor (or multi sensor) scan* will be found in any region S. The belief-mass function in (B.3) provides the global likelihood density.

B.4 Belief-mass function of target motion model

The probability mass $p(S|x_k) = Pr(X_{k+1} \in S)$, which is the probability that the target state X_{k+1} will be found in the region S conditioned on previous state x_k, provides the statistical measure of the target dynamics. For a randomly varying finite length set Γ_{k+1}, the sufficient statistics is captured in the belief mass function given by

$$\beta_{\Gamma_{k+1}}(S|X_k) = Pr(\Gamma_{k+1} \subseteq S). \tag{B.4}$$

The Markov transition density is calculated from the belief-mass measure in (B.4).

B.5 Basics of FISST mathematics

Unlike probability mass, belief mass functions are non-additive measures. In general, if $S_1 \cap S_2 = \phi$, while joint probability mass $p(S_1 \cup S_2|x) = p(S_1|x) + p(S_2|x)$, the joint belief mass $\beta(S_1 \cup S_2|x) \geq \beta(S_1|x) + \beta(S_2|x)$. It is also noted in Mahler (2004b) that the belief-mass measure behaves like probability mass on certain abstract topological spaces. This additional property introduces some additional complexities in calculating densities from belief mass functions.

In general, the relation between the belief-mass measure of a certain randomly varying set (the probability of the random set being in the region S) is given by

$$\beta(S|B) = Pr(A \in S|B) \int_S f(A|B)\delta A, \tag{B.5}$$

where A denotes either the target dynamic set Γ_k or the sensor observation set Σ. The integration in (B.5) refers to the sum of densities or likelihoods of all possibilities suggested by the random set.

B.6 Set integral rule

The integration in (B.5) follows FISST "set integeral" rule. An illustrative example (following Mahler, 2004b), will be useful to demonstrate the "set integral."

Assuming a function $F(Y)$ is given for a finite set variable Y, $F(Y)$ can have the following forms:

$$F(\phi) = \text{probability that } Y = \phi,$$
$$F(\{y\}) = \text{likelihood that } Y = \{y\},$$
$$F(\{y_1, y_2\}) = \text{likelihood that } Y = \{y_1, y_2\},$$
$$\vdots$$
$$F(\{y_1, y_2, \ldots, y_j\}) = \text{likelihood that } Y = \{y_1, y_2, \ldots, y_j\}. \tag{B.6}$$

In general, this $F(Y)$ can be a likelihood $F(Z) = f(Z|X)$ or Markov density $F(X) = f_{k+1|k}(X|X_k)$ and the forms refer to the possible sets. The "set integral" suggests that

$$\int_S F(Y)\delta Y = F(\phi) + \sum_{j=1}^{\infty} \frac{1}{j!} \underbrace{\int_{S \times \cdots \times S}}_{j \text{ times}} F(\{y_1, y_2, \ldots, y_j\}) dy_1 \cdots dy_j$$

$$= F(\phi) + F_S(1) + F_S(2) + \cdots, \tag{B.7}$$

where

$$F_S(j) = \frac{1}{j!} \underbrace{\int}_{S \times \cdots \times S \; j \text{ times}} F(\{y_1, y_2, \ldots, y_j\}) dy_1 \cdots dy_j$$

denotes the total probability that Y contains j elements.

B.7 Set derivative rule

In (B.5), a procedure is suggested to build the belief mass measure of a finite set from its density. For building density or likelihood from belief-mass functions, an operation opposite to "set integral" is needed. This operation is termed as "set derivative" and also follows FISST rules.

If $Y = \{y_1, y_2, \ldots, y_m\}$, the set derivatives are defined as follows:

$$\frac{\delta \beta}{\delta y_j}(S) = \frac{\delta}{\delta y_j} \beta(S) = \lim_{\lambda E_y} \frac{\beta(S \cup E_y) - \beta(S)}{\lambda(E_z)},$$

$$\frac{\delta \beta}{\delta Y}(S) = \frac{\delta^m}{\delta y_1 \cdots \delta y_m} \beta(S) = \frac{\delta}{\delta y_1} \cdots \frac{\delta}{\delta y_m} \beta(S),$$

$$\frac{\delta \beta}{\delta \phi}(S) = \beta(S).$$

B.8 Calculating likelihoods and Markov densities

According to the "set derivative" and "set integral" rules, the belief-mass measure and densities are related to each other as

$$\beta(S) = \int_S \frac{\delta \beta}{\delta X}(\phi) \delta X, \tag{B.8}$$

$$F(X) = \left[\frac{\delta}{\delta X} \int_S F(Y) \delta Y \right]_{S=\phi}. \tag{B.9}$$

The expressions in (B.8) and (B.9) are the key to calculating multi-target likelihoods and Markov densities under random set formalism.

- The true likelihood $f(Z|X)$ is given by

$$f(Z|X) = \frac{\delta \beta}{\delta Z}(\phi|X). \tag{B.10}$$

- The true Markov density is given by

$$f_{k+1|k}(X_{k+1}|X_k) = \frac{\delta \beta_{k+1|k}}{\delta X_{k+1}}(\phi|X_k). \tag{B.11}$$

Based on the target dynamic model and sensor model, the densities can be calculated after constructing the belief-mass measure of the appropriate sets. After the likelihood and

Markov density are obtained, standard Bayesian recursion updates the target state in the usual manner.

B.9 Standard rules of FISST calculus

Like ordinary differential calculus, FISST calculus follows some rules. These are summarized below for reference (for details, see Mahler, 2004b, p. 31).

B.9.1 Sum rule

$$\frac{\delta}{\delta Z}[a_1 \beta_1(S) + a_2 \beta_2(S)] = a_1 \frac{\delta \beta_1}{\delta Z}(S) + a_2 \frac{\delta \beta_2}{\delta Z}(S),$$

$$\int [a_1 F_1(S) + a_2 F_2(S)] \delta Z = a_1 \int F_1(S) \delta Z + a_2 \int F_2(S) \delta Z.$$

B.9.2 Product rule

$$\frac{\delta}{\delta z}[\beta_1(S)\beta_2(S)] = \frac{\delta \beta_1}{\delta z}(S)\beta_2(S) + \beta_1(S)\frac{\delta \beta_2}{\delta z}(S),$$

$$\frac{\delta}{\delta Z}[\beta_1(S)\beta_2(S)] = \sum_{W \subseteq Z} \frac{\delta \beta_1}{\delta W}(S) \frac{\delta \beta_2}{\delta (Z - W)}(S).$$

B.9.3 Constant rule

$$\frac{\delta}{\delta Z} K = 0.$$

B.9.4 Chain rule

$$\frac{\delta}{\delta z} f(\beta(S)) = \frac{df}{dx}(\beta(S))\frac{\delta \beta}{\delta z}(S),$$

$$\frac{\delta}{\delta z} f(\beta_1(S) \cdots \beta_n(S)) = \sum_{i=1}^{n} \frac{\partial f}{\partial x_i}(\beta_1(S) \cdots \beta_n(S))\frac{\delta \beta_i}{\delta z}(S).$$

B.9.5 Power rule

Let $Z = \{z_1, \ldots, z_k\}$ and let $n \geq 0$ be an integer. Let $p(S)$ be a probability mass function with density function $f_p(z)$. Then,

$$\frac{\delta}{\delta Z} p(S)^n = \begin{cases} \frac{n!}{(n-k)!} p(S)^{n-k} f_p(z_1) \ldots f_p(z_k), & \text{if } k \leq n, \\ 0, & \text{if } k > n. \end{cases}$$

Appendix C

Pseudo-functions in object tracking

C.1 Kalman filter prediction

$$\left[\hat{\mathbf{x}}_{k|k-1}, \mathbf{P}_{k|k-1}\right] = \text{KF}_\text{P} \left[\hat{\mathbf{x}}_{k-1|k-1}, \mathbf{P}_{k-1|k-1}, \mathbf{F}, \mathbf{Q}\right],$$

which is defined by

$$\hat{\mathbf{x}}_{k|k-1} = \mathbf{F}\hat{\mathbf{x}}_{k-1|k-1},$$
$$\mathbf{P}_{k|k-1} = \mathbf{F}\mathbf{P}_{k-1|k-1}\mathbf{F}^T + \mathbf{Q}.$$

C.2 Measurement prediction

$$\left[\hat{\mathbf{y}}_{k|k-1}, \mathbf{S}_k\right] = \text{MP} \left[\hat{\mathbf{x}}_{k|k-1}, \mathbf{P}_{k|k-1}, \mathbf{H}, \mathbf{R}\right].$$

The measurement prediction is defined by

$$\hat{\mathbf{y}}_{k|k-1} = \mathbf{H}\hat{\mathbf{x}}_{k|k-1},$$
$$\mathbf{S}_k = \mathbf{H}\mathbf{P}_{k|k-1}\mathbf{H}^T + \mathbf{R}.$$

C.3 Kalman filter estimation

$$\left[\hat{\mathbf{x}}_{k|k}, \mathbf{P}_{k|k}\right] = \text{KF}_\text{E} \left[\mathbf{y}, \hat{\mathbf{x}}_{k|k-1}, \mathbf{P}_{k|k-1}, \mathbf{H}, \mathbf{R}\right],$$
$$\left[\hat{\mathbf{y}}_{k|k-1}, \mathbf{S}_k\right] = \text{MP} \left[\hat{\mathbf{x}}_{k|k-1}, \mathbf{P}_{k|k-1}, \mathbf{H}, \mathbf{R}\right],$$
$$\mathbf{K}_k = \mathbf{P}_{k|k-1}\mathbf{H}^T\mathbf{S}_k^{-1},$$
$$\hat{\mathbf{x}}_{k|k} = \hat{\mathbf{x}}_{k|k-1} + \mathbf{K}_k \left(\mathbf{y} - \hat{\mathbf{y}}_{k|k-1}\right),$$
$$\mathbf{P}_{k|k} = (\mathbf{I} - \mathbf{K}_k\mathbf{H})\mathbf{P}_{k|k-1},$$

where $\mathbf{I}$ denotes the identity matrix.

C.4 Gaussian mixture

$$\left[\hat{\mathbf{x}}_{k|k}, P_{k|k}\right] = \text{GMix}\left[\{\hat{\mathbf{x}}_{k|k}(i), P_{k|k}(i), \mu_{k|k}(i)\}_i\right],$$

$$\hat{\mathbf{x}}_{k|k} = \sum_{i=1}^{d} \hat{\mathbf{x}}_{k|k}^i \mu_{k|k}(i),$$

$$P_{k|k} = \sum_{i=1}^{d} \mu_{k|k}(i)\{P_{k|k}^i + [\hat{\mathbf{x}}_{k|k}^i - \hat{\mathbf{x}}_{k|k}][\hat{\mathbf{x}}_{k|k}^i - \hat{\mathbf{x}}_{k|k}]^T\}.$$

C.5 Single-object tracking data association

$$\left[p(\chi_k|\mathbf{Y}^k), \{\boldsymbol{\beta}_k(i)\}_{i=0}^{m_k}\right] = \text{ISTDA}\left[p(\chi_k|\mathbf{Y}^{k-1}), \{p_k(i)\}_{i=1}^{m_k}\right],$$

where

$$\boldsymbol{\beta}_k(i) \overset{\triangle}{=} p(\theta_k(i)|\chi_k, \mathbf{Y}^k) = \frac{1}{\Lambda_k}\begin{cases} 1 - P_D P_G, & i = 0, \\ P_D P_G \dfrac{p_k(i)}{\rho_k(i)}, & i > 0, \end{cases}$$

$$p(\chi_k|\mathbf{Y}^k) = \frac{\Lambda_k p(\chi_k|\mathbf{Y}^{k-1})}{1 - (1 - \Lambda_k)p(\chi_k|\mathbf{Y}^{k-1})},$$

$$\Lambda_k = 1 - P_D P_G + P_D P_G \sum_{i=1}^{m_k} \frac{p_k(i)}{\rho_k(i)},$$

where Λ_k is the measurement likelihood ratio at time k.

C.6 Multiple-object tracking data association

$$\left[\{p(\chi_k^\tau|\mathbf{Y}^k), \{\boldsymbol{\beta}_k^\tau(i)\}_{i\geq0}\}_\tau\right] = \text{JMTDA}\left[\{p(\chi_k^\tau|\mathbf{Y}^{k-1}), \{p_k^\tau(i)\}_{i>0}\}_\tau\right],$$

$$p(\varepsilon|\mathbf{Y}^k) = c_k^{-1} \prod_{\tau \in T_0(\varepsilon)} \left(1 - P_D^\tau P_G^\tau P\{\chi_k^\tau|\mathbf{Y}^{k-1}\}\right)$$

$$\times \prod_{\tau \in T_1(\varepsilon)} \left(P_D^\tau P_G^\tau P\{\chi_k^\tau|\mathbf{Y}^{k-1}\}\frac{p_k^\tau(i(\tau, \varepsilon))}{\rho_k(i(\tau, \varepsilon))}\right),$$

where the normalization constant c_k is calculated by utilizing the fact that feasible joint events are mutually exclusive and that they form an exhaustive set

$$\sum_\varepsilon p(\varepsilon|\mathbf{Y}^k) = 1.$$

Note that product operation on an empty set equals 1.

$$p\big(\theta_k^\tau(0)|\mathbf{Y}^k\big) = \sum_{\varepsilon \in \Xi(\tau,0)} p\big(\varepsilon|\mathbf{Y}^k\big),$$

$$p\big(\chi_k^\tau, \theta_k^\tau(i)|\mathbf{Y}^k\big) = \sum_{\varepsilon \in \Xi(\tau,i)} p\left(\varepsilon|\mathbf{Y}^k\right),$$

$$p\big(\chi_k^\tau, \theta_k^\tau(0)|\mathbf{Y}^k\big) = \frac{\big(1 - P_D^\tau P_G^\tau\big) p\left(\chi_k^\tau|\mathbf{Y}^{k-1}\right)}{1 - P_D^\tau P_G^\tau\, p\left(\chi_k^\tau|\mathbf{Y}^{k-1}\right)} p\big(\theta_k^\tau(0)|\mathbf{Y}^k\big),$$

$$p\big(\chi_k^\tau|\mathbf{Y}^k\big) = \sum_{i=0}^{m_k} p\big(\chi_k^\tau, \theta_k^\tau(i)|\mathbf{Y}^k\big),$$

$$\beta_k^\tau(i) \triangleq p\big(\theta_k^\tau(i)|\chi_k^\tau, \mathbf{Y}^k\big) = \frac{p\left(\chi_k^\tau, \theta_k^\tau(i)|\mathbf{Y}^k\right)}{p\left(\chi_k^\tau|\mathbf{Y}^k\right)}, \quad i \geq 0.$$

C.7 IMM mixing step

$$\left[\big\{\mu_{k|k-1}(i), \hat{\mathbf{x}}_{k|k-1}^i, \mathbf{P}_{k|k-1}^i\big\}_i\right]$$

$$= \mathrm{IMM}_{\mathrm{MP}}\big[\big\{\mu_{k-1|k-1}(i), \hat{\mathbf{x}}_{k-1|k-1}^i, \mathbf{P}_{k-1|k-1}^i, \mathbf{F}_i, \mathbf{Q}_i\big\}_i, \mathbf{\Gamma}\big],$$

$$\mu_{k|k-1}(i) = p(r_k = i|\mathbf{y}^{k-1})$$

$$= \sum_{j=1}^{d} p(r_k = i|r_{k-1} = j, \mathbf{y}^{k-1}) p(r_{k-1} = j|\mathbf{y}^{k-1})$$

$$= \sum_{j=1}^{d} \Gamma_{ji}\, p(r_{k-1} = j|\mathbf{y}^{k-1})$$

$$= \sum_{j=1}^{d} \Gamma_{ji}\, \mu_{k-1|k-1}(j),$$

$$\hat{\mathbf{x}}_{k|k-1}^i = \mathbf{F}_i \hat{\mathbf{x}}_{k-1|k-1}^{0i} + \mathbf{u}_i,$$

$$\mathbf{P}_{k|k-1}^i = F_i \mathbf{P}_{k-1|k-1}^{0i} \mathbf{F}_i^T + \mathbf{Q}_i.$$

References

Ackerson, G. A. and Fu, K. S. (1970) On state estimation in switching environments, *IEEE Transactions on Automatic Control*, **15**(1), 10–17.

Alspach, D. L. and Sorenson, H. W. (1971) Nonlinear Bayesian estimation using Gaussian sum approximations, *IEEE Transactions on Automatic Control*, **17**(4), 439–448.

Anderson, B. D. O. and Moore, J. B. (1979) *Optimal Filtering*. New Jersey: Prentice-Hall.

Bar-Shalom, Y. (1978) Tracking methods in a multitarget environment, *IEEE Transactions on Automatic Control*, **23**(4), 618–626.

Bar-Shalom, Y. (2000) Update with out-of-sequence measurements in tracking: exact solution, in *Proc. SPIE: Signal and Data Processing of Small Targets 2000*, O. E. Drummond (ed.), Vol. 4048, pp. 541–556.

Bar-Shalom, Y. and Birmiwal, K. (1982) Variable dimension filter for maneuvering target tracking, *IEEE Transactions on Aerospace and Electronic Systems*, **18**(5), 621–629.

Bar-Shalom, Y. and Fortmann, T. E. (1988) *Tracking and Data Association*. New York: Academic Press.

Bar-Shalom, Y. and Li, X. R. (1993) *Estimation and Tracking: Principles, Techniques, and Software*. MA: Artech House.

Bar-Shalom, Y. and Li, X. R. (1995) *Multitarget-Multisensor Tracking: Principles and Techniques*. Storrs, CT: YBS Publishing.

Bar-Shalom, Y. and Tse, E. (1975) Tracking in a cluttered environment with probabilistic data association, *Automatica*, **11**(5), 451–460.

Bar-Shalom, Y., Challa, S. and Blom, H. A. P. (2005) IMM estimator versus optimal estimator for hybrid systems, *IEEE Transactions on Aerospace and Electronic Systems*, **41**(3), 986–991.

Bar-Shalom, Y., Chang, K. C. and Blom, H. A. P. (1989) Automatic track formation in clutter with a recursive algorithm, in *Proc. 28th Conference on Decision and Control*, Florida, December, pp. 1402–1408.

Bar-Shalom, Y., Chang, K. C. and Blom, H. A. P. (1990) Automatic track formation in clutter with a recursive algorithm, in *Multitarget-Multisensor Tracking: Advanced Applications*, Y. Bar-Shalom (ed.). MA: Artech House, pp. 25–42.

Bar-Shalom, Y. (ed.) (1990) *Multitarget-Multisensor Tracking: Advanced Applications*. MA: Artech House.

Bar-Shalom, Y., Li. X. R. and Kirubarajan, T. (2001) *Estimation with Applications to Tracking and Navigation: Theory Algorithms and Software*. New York: Wiley.

Bayes, T. (1764) Essay towards solving a problem in the doctrine of chances, *Philosophical Transactions of the Royal Society of London*, **53**, 370–418.

Berger, J. O. (1985) *Statistical Decision Theory and Bayesian Analysis*. New York: Springer.

Bergman, N. (2001) Posterior Cramér–Rao bounds for sequential estimation, in *Sequential Monte Carlo Methods in Practice*, A. Doucet, N. de Freitas and N. Gordon (eds.). New York: Springer, pp. 321–338.

Bernardo, J. M. and Smith, A. F. M. (1994) *Bayesian Theory*. New York: Wiley.

Berry, D. A. (1996) *Statistics: A Bayesian Perspective*. Belmont: Duxbury.

Blackman, S. (1986) *Multiple Target Tracking with Radar Applications*. MA: Artech House.

Blackman, S. and Popoli, R. (1999) *Design and Analysis of Modern Tracking Systems*. MA: Artech House.

Blair, W. D. and Watson, G. A. (1992) Interacting multiple bias model algorithm with application to tracking maneuvering targets, in *IEEE Proc. 31st Conference on Decision and Control*, Tucson, AZ, December, pp. 3790–3795.

Blair, W. D. and Watson, G. A. (1994) IMM Algorithm for solution to benchmark problem for tracking maneuvering targets, in *Proc. SPIE Symposium on Acquisition, Tracking and Pointing*, Orlando, FL.

Blair, W. D., Watson G. A. and Hoffman, S. A. (1993) Second order interacting multiple model algorithm for tracking maneuvering targets, in *Proc. SPIE: Signal and Data Processing of Small Targets 1993*, O. E. Drummond (ed.), vol. 1954, pp. 518–529.

Blair, W. D., Watson, G. A. and Rice, T. R. (1991) Interacting multiple model filter for tracking maneuvering targets in spherical coordinates, in *IEEE Proceedings of SOUTHEASTCON' 91*, Williamsburg, VA, vol. 2, 1991, pp. 1055–1059.

Blom, H. A. P. (1984a) A sophisticated tracking algorithm for ATC surveillance data, in *Proc. International Radar Conference*, Paris, May.

Blom, H. A. P. (1984b) An efficient filter for abruptly changing systems, in *Proc. 23rd IEEE Conference on Decision and Control*, Las Vegas, NV, December, pp. 656–658.

Blom, H. A. P. and Bar-Shalom, Y. (1988) The interacting multiple model algorithm for systems with Markovian switching coefficients, *IEEE Proceedings on Auto Control AC*, **33**(8), 780–783.

Blom, H. A. P. and Bloem, E. (2002) Combining IMM and JPDA for tracking multiple maneuvering targets in clutter, in *Proc. 5th International Conference on Information Fusion, Fusion 2002*, Annapolis, MD, July, vol. 1, pp. 705–712.

Bochardt, O., Calhoun, R., Uhlmann, J. K. and Julier, S. J. (2006) Generalized information representation and compression using covariance union, in *Proc. 9th International Conference on Information Fusion, Fusion 2006*, Florence, Italy, July, pp. 1–7.

Boult, T. (1998) Frame-rate multi-body tracking for surveillance, in *Proc. DARPA Image Understanding Workshop*, Monterey, CA, pp. 305–308.

Box, G. P. A. and Tiao, G. C. (1973) *Bayesian Inference in Statistical Analysis*. New York: Wiley.

Bucy, R. and Senne, K. (1971) Digital synthesis of non-linear filters, *Automatica*, **7**, 287–298.

Chakravorty, R. and Challa, S. (2004) Fixed lag smoothing technique for track maintenance in clutter, in *Proc. Intelligent Sensors Sensor Networks and Information Processing*, pp. 119–124.

Chakravorty, R. and Challa, S. (2005) Smoothing framework for automatic track initiation in clutter, in *Proc. 8th International Conference on Information Fusion*, Philadelphia, PA, July, pp. 54–61.

Chakravorty, R. and Challa, S. (2006) Augmented state integrated probabilistic data association smoothing for multiple target tracking in clutter, *Journal of Advances in Information Fusion*, **1**(1), 63–74.

Chakravorty, R. and Challa, S. (2009) Multitarget tracking algorithm: joint IPDA and Gaussian mixture PHD filter, in *Proc. 12th International Conference on Information Fusion, Fusion 2009*, Seattle, WA, July, pp. 316–323.

Challa, S. (1998) Nonlinear state estimation and filtering with applications to target tracking problems. Unpublished PhD thesis, Queenland University of Technology.

Challa, S., Bar-Shalom, Y. and Krishnamurthy, V. (2000) Nonlinear filtering via generalized Edgeworth series and Gauss–Hermite quadrature, *IEEE Transactions on Signal Processing*, **48**(6), 1816–1820.

Challa, S., Evans, R. J. and Mušicki, D. (2002c) Target tracking – a Bayesian perspective, in *Proc. 14th International Conference on Digital Signal Processing*, Vol. 1, pp. 437–440.

Challa, S., Evans, R. and Wang, X. (2002a) A Bayesian solution to the OOSM problem, Submitted to *IEEE Transactions on Aerospace and Electronic Systems*.

Challa, S., Evans, R. and Wang, X. (2002b) Target tracking in clutter using time-delayed out-of-sequence measurements, in *Proc. Defence Applications of Signal Processing (DASP)*, July.

Challa, S., Evans, R. J. and Wang, X. (2003) A Bayesian solution and its approximations to out-of-sequence measurement problems, *Information Fusion*, **4**(3), 185–199.

Challa, S. and Koks, K. (2004) Bayesian and Dempster–Shafer fusion, in *Sadhana*, Vol. 29, Part 2, pp. 145–176.

Challa, S. and Koks, D. (2005) An introduction to Bayesian and Dempster–Shafer data fusion, DSTO Technical Report 1436.

Challa, S., Vo, B. and Wang, X. (2002c) Bayesian approaches to track existence – IPDA and random sets, in *Proc. 5th International Conference on Information Fusion, Fusion 2002*, Annapolis, MD, July, Vol. 2, pp. 1228–1235.

Chan, Y. T., Hu, A. G. and Plant, J. B. (1979) A Kalman filter based tracking scheme with input estimation, *IEEE Transactions on Aerospace and Electronic Systems*, **15**(2), 237–244.

Collins, R. T., Lipton, A. J., Kanade, T., Fujiyoshi, H., Duggins, D., Tsin, Y., Tolliver, D., Enomoto, N. and Hasegawa, O. (2000) A system for video surveillance and monitoring, Carnegie Mellon University, Pittsburgh, PA, Technical Report, CMU-RI-TR-00-12.

Cover, T. M. and Thomas, J. A. (2006) *Elements of Information Theory*. New York: Wiley.

Cutaia, N. J. and O'Sullivan, J. A. (1995) Identification of maneuvering aircraft using class dependent kinematic models, ESSRL-95-13, May.

Daum, F. E. (1986) Exact finite-dimensional nonlinear filters, *IEEE Transactions on Automatic Control*, **31**(7), 616–622.

Davies, A. C., Yin, J. H. and Velastin, S. A. (1995) Crowd monitoring using image processing, *Electronic and Communication Engineering Journal*, **7**(1), 37–47.

de Laplace, P. S. (1812) *Théorie analytique des probabilités*, Paris: Courcier Imprimeur.

Erdinc, O., Willett, P. and Bar-Shalom, Y. (2005) Probability hypothesis density filter for multitarget multisensor tracking, in *Proc. 8th International Conference on Information Fusion*, Philadelphia, PA, July, pp. 146–153.

Farina, A. and Studer, F. A. (1985) *Radar Data Processing, Vol. I: Introduction and Tracking, Vol. II: Advanced Topics and Applications*. Hertfordshire, England: Research Studies Press and New York: Wiley.

Farooq, M., Bruder, S., Quach, T. and Lim, S. S. (1992) Adaptive filtering techniques for manoeuvring targets, in *Proc. 34th Midwest Symposium on Circuits and Systems*, vol. 1, Monterey, CA, USA, pp. 31–34.

Gelman, A., Carlin, J. B., Stern, H. S. and Rubin, D. B. (1995) *Bayesian Data Analysis*. London: Chapman and Hall.

Gholson, N. H. and Moose, R. L. (1977) Maneuvering target tracking using adaptive state estimation, *IEEE Transactions on Aerospace and Electronic Systems*, **13**(3), 310–317.

Goodman, I. R., Mahler, R. and Nhuyen, H. T. (1997) *Mathematics of Data Fusion*. Amsterdam: Kluwer.

Haritaoglu, I., Harwood, D. and Davis, L. S. (2004) W^4: Real-time surveillance of people and their activities, *IEEE Transactions on Pattern Analysis and Machine Intelligence*, **22**(8), 809–830.

Helmick, R. E., Blair, W. D. and Hoffman, S. A. (1993) Interacting multiple-model approach to fixed-interval smoothing, in *Proc. 32nd IEEE Conference on Decision and Control*, San Antonio, TX, December, pp. 3052–3057.

Helmick, R. E., Blair, W. D. and Hoffman, S. A. (1995) Fixed-interval smoothing for Markovian switching systems, *IEEE Transactions on Information Theory*, **41**(6), 1845–1855.

Helmick, R. E., Blair, W. D. and Hoffman, S. A. (1996) One-step fixed-lag smoothers for Markovian switching systems, *IEEE Transactions on Automatic Control*, **41**(7), 1051–1056.

Hernandez, M., Farina, A. and Ristic, B. (2006) PCRLB for tracking in cluttered environments: measurement sequence conditioning approach, *IEEE Transactions on Aerospace and Electronic Systems*, **42**(2), 680–704.

Hernandez, M., Marrs, A. D., Gordon, N. J., Maskell, S. R. and Reed, C. M. (2002) Cramér–Rao bounds for non-linear filtering with measurement origin uncertainty, in *Proc. 5th International Converence on Information Fusion*, Fusion 2002, Annapolis, MD, July, Vol. 2, pp. 18–25.

Hernandez, M., Ristic, B. and Farina, A. (2005) A performance bound for manoeuvring target tracking using best-fitting Gaussian distributions, in *Proc. 8th International Conference on Information Fusion*, Philadelphia, PA, July, pp. 1–8.

Hilton, R. D., Martin, D. A. and Blair, W. D. (1993) Tracking with time delayed data in multisensor system, Technical Report, NSWCDD/TR-93/351, Dahlgren, VA.

Ho, Y. C. and Lee, R. C. K. (1964) A Bayesian approach to problems in stochastic estimation and control, *IEEE Transactions on Automatic Control*, **9**(4), 333–339.

Hoffman, J. R. and Mahler, R. (2002) Multitarget miss distance and its applications, in *Proc. 5th International Conference on Information Fusion*, Fusion 2002, Annapolis, MD, July, Vol. 2, pp. 149–155.

Hogg, R. V. and Craig, A. T. (1995) *Introduction to Mathematical Statistics*. New Jersey: Prentice-Hall.

Houles, A. and Bar-Shalom, Y. (1989) Multisensor tracking of a maneuvring target in clutter, *IEEE Transactions on Aerospace and Electronic Systems*, **25**(2), 176–189.

Hu, W., Tan, T., Wang, L. and Maybank, S. (2004) A survey on visual surveillance of object motion and behaviors, *IEEE Transactions on Systems, Man, and Cybernetics – Part C: Applications and Reviews*, **34**(3), pp. 334–352.

Hwang, I., Balakrishnan, H., Roy, K. and Tomlin, C. (2004) Multiple-target tracking and identity management in clutter, with application to aircraft tracking, in *Proc. 2004 American Control Conference*, Boston, MA, June 30–July 2, pp. 3422–3428.

Ito, K. and Xiong, K. (2000) Gaussian filters for nonlinear filtering problems, *IEEE Transactions on Automatic Control*, **45**(5), 910–927.

Jaffer, A. G. and Gupta, S. C. (1971) Recursive Bayesian estimation with uncertain observation, *IEEE Transactions on Information Theory*, **17**, 614–616.

Jazwinski, A. H. (1970) *Stochastic Processes and Filtering Theory*. New York: Academic Press.

Julier, S. J., Uhlmann, J. K. and Durrant-Whyte, H. (2000) A new method for the nonlinear transformation of means and covariances in filters and estimators, *IEEE Transactions on Automatic Control*, **45**(3), 477–482.

Kalman, R. E. (1960) A new approach to linear filtering and prediction problems, *Transactions of the ASME, Journal of Basic Engineering*, March, 35–45.

Kingman, J. F. C. (1992) *Poisson Processes*. Oxford: Oxford University Press.

Kramer, S. C. and Sorenson, H. W. (1988) Bayesian parameter estimation, *IEEE Transactions on Automatic Control*, **33**(2), 217–222.

Krause, S. S. (1995) *Avoiding Mid-air Collisions*. TAB Books.

Kudryavtsev, L. D. (2001) Implicit function, in *Encyclopedia of Mathematics*, M. Hazewinkel (ed.). New York: Springer.

Lerner, U. N. (2002) *Hybrid Bayesian Networks for Reasoning about Complex Systems*. Stanford, CA: Stanford University Press.

Lerro, D. and Bar-Shalom, Y. (1990) Automated tracking with target amplitude information, in *Proc. 1990 American Control Conference*, San Diego, CA, May, pp. 2875–2880.

Lerro, D. and Bar-Shalom, Y. (1993) Interactive multiple model tracking with target amplitude feature, *IEEE Transactions on Aerospace and Electronic Systems*, **29**(2), 494–509.

Li, X. R. (1994) Multiple-model estimation with variable structure: some theoretical considerations, in *Proc. 33rd IEEE Conference on Decision and Control*, Orlando, FL, December, pp. 1199–1204.

Li, X. R. (2000) Multiple-model estimation with variable structure – part II: model-set adaptation, *IEEE Transactions on Automatic Control*, **45**(11), 2047–2060.

Li, X. R. and Bar-Shalom, Y. (1992) Mode-set adaptation in multiple-model estimators for hybrid systems, in *Proc. 1992 American Control Conference*, Chicago, IL, June, pp. 1794–1799.

Li, X. R. and Bar-Shalom, Y. (1996) Multiple-model estimation with variable structure, *IEEE Transactions on Automatic Control*, **41**(4), 478–493.

Li, X. R. and Bar-Shalom, Y. (1997) Intelligent PDAF: refinement of IPDAF for tracking in clutter, in *Proc. 29th SSST*, March, pp. 133–137.

Li, X. R. and He, C. (1999) 2M-PDAF: an integrated two-model probabilistic data association filter, in *Proc. SPIE: Signal and Data Processing of Small Targets 1999*, O. E. Drummond (ed.), pp. 384–395.

Li, X. R. and Zhang, Y. (2000) Multiple-model estimation with variable structure–part V: likely-model set algorithm, *IEEE Transactions on Aerospace and Electronic Systems*, **36**(2), pp. 448–466.

Li, X. R., Zhi, X. and Zhang, Y. (1999) Multiple-model estimation with variable structure part III: model-group switching algorithm, *IEEE Transactions on Aerospace and Electronic Systems*, **35**(1), 225–240.

Lin, L., Bar-Shalom, Y. and Kirubarajan, T. (2006) Track labeling and PHD filter for multitarget tracking. *IEEE Transactions on Aerospace and Electronic Systems*, **42**(3), 778–795.

Lindley, D. V. (1972) *Bayesian Statistics, A Review*. Philadelphia: Society for Industrial and Applied Mathematics.

Lipton, A. J., Fujiyoshi, H. and Patil, R. S. (1998) Moving target classification and tracking from real-time video, in *Proc. 4th IEEE Workshop on Applications of Computer Vision*, pp. 8–14.

Mahalanabis, A. K., Zhou, B. and Bose, N. K. (1990) Improved multi-target tracking in clutter by PDA smoothing, *IEEE Transactions on Aerospace and Electronic Systems*, **26**(1), 113–121.

Mahler, R. (1997) Multisensor-multitarget statistics, in *A Unified Approach to Data Fusion – Proceedings of 7th Joint Data Fusion Symposium*, F. A. Sadjadi (ed.), pp. 154–174.

Mahler, R. (2000) Approximate multisensor-multitarget joint detection, *IEEE Transactions on Aerospace and Electronic Systems*, to appear.

Mahler, R. (2003a) Multitarget Bayes filtering via first-order multitarget moments, *IEEE Transactions on Aerospace and Electronic Systems*, **39**(4), 1152–1178.

Mahler, R. (2003b) Objective functions for Bayesian control-theoretic sensor management, I: Multitarget first-moment approximation, in *Proc. IEEE Conference on Aerospace and Electronics Systems*, vol. 4, pp. 1905–1923.

Mahler, R. (2004a) "Statistics 101" for multisensor, multitarget data fusion, *IEEE Magazine of Aerospace and Electronic Systems*, **19**(1), 53–64.

Mahler, R. (2004b) An introduction to multisource-multitarget statistics and its applications, Technical report, Lockheed Martin.

Mahler, R. (2007) PHD filters of higher order in target number, *IEEE Transactions on Aerospace and Electronic Systems*, **43**(4), 1523–1543.

Mallick, M., Coraluppi, S. and Bar-Shalom, Y. (2001b) Comparison of out-of-sequence measurement algorithms in multi-platform target tracking, in *Proc. 4th International Conference on Information Fusion*, Fusion 2001, Montreal, Quebec, August, Vol. II, pp. ThB1-11–18.

Mallick, M., Coraluppi, S. and Carthel, C. (2001a) Advances in asynchronous and decentralized estimation, in *Proc. 2001 IEEE Aerospace Conference*, Big Sky, MT, March.

Marcus, G. D. (1979) Tracking with measurements of uncertain origin and random arrival times. Unpublished thesis, University of Connecticut.

Mazor, E., Averbuch, A., Bar-Shalom, Y. and Dayan, J. (1998) Interacting multiple model methods in target tracking: a survey, *IEEE Transactions on Aerospace and Electronic Systems*, **34**(1), 103–123.

McGinnity, S. and Irwin, G. (2001) Manoevring target tracking using a multiple-model bootstrap filter, in *Sequential Monte Carlo Methods in Practice*, A. Doucet, N. de Freitas and N. Gordon (eds.), New York: Springer, pp. 479–496.

Moose, R. L. (1975) An adaptive state estimation solution to the maneuvering target tracking problem, *IEEE Transactions on Automatic Control*, **20**, 359–362.

Moose, R. L., Vanlandingham, H. F. and McCabe, D. H. (1979) Modeling and estimation for tracking maneuvering targets, *IEEE Transactions on Aerospace and Electronic Systems*, **15**(3), 448–456.

Mori, S., Chong, C. Y., Tse, E. and Wishner, R. P. (1986) Tracking and classifying multiple targets without a priori identification, *IEEE Transactions on Automatic Control*, **31**(5), 401–408.

Munir, A. and Atherton, D. P. (1994) Maneuvering target tracking using an adaptive interacting multiple model algorithm, in *Proc. 1994 American Control Conference*, vol. 2, Baltimore, MD, pp. 1324–1328.

Mušicki, D. (1994) Automatic tracking of maneuvering targets in clutter using IPDA. Unpublished PhD dissertation, University of Newcastle, New South Wales, Australia.

Mušicki, D. and Evans, R. J. (1995) Integrated probabilistic data association – finite resolution, *Automatica*, **31**(4), pp. 559–570.

Mušicki, D. and Evans, R. J. (2002) Joint integrated probabilistic data association JIPDA, in *Proc. 5th International Conference on Information Fusion, Fusion 2002*, Annapolis, MD, July, pp. 1120–1125.

Mušicki, D. and Evans, R. J. (2004a) Clutter map information for data association and track initialization, *IEEE Transactions on Aerospace and Electronic Systems*, **40**(2), 387–398.

Mušicki, D. and Evans, R. J. (2004b) Joint integrated probabilistic data association – JIPDA, *IEEE Transactions on Aerospace and Electronic Systems*, **40**(3), 1093–1099.

Mušicki, D. and Evans, R. J. (2008) Multi-scan multi-target tracking in clutter with ITS, *IEEE Transactions on Aerospace and Electronic Systems*, to appear.

Mušicki, D. and La Scala, B. F. (2005) Limits of linear multitarget tracking, in *Proc. 8th International Conference on Information Fusion, Fusion 2005*, Philadelphia, PA, July, pp. 205–210.

Mušicki, D. and La Scala, B. F. (2008) Multi-target tracking in clutter without measurement assignment, *IEEE Transactions on Aerospace and Electronic Systems*, **44**(3), 877–896.

Mušicki, D. and Morelande, M. (2004) Gate volume estimation for target tracking, in *Proc. 7th International Conference on Information Fusion, Fusion 2004*, Stockholm, Sweden, June 28–July 1, pp. 455–462.

Mušicki, D. and Suvorova, S. (2004) Target tracking initiation comparison and optimisation, in *Proc. 7th International Conference on Information Fusion, Fusion 2004*, Stockholm, Sweden, June 28–July 1, pp. 28–32.

Mušicki, D. and Suvorova, S. (2008) Tracking in clutter using IMM-IPDA based algorithms, *IEEE Transactions on Aerospace and Electronic Systems*, **44**(1), 111–126.

Mušicki, D. and Wang, X. (2004) Reliability of PDA based target tracking in clutter, in *Proc. 7th International Conference on Information Fusion, Fusion 2004*, Stockholm, Sweden, June 28–July 1, pp. 1257–1262.

Mušicki, D., Challa, S. and Suvorova, S. (2004a) Automatic track initiation of maneuvering target in clutter, in *Asian Control Conference, ASCC 2004*, Melbourne, Australia, July, pp. 1008–1014.

Mušicki, D., Evans, R. J. and La Scala, B. F. (2003) Integrated tracking splitting suite of target tracking filters, in *Proc. 6th International Conference on Information Fusion, Fusion 2003*, Cairns, Australia, July, pp. 1039–1047.

Mušicki, D., Evans, R. J. and Stanković, S. (1994) Integrated probabilistic data association, *IEEE Transactions on Automatic Control*, **39**(6), 1237–1241.

Mušicki, D., La Scala, B. F. and Evans, R. J. (2004b) Integrated track splitting filter for manoeuvring targets, in *Proc. 7th International Conference on Information Fusion, Fusion 2004*, Stockholm, Sweden, June 28–July 1, pp. 146–152.

Mušicki, D., La Scala, B. F. and Evans, R. J. (2007) The integrated track splitting filter – efficient multi-scan single target tracking in clutter, *IEEE Transactions on Aerospace and Electronic Systems*, **43**(4), 1409–1425.

Mušicki, D., Mallick, M., La Scala, B. F., Strange, S. and Evans, R. (2005a) LMITS as an efficient MHT, in *Proc. SPIE: Signal and Data Processing of Small Targets 2005*, O. E. Drummond (ed.), pp. OVI–OVI2.

Mušicki, D., Suvorova, S., Morelande, M. and Moran, W. (2005b) Clutter map and target tracking, in *Proc. 8th International Conference on Information Fusion, Fusion 2005*, Philadelphia, PA, July, pp. 69–76.

Olson T. and Brill, F. (1997) Moving object detection and event recognition algorithms for smart cameras, in *Proc. DARPA Image Understanding Workshop*, pp. 159–175.

Panta, K., Vo, B. and Singh, S. (2005) Improved PHD filter for multi-target tracking, in *Proc. International Conference on Intelligent Sensing and Information Processing*, pp. 213–218.

Rapoport, I. and Oshman. Y. (2004) Recursive Weiss–Weinstein lower bounds for discrete-time nonlinear filtering, in *Proc. 43rd IEEE Conference on Decision and Control*, Atlantis, Paradise Island, Bahamas, December, pp. 2662–2667.

Rasmussen, C. and Hager, G. D. (2001) Probabilistic data association methods for tracking complex visual objects, *IEEE Transactions on Pattern Analysis and Machine Intelligence*, **23**(6), 560–576.

Ravn, O., Larsen, T. D., Andersen, N. A. and Poulsen, N. K. (1998) Incorporation of time delayed measurements in a discrete time Kalman filter, in *Proc. 37th IEEE Conference on Decision and Control*, Tampa, FL, December, pp. 3972–3977.

Regazzoni, C. S. and Tesei, A. (1996) Distributed data fusion for real time crowding estimation, *Signal Processing*, **53**(1), 47–63.

Regazzoni, C. S., Tesei, A. and Munro, V. (1993) A real time vision system for crowding monitoring, *Proc. IECON*, pp. 1860–1864.

Reid, D. B. (1979) An algorithm for tracking multiple targets, *IEEE Transactions on Automatic Control*, **24**(6), 843–854.

Ricker, G. G. and Williams, J. R. (1978) Adaptive tracking filter for maneuvering targets, *IEEE Transactions on Aerospace and Electronic Systems*, **14**(1), 185–193.

Ristic, B. and Morelande, M. (2007) Comments on: Cramer–Rao lower bound for tracking multiple targets, in *Proceedings of the IET Radar Sonar Navigation*, **1**(1), 74–76.

Ristic, B., Farina, A. and Hernandez, M. (2004) Cramer–Rao lower bound for tracking multiple targets, in *Proceedings of the IET Radar Sonar Navigation*, **151**(3), 129–134.

Ross, S. M. (2003) *Introduction to Probability Models*. Orlando, FL: Academic Press.

Roumeliotis, S. I. and Bekey, G. A. (1997) An extended Kalman filter for frequent local and infrequent global sensor data fusion, in *Proc. SPIE (Sensor Fusion and Decentralized Control in Autonomous Robotic Systems)*, Pittsburgh, PA, October 14–19, pp. 11–22.

Salmond, D. (1990) Mixture reduction algorithms for target tracking in clutter, *Proc. SPIE: Signal and Data Processing of Small Targets, 1990*, O. E. Drummond (ed.), pp. 434–445.

Sidenbladh, H. (2003) Multi-target particle filtering for the probability hypothesis density, in *Proc. 6th International Conference on Information Fusion*, Fusion 2003, Cairns, Australia, July, pp. 800–806.

Singer, R. A., Sea, R. G. and Housewright, K. B. (1974) Derivation and evaluation of improved tracking filters for use in dense multitarget environments, *IEEE Transactions on IT*, **IT-20**(4), 423–432.

Stigler, S. M. (1986) *The History of Statistics, The Measurement of Uncertainty before 1900*. The Belknap Press of Harvard University Press.

Sworder, D. D. and Boyd, J. E. (1999) *Estimation Problems in Hybrid Systems*. New York: Cambridge University Press.

Tan, T. N., Sullivan, G. D. and Baker, K. D. (1998) Model-based localization and recognition of road vehicles, *International Journal of Computer Vision*, **27**(1), 5–25.

Thomopoulos, S. C. A. and Zhang, L. (1994) Decentralized filtering with random sampling and delay, *Information Sciences*, **81**(1), 117–131.

Tichavský, P., Muravchik, C. H. and Hehorai, A. (1998) Posterior Cramér–Rao bounds for discrete-time nonlinear filtering, *IEEE Transactions on Signal Processing*, **46**(5), 1386–1396.

Tugnait, J. K. (1982) Detection and estimation for abruptly changing systems, *Automatica*, **18**(5), 607–615.

Van Trees, H. L. (1968) *Detection, Estimation, and Modulation Theory, Part 1*. New York: Wiley.

Vo, B. and Ma, W. (2006) The Gaussian mixture probability hypothesis density filter, *IEEE Transactions on Signal Processing*, **54**(11), 4091–4104.

Vo, B., Singh, S. and Doucet, A. (2005) Sequential Monte Carlo methods for multi-target filtering with random finite sets, *IEEE Transactions on Aerospace and Electronic Systems*, **41**(4), 1224–1245.

Vo, B. T., Vo, B. N. and Cantoni, A. (2006) The cardinalized probability hypothesis density filter for linear Gaussian multi-target models, in *Proc. 40th Annual Conference on Information Sciences and Systems*, pp. 681–686.

Wang, X. and Challa, S. (2003) Augmented state IMM-PDA for OOSM solution to maneuvering target tracking in clutter, in *Proc. 2003 International Radar Conference*, pp. 479–485.

Wang, X. and Mušicki, D. (2007) Low elevation sea-surface target tracking, using IPDA type filters, *IEEE Transactions on Aerospace and Electronic Systems*, **43**(2), 759–774.

Wang, X., Challa, S., Evans, R. J. and Li, X. R. (2003) Minimal submodel-set algorithm for maneuvering target tracking, *IEEE Transactions on Aerospace and Electronic Systems*, **39**(4), 1218–1231.

Wang, X., Mušicki, D., Richard, E. and Fletcher, F. (2008) Efficient and enhanced multi-target tracking with Doppler measurements, *IEEE Transactions on Aerospace and Electronic Systems*, to appear.

Williams, J. L. and Mayback, P. S. (2003) Cost-function-based Gaussian mixture reduction for target tracking, in *Proc. 5th International Conference on Information Fusion, Fusion 2003*, Cairns, Australia, July, pp. 1047–1054.

Wren, C. R., Azarbayejani, A., Darrell, T. and Pentland, A. P. (1997) Pfinder: real-time tracking of the human body, *IEEE Transactions on Pattern Analysis and Machine Intelligence*, **19**(7), 780–785.

Zajic, T. and Mahler, R. (2003) A particle-systems implementation of the PHD multi-target tracking filter, in *Proc. SPIE Signal Process, Sensor Fusion Target Recognition XII*, pp. 291–299.

Index

Printed in the United States
By Bookmasters